Fire Litigation Handbook

Dennis J. Berry
of the Massachusetts Bar

National Fire Protection Association
Batterymarch Park
Quincy, MA 02269

Cl-2120650
First Printing November, 1984

Editor: Gene A. Moulton
Assistant Editor: Marion Cole
Art Coordinators: Eileen Harrington
                 Nancy Maria
Composition: Louise Grant
             Geoffrey Stevens
Production Coordinators: Elizabeth Carmichael
                         Donald McGonagle
                         Debra Rose

NFPA No. SPP-79
ISBN 0-87765-280-5
Library of Congress Card No. 84-60034
Printed in the United States of America

Table of Contents

Foreword

In the litigatious society in which we live, injury and damage from fire is increasingly the subject of courtroom confrontation. This handbook has been written to assist the practicing lawyer represent either the plaintiff or the defendant in the preparation and trial of such a case.

Litigation is not the solution to the problem of fire in our country. Yet those who have taken the proper steps to avoid and suppress fire are entitled to have these facts forcefully brought out in an appropriate forum.

It is our intent that this book improve the quality of fire litigation for both sides and in so doing enable us to take another step toward furthering the cause of life safety from fire.

As a public advocate for firesafety, we believe that this in itself is a worthy goal.

Robert W. Grant
President
National Fire Protection Association

Preface

Injury and damage from fire have been a fact of life for centuries, but only recently have large numbers of civil cases been filed seeking redress against those who are allegedly reponsible for the losses. Fire litigation is clearly a growing area of concern for both plaintiffs and defendants. However, unlike many other aspects of civil litigation which have come to the fore in recent years, fire is a phenomenon well within the experience of every judge, juror, and witness. Unless the advocate has a full command of the facts and the law in a fire case, he or she runs the risk of the judge or jury making too many assumptions based, not on the evidence, but on their own experience. The purpose of this handbook is to aid the practicing attorney in the planning, preparation and presentation of a case involving a fire incident. This handbook will not replace basic legal research for the full development of any theory, nor will it make an attorney a qualified expert on the technical aspects of fire. What it does do is bring together in one volume arranged for practical use, basic information on law and science needed for a fire case.

In order to organize the material in the most practical fashion, the handbook is divided into two sections. Section I deals with the legal aspects of a fire case and Section II discusses many technical aspects associated with fire. By dividing the book this way, reference can easily be made to a specific area whenever a question arises.

Acknowledgments

Few books with as wide a scope as the *Fire Litigation Handbook* can be credited to a single person. This handbook required the cooperative efforts of many individuals to carry through the writing, editing, and production.

First among these is the law firm of Cozen, Begier & O'Connor of Philadelphia. Through its attorneys, this firm—one of the largest in the United States devoted extensively to fire- and property-related losses—served as the sole consultant to the NFPA for this book. The knowledge, wisdom and cooperation of the attorneys involved in helping to produce Section I, concerned with legal aspects of fire cases, is greatly appreciated.

Specifically, I would like to acknowledge the efforts of Joseph A. Gerber, who participated in the early planning of the work and the first draft, as well as of Michael J. Izzo, Jr., who chaired subsequent revisions leading to the final draft. Capable assistance was provided by Susan M. Danielski of the firm's Legal Research Department.

The chairmen of the three principal operating departments of the firm who also participated actively in the preparation of materials are Richard C. Glazer, Chairman of the Subrogation Unit; Robert R. Reeder, Chairman of the Defense Unit; and David R. Strawbridge, Chairman of the Arson and Fraud Defense Unit.

Many other partners and associates also contributed to the development of this handbook. With great appreciation, I note the support provided to this project by the leadership of the firm: Stephen A. Cozen, Harry P. Begier, Jr., and Patrick J. O'Connor.

Additionally, the NFPA commends the decision of Cozen, Begier & O'Connor to donate its potential compensation and royalties from this work to a consortium of burn center hospitals and clinics in the metropolitan Philadelphia area.

Working closely with me to produce some of the chapters in Section II was Ann M. McNamara, a former member of the NFPA's Public Affairs Division and a student at Suffolk University Law School, Boston. Gene A. Moulton was the editor who had full charge, not only of turning the complete text into coherent prose but also of overseeing production and of reminding recalcitrant authors of impending deadlines. Marion Cole was most helpful in copy editing and polishing much of Section II.

The technical strength of this book is based largely on the unmatched depth of expertise within the NFPA in the whole subject area of fire. Particular assistance was given by those members of the Engineering Services, Fire Analysis, and Fire Investigation staffs who reviewed the text and made helpful suggestions all along the way. Among this group are Richard E. Stevens, recently-retired Vice President and Chief Engineer, along with John R. Anderson, Richard L. Best, John K. Bouchard, John M. Caloggero, Ron Coté, Martin F. Henry, Thomas J. Klem, James K. Lathrop, and Wilbur L. Walls—all active or recently-retired members of the NFPA technical staff. Chapter 9 is based entirely on work by Martha H. Curtis, Michael J. Karter, Jr., Donald J. Redding, and Arthur E. Washburn of the Fire Analysis staff. A posthumous note of thanks is extended to Dr. Louis Derry, former head of the Fire Analysis Division, for his leadership in obtaining the data and developing the method of statistical analysis that has kept the NFPA in the forefront of evaluating the nation's fire problem.

The job of putting all the words onto paper in a readable fashion fell, in the first instance, to my secretary, Virginia M. Glass. After that, the many editorial changes were ably executed by Elizabeth K. Carmichael and her Standards and Text Processing staff.

Without the leadership of NFPA management under President Robert W. Grant and Vice President Anthony R. O'Neill, this work would not have been possible. The inception of this book can be traced directly to questions posed to me by Bob Grant at a breakfast meeting more than two years ago. Kathleen A. McGloin, my direct supervisor, allowed me sufficient time to do all the research, writing, and rewriting necessary for a finished product. Assistant Vice President Richard T. Winn and Celia Wolf, Director of Product Development and Marketing, consistently helped this project weather the rocky shoals of growth through which any new book passes as basic ideas are translated into final chapters.

To all of these people I extend a deep and heartfelt thanks. I hope that the work I have done properly reflects their ever-present high level of support.

Dennis J. Berry
September, 1984

Introduction to Section I

In today's world, fire is a commonplace occurrence with which many prospective jurors have had some experience. Fires may involve relatively simple facts or may arise in complex circumstances involving sophisticated chemistry or machinery. By its nature, fire destroys and, therefore, proper assembly of proofs in a fire case may present considerable obstacles to a practicing lawyer who is unfamiliar with fire science and investigation. Development of persuasive proofs in a fire case will confront the practicing attorney with the challenge of gathering the necessary factual information and integrating the information with scientific and legal theory to properly represent a client.

The focus of chapters 1-5, which comprise Section I, is on the practical and legal considerations in handling a fire case from the initial investigation through the development of legal theory. How to develop relevant facts and how to use scientific and circumstantial evidence are among the many pertinent topics covered in Section I. The purpose of these chapters is to provide counsel, or others involved with fire litigation, with appropriate information and insight to assist in the overall preparation of a fire case.

The handbook format was chosen because several basic assumptions were made in the writing of these chapters. First, the handbook is designed to be a guide for everyday reference to those involved with fire litigation, rather than an academic or legal treatise. Numerous scholarly reference materials exist which discuss legal issues unique to a particular jurisdiction. This work does not attempt to treat litigation or other legal problems in general. It is assumed that a practicing attorney will consult these other reference materials with respect to general legal problems. The treatment given to legal or factual issues here is limited to those which have specific or unique application to cases involving fires.

Finally, this presentation is not intended to be adversarial in nature but, rather, it is designed to discuss those considerations which are important and common to all types of fire cases. It remains the sole responsibility of the attorney to recognize and advance the proper legal issues and positions of a client and to select the proper application of the general principles discussed in this handbook.

Section I Table of Contents (detailed)

CHAPTER 1

THE INVESTIGATION — OBJECTIVES AND METHODS

§1.1 Introduction — Know What You Are Looking For

It is important to recognize that a fire investigation differs substantially from many other types of investigations in which counsel may be involved. After an intersection accident, for example, the intersection remains available to be drawn, photographed, measured and seen long after the occurrence of the accident. Fires are destructive and, often, in destroying a structure or causing an injury, important evidence may also be eliminated in the process. Important evidence may also be discarded by the fire department during cleanup or overhaul operations. Because of this, the need for a thorough and complete investigation at the earliest possible time is of extreme importance. An appreciation of the importance of the use of proper investigative techniques and scientific experts and of the need to gather and preserve relevant physical evidence is essential.

Before a meaningful investigation can be undertaken, the attorney involved in a fire case must have a clear understanding of the investigative objective. Without proper direction and objectives, the investigation will become a fruitless search and its results next to worthless for trial preparation. Whether the investigation is commenced immediately after the fire is extinguished or at a much later point in time, its *primary focus must be causation*, or what fire science experts and investigators term *"Cause and Origin."*

The determination of cause and origin provides the foundation for the development of the factual and legal theory underlying the case. In turn, effective analysis of legal rights and duties depends upon this determination. Since many legal rights and duties pertain not only to what caused the fire itself but what caused the specific injuries at issue, the cause and origin investigation must often go further into an analysis of why the fire spread from its origin point to other areas. All these factors emphasize that the preliminary determination of cause and origin itself is an essential fact in any legal analysis. This first step leads to a complete formulation of liability and causation which will ultimately be presented at trial.

§1.2   Cause, Origin and Spread

When knowledgeable fire investigators speak of "cause and origin," they are denoting source of ignition and geographic location of where the fire began. It can be easily understood that before one attempts to determine what caused the fire, one must try to ascertain, by proper investigative technique, precisely where the fire began or had its origin. Once this is determined, the investigation can proceed further to determine the likely sources of ignition.

When the geographic area or point of origin of the fire is determined, two separate levels of causation investigation should be pursued. The first deals with the actual start of the fire. What was the ignition source within the area of origin which started the fire? This analysis involves a consideration and evaluation of all heat- or energy-producing sources located in the defined area of origin. The second level of analysis deals with the precise cause of personal injuries or property damage which resulted from the fire. This analysis may be entirely different from an examination of the instrumentality which caused the fire itself. For example, smoke detectors, alarms, panic doors, and fire escapes are maintained as life safety measures. A failure of any of these types of equipment may be causally related to injuries or death, irrespective of the specific cause of the fire. With respect to property damage, for example, certain types of occupancies require, by law, properly designed and functioning sprinkler systems, or chemical extinguishers. The failure of extinguishing or suppression equipment to eliminate or retard an incipient fire may be actionable as the cause of enhanced damages.

As a final example, certain building materials known as polyiso-cyanurate foam insulation (polyurethane), when not properly installed and protected, can cause uncontrolled and rapid fire acceleration accompanied by highly toxic fumes. Great acceleration or an inability to extinguish the fire may result from improper installation or product defects in these building materials. Spread of fire may also be caused by the absence of required fire walls, fire stops, or by other deficiencies in construction or design of buildings which allow fire to spread laterally or vertically. From the standpoint of legal theories of causation, the reason for fire spread may be equally as important as the determination of the cause of the fire. Accordingly, in such cases the fire investigation must have an expansive focus in determining these three elements: (1) origin, (2) cause, and (3) spread.

Skilled fire investigators and scientific experts are trained to reconstruct the events leading to a fire and determine cause and origin, as well as spread, from a number of available sources. These include photographs, physical evidence retained by others, statements and interviews of witnesses, and public and private records. Proper determination of the type of occupancy involved will refer counsel to numerous codes and standards which may be relevant to the later development of legal theories of liability and defense. At this point it is sufficient to emphasize that a thorough and meticulous investigation, which concentrates on the objectives of determining origin, cause, and the reasons for spread, will provide the basic factual framework within which legal theories and positions can be developed.

§1.3   *Fire Cause and Origin Investigators*

The attorney or other person responsible for supervising the investigation in a fire case must ensure that the proper personnel and investigation procedures are coordinated in a comprehensive effort to determine cause and origin and a viable legal approach. The fire investigation expert is the critical person in any investigation concerning fire causation.

Wherever the fire occurs, there is usually a governmental official who has a legal duty to investigate fires. Townships, municipalities, cities, counties, state governments, and the federal government may have employees who respond to a fire and whose official duties include the determination of cause and origin. The specific role of these "fire marshals" or "arson investigators" will be discussed in subsequent sections. It is important, however, to note here that a first step in the investigation should be to determine the existence and identity of fire marshals who may have jurisdiction over the fire scene.

In many cases it is quite common for local, state, and federal agencies to cooperate in a fire investigation. These government officials should be contacted to obtain the results of their on-site investigation into causation. Public officials, such as fire marshals, not only receive formal schooling in fire investigative technique, but have extensive opportunities to apply this schooling in numerous on-site investigations.

Generally, the results of the fire marshal's investigation are considered public information and are available upon request. On the practical side, where the official investigation indicates an accidental cause, the reports of public fire marshals are readily provided upon written request. However, procedures necessarily vary in different juristictions — some require the issuance of a subpoena whereas others require only a letter and payment of copy fees. In some limited situations — particularly those involving criminal investigations — the information may be unavailable, even by subpoena, until formal criminal proceedings are commenced or concluded. Where the official investigation results in the submission of evidence to a grand jury, that evidence is generally protected from discovery except in unusual or compelling circumstances. This can prove to be a hinderance to information gathering in civil cases; however, there is some recent support for relaxing strict rules of grand jury secrecy where the policies for secrecy are not furthered by protecting the information. As a practical matter, most significant information exchanges between official fire investigators and private counsel occur before submission of cases to a grand jury and, therefore, this type of restriction may be more imaginary than real. Arson reporting immunity statutes also facilitate the exchange of public and private information (Sec §5.5.9 *infra*). At some point, however, this valuable information should become available for use both in either a civil or criminal action.

The thoroughness and detail of a public fire marshal's investigative efforts are often a function of the scope of official duty, the size and sophistication of the fire department, and its budget for schooling and specialized equipment. Obviously, record keeping and evidence preservation may vary substantially from a small town to a large city.

In some instances, the official governmental fire investigator's duty will be limited by statute or ordinance to a broad determination that fire is either accidental or incendiary, i.e., intentionally set. In some cases, only incendiary fires are fully investigated in order to fulfill the requirements of the criminal justice system, whereas accidental fires may not be analyzed in detail. While the conclusion that the fire was "accidental" in an official report may be helpful and valid, it has limitations. In a civil case, more information will be necessary to develop the needed proofs for liability or defense. Accordingly, while it is helpful to rely upon the official fire investigator for preliminary information, if the investigation stops there the investigation will be incomplete and the overall objectives may not be fulfilled.

Because of these limitations, the job of making more specific findings as to cause and origin is often given by the attorney to private, independent investigation experts who specialize in determining fire causation. Some of these investigators are former fire marshals, some are not. Although a private fire investigation expert need not have served in a governmental capacity, experience teaches that, in general, persons with official fire investigation backgrounds have proven themselves to be competent and reliable on-scene investigators.

It is important also to note a further caveat with regard to cause and origin investigators. While the fire marshal or private fire investigator may be experienced in determining cause and origin of fires and be able to recognize and identify both simple as well as complex causes, the explanation of electrical or chemical causation may require skills beyond mere identification. Such conclusions often require a high degree of scientific or engineering ability and, therefore, the need for additional expert advice. Fires occur in an endless variety of ways, and often a complete and plausible explanation of the chain of events and combination of circumstances requires the aid of scientific and engineering disciplines. Thus, depending upon the circumstances of an individual case, it may be necessary to have not only the fire marshal and an independent private fire investigation expert evaluate the facts, but to call in a chemist, metallurgist, electrical engineer, physicist or other scientific expert in order to reach a thoroughly explained conclusion as to causation. The selection of such experts is discussed in §3.3, *infra*.

§1.4   Development and Preservation of Evidence

§ 1.4.1   Introduction — Investigative Method

Since the central focus of the fire investigation is on cause and origin, this section will discuss ways to develop and preserve cause and origin evidence for later use at trial. Because a fire scene is subject to change due to clean-up or overhaul operations, demolition, or repairs, evidence gathered at the earliest point in time after extinguishment will be of the greatest probative value in determining causation.

Naturally, a more comprehensive investigation will result if the opportunity is afforded to conduct it promptly after the occurrence of the fire or at some time before the remains of the physical scene have been disturbed. If this opportunity exists, an investigative team should be promptly organized to conduct an on-site investigation. Obviously, the opportunity to be on site will not always be present. Experience shows that a late investigation will create practical problems for counsel. First, the opportunity to view the entire scene may be lost. As a result, reliance must be placed on the judgment or competence of others to select the relevant areas to examine, preserve or photograph. Second, cross examination of the latecomer is always easier as opposing counsel can always point to areas not seen or examined personally. Finally, the earliest investigation provides the best opportunity to develop complete information because recollections are fresh. However, meaningful investigations can still be done at later points in time if focused on assembling information about the earliest conditions under which the fire took place.

A variety of investigative techniques and tools come into play in a cause and origin investigation. In addition to interviewing any available eyewitnesses, and documenting the scene with photographs or video film, the general investigative method involves a meticulous examination of the structure and fire debris for an analysis of "burn patterns." Fire science has shown that a fire will typically burn in an upward direction as it consumes fuel and seeks oxygen. In doing so, a fire path will usually emerge and leave behind traceable burn patterns on walls, floors and other areas. Where total destruction does not occur, the traceable patterns of burning and charring will be more evident and will more clearly identify the path of the fire. In analyzing the visible burn patterns, fire investigators will generally seek V-shaped patterns on damaged areas of a structure or objects located at the scene. The V-pattern is then traced to its lowest point; this point is generally indicative of the point of origin. Although this method seems simplistic, it is logical and is supported by well-established principles of fire science.

Obviously, where a structure is totally destroyed the application of the general investigative method will be more difficult. Nevertheless, the same basic method is also utilized in examining totally destroyed structures. Although the burn patterns may not be as physically evident, the lowest points of fire may be traced through a physical examination of the areas of deepest charring in wood or those areas showing the most severe fire damage. These areas are generally indicative of the place where the fire has burned the longest. Again, although there may be exceptions to the general rule, such areas are also generally indicative of fire origin.

Fire, generally, burns in an upward direction. The notable exception to this general rule is the situation where a flammable liquid has been used to accelerate a fire. At its inception, a flammable liquid fire will burn downward, following the path of the flammable liquid which may be absorbed into the structure of a building or which flows downward as the result of gravity. After the accelerant has been exhausted the fire will then burn in an upward direction.

As noted, the general investigative method involves analyzing the burn patterns and destruction with the objective of finding the geographic point of origin. Thereafter, determining the cause becomes the focus of the investigation. The cause is generally some source of heat energy which provides the ignition of the fire in the area of origin.

Obviously, some sources of ignition, such as a carelessly discarded cigarette or match, will never be found in the fire debris even though the point of origin is precisely identified. In such cases, the investigative method should focus on an analysis of the overall area of origin and the elimination of potential causes either by physical examination or through the use of scientific proofs.

Other potential ignition sources, such as electrical equipment, wiring, motors, appliances, or chemical residues, leave identifiable evidence of causation. Such evidence can be photographed and physically preserved for further analysis. For example, failures in electrical equipment and wiring produce certain physical manifestations of the failure itself such as beading of copper wires or massive melting of motor windings. Certain flammable liquids produce patterns of downward burning and irregular patterns of burning which are not only identifiable, but which leave recoverable traces of their existence in the burned material. It is not sufficient merely to preserve this evidence (although its preservation is essential). The investigative method must go further and make distinctions between cause and effect. In the above example involving electrical equipment, the investigation must determine that the electrical evidence represents the cause of the fire, rather than evidence of a condition which resulted because of the fire damage. In the example of the flammable liquids, the investigation must go further to determine if the presence of flammable liquids is ordinary or unusual to the occupancy of the fire premises. Obviously, different conclusions can be drawn as to the accidental or intentional nature of the fire depending on the result of this inquiry.

The foregoing illustrates the importance of a complete and expansive investigation. One which is too narrowly focused may result in incorrect or misleading conclusions. First findings as to causation may, indeed, be accurate; however, they must be verified empirically within the confines of known scientific rules. The completeness of the investigation and the ability to tie the circumstances together into a cohesive set of consistent proofs, bearing upon the theory of liability or defense, is the ultimate objective of the overall investigative effort.

§ 1.4.2   On-Site Inspections

From the foregoing, it is apparent that an on-site inspection by an experienced fire investigator is extremely desirable and important. It is also important for the attorney to be present at an on-site inspection if at all possible. There is no substitute for a firsthand observation of the fire scene and the investigative process itself. Moreover, a firsthand experience with the fire scene is of invaluable assistance to the attorney in the preparation of lines of direct testimony and cross-examination. Counsel who has been present at the scene has a distinct advantage over an opponent who has never been there. The attorney's factual observations of the scene do not constitute trial evidence. However, a sensitivity to the facts will be developed that aid immeasurably to meaningful examination of witnesses at trial.

As previously noted, it is important to make an on-site inspection before the physical scene is disturbed. Every effort should be made to ensure that the scene is not disturbed until it can be thoroughly examined by an experienced investigator. Police or fire marshals conducting an official investigation generally ensure that the fire scene remains undisturbed until the investigation is complete. Where no official investigation is conducted, hiring a private guard service is one way to keep the scene relatively undisturbed until an investigation can be conducted. Often investigative objectives must give way to safety considerations.

Limited demolition is sometimes necessary to eliminate dangerous conditions. In such cases the investigation will be made more difficult — but certainly not impossible. Other methods of reconstruction of events will then become more significant. Obviously, every effort should be made to obtain an early site inspection.

Although the development of cause and origin information has its primary focus on identifying and isolating the precise point of origin and cause, the relevance of general information should not be overlooked. Much general information becomes highly relevant as a case develops; the more detailed information that is gathered, the better prepared the case will be.

There are several kinds of general information that should be noted by the fire investigator and studied by the attorney. Initially, detailed measurements should be taken at the site to determine the square footage, size, height and construction of the building. Although these measurements may at first appear innocuous, they often assume great importance. For example, given the materials from which the building is constructed and the fuel loads of contents, the area or square footage will be important in determining the rate of fire spread. These measurements may also be significant to determine code-required clearances between combustible surfaces and heat-producing objects such as chimneys, woodstoves, flue pipes and similar installations. Measurements concerning the height of walls, ceilings and structural members may disclose the absence of required fire stops between rooms, in pipe chases or in utility openings between floors. All this may be of significance in explaining the spread of fire and the enhancement of property damage or loss of life.

During the on-site inspection, an analysis should be made of the materials of construction of the building itself. Numerous building materials carry designated fire ratings and have been tested by independent agencies. Identification of the specific construction materials may provide the basis for theories of liability concerning fire spread or may simply provide basic factual information which will explain the sequence of events of the fire from first discovery until extinguishment.

A second focus on general information should involve the type of occupancy or use of the premises. Different rules and regulations apply to residential buildings, multistory apartment buildings, commercial buildings, warehouses, and various manufacturing operations.

Each of these types of occupancy is required to conform to different standards with respect to firesafety or life safety. For example, schools, hospitals and multistory dwellings often require the installation of sprinkler systems, smoke detectors, direct-tie fire alarm systems, and specified numbers and types of fire exit doors. Depending upon the circumstances of a particular case, an on-site inspection may lead to discovery of violations of life safety features specified for certain occupancies. Numerous manufacturing occupancies are classified as ultrahazardous locations. In addition to required sprinklers and alarms, they must possess specialized explosion-proof electrical equipment, dust or fiber removal systems, or other specialized equipment. The significance of properly functioning code-required equipment is enhanced by the hazardous environment in such occupancies.

In addition to the premises themselves, an examination should also be made of the materials stored in a particular occupancy. For example, certain warehousing operations are subject to rules and regulations specifying storage arrangements and other requirements relevant to firesafety. These rules and regulations are contained in private and public fire codes, such as the NFPA codes. Also, contractual arrangements between a warehouser and storage interests often contain firesafety requirements. Accordingly, a determination of the nature and method of storage in a warehouse occupancy may lead to a later conclusion that improper storage of materials was connected to the cause, origin or spread of the fire.

All on-site general determinations, such as measurements, distances, location of materials, equipment, machinery, and furnishings, should be documented as fully as possible by photographs or through the use of blueprints, drawings, maps, sketches, and diagrams. All of this information can thereafter be critically evaluated both by fire investigative experts and scientific experts in development of the theory of liability or defense in the case.

In addition to on-site inspections conducted by an experienced fire investigator and, where appropriate, by scientific experts, it is important to develop information concerning the prefire conditions from your client, employees of the client, or other persons familiar with the premises. A comparison of prefire conditions with those which exist after the fire is extinguished will help to explain the identity and location of seriously damaged contents or specific parts of the structure after the building has collapsed or been seriously damaged.

§ 1.4.3   Witnesses

An important aspect of any investigation is locating and identifying important witnesses and preserving their observations and testimony for later use at trial. Witnesses fall into several categories. The first of these are eyewitnesses. It is always important to identify and locate any eyewitnesses to the actual occurrence of the fire. Eyewitness accounts can sometimes direct the origin investigation to a specific location or piece of equipment. These observations can later be verified and documented by physical examination of the site, damage and debris.

The earliest eyewitness observations are extremely significant and should be documented. These include first visual observations of smoke, the fire itself, its speed, direction, and spread. For example, the color of smoke may, at first, seem insignificant; however, it may later be an important circumstantial fact in identifying the type of fire. For example, observations of heavy black smoke at the incipient stages of a fire may indicate the burning of petroleum-based products. At a refinery or other place where petroleum products are usually stored this observation may not be significant. However, the same observation may have extreme significance if petroleum-based products are burning in the living room of a residence or on the dance floor of a bar-restaurant which is the subject of a foreclosure action. First observations of the color of fire are also significant; the color of flames may indicate the burning of natural gas, paper, plastics or other substances. Finally, it is also important to document other sensory perceptions such as odors, sounds and heat intensity on the part of eyewitnesses. Gasoline and other petroleum-based products have distinct odors which are recognized by most people. The pungent odor of burning wire insulation or the distinctive whooshing sound of a vapor flash fire are also important sensory perceptions worthy of note and placement in the factual framework of the fire.

Finally, eyewitness accounts should contain a detailed description of all actual observations. These should be documented as to date, time, and physical location. A helpful device for reconstructing eyewitness accounts is to have a scale drawing or blueprint of the relevant fire scene available when the witness is interviewed. Each witness can then place himself or herself at a specific geographic location at the fire scene. Furthermore, use of an accurate diagram in conjunction with a witness' statement is a most useful tool to refresh recollection and re-orient the witness to the geography of a structure when testimony occurs long after the structure has been destroyed. Similarly, the use of charts and diagrams to illustrate witnesses' testimony is also extremely helpful in assisting the jury to understand a completely unfamiliar location. Lay witness testimony, if obtained early in the investigation, can become extremely reliable circumstantial evidence in a case, especially where such witnesses are unrelated to parties of the case.

In addition to eyewitnesses, there are other significant witnesses whose information is of extreme importance to a proper fire investigation. Among these are those witnesses who have familiarity with the prefire conditions of a damaged building. They should be extensively interviewed to accomplish several objectives. First, the prefire physical condition should be fully outlined. Everything within a building which explains or somehow fits into the overall investigation should be identified and located. This includes the location of machinery, equipment, furnishings, staircases, room dividers, fire walls, electric service panels, portable equipment, appliances, etc.

In interviewing witnesses as to prefire conditions it is useful to have them make drawings, sketches or diagrams or produce blueprints or other plans precisely locating each item of significance to the investigation. Documentation of prefire conditions in this form is extremely useful in reconstructing the damaged structure and identifying and locating its contents.

Another area of importance to investigate is prefire events and relationships: these should be examined in support of such witnesses' statements concerning the building and its equipment. These should include a compilation of all relevant contracts and records pertaining to the building and its equipment, such as purchase agreements, leases, service contracts for heating equipment, warehousing and storage agreements, building plans and construction contracts, purchase invoices, service or repair invoices, product literature and advertising, and all other documentation which pertains to liability or defense issues arising out of the cause and origin investigation. Each category of documents may become relevant to a liability theory or defense position at a later time. Accordingly, comprehensive development of this information may later provide a critical missing element in the preparation of the fire case.

If the cause and origin investigation points to the failure or malfunction of specific industrial equipment or manufacturing machinery, plant engineers or other persons responsible for the operation and maintenance of this equipment should be interviewed in detail to determine the proper procedures and functioning of this equipment in the manufacturing process. Often, running logs are kept which show the operating hours of equipment and its maintenance history. Maintenance personnel or plant engineers are often highly skilled and conversant with actual operation of the equipment and can be most valuable in assisting a scientific expert in confirming necessary facts to prove out the cause and origin theory.

Fire department personnel are another source of important witness information. Fire department personnel have been trained to make and record observations made at the fire scene. Because of experience and training, fire fighters are more apt than lay witnesses to look for facts and circumstances which are relevant to a determination of cause and origin. Generally, reports made by responding fire companies record the following information: the time alarms are received; the time of arrival; the use of specialized fire fighting equipment; observed weather conditions; and the names of all responding fire fighters. The development of the sequence of events which occurs at a fire scene can be relevant to a number of issues concerning cause, origin, and spread.

Fire department witnesses and records are an important source of this type of information. A fire department's recorded data provides a framework within which to develop a reliable chronology for theories of liability and defense.

In a particular case, it may be important to interview radio or dispatching personnel who receive and, in some circumstances, record incoming telephone alarms. Often, information as to the first-recorded observations can be derived from these sources.

Personnel in a first-arriving fire company are usually the first trained persons to make observations of the scene of the fire from the exterior. More significantly, they are often the first eyewitnesses to fire conditions on the inside of a building. Observations of these trained professionals can assist greatly in isolating the area of origin or explaining the propagation of the fire and the fuels located in the premises.

The mechanics of obtaining interviews or statements from fire department personnel differ from jurisdiction to jurisdiction. Generally, fire department "run reports" are available upon request. Certain fire departments in larger cities have fixed procedures for interviewing fire marshals, fire fighters, and other investigative personnel.

As a final comment the attorney should remember that interviewing witnesses is but a single aspect of the overall investigative process. Clearly, the well-prepared attorney will have interviewed and documented the testimony of relevant witnesses as an integral part of a cause and origin investigation. However, the statements made by witnesses must be carefully compared to the results of the physical investigation in order to weigh their accuracy and probative value.

§1.4.4 Photographs and Other Visual Aids

The old adage that a picture is worth a thousand words rings especially true in a fire case. Because the scene of the fire is often demolished and proof of liability or defense is based on reconstruction of the facts through circumstantial evidence, photographs and other visual aids play an important role. In addition, the expectations of many jurors in this television age demand that attorneys present videotapes, photographs or slides as evidence in a case.

The first objective of photographic evidence in a fire case is to "preserve" the scene of a fire and relevant evidence of the cause and origin. It is suggested that one can never have too many photographs of the fire scene. The initial investment in film and development pales in relation to the importance of a graphic color photograph or film of a critical area or object in constructing a solid case. From an investigative standpoint, the objective is to document as many relevant facts as to cause and origin as possible. From the trial lawyer's standpoint, jurors should not have to guess about critical locations or instrumentalities when they can be shown a highly demonstrative photograph or film.

One can never have too many photographs of a fire scene, from an investigative standpoint, because when theories of causation are being developed and evaluated during the initial stages of the investigation, not every relevant or critical part of the physical scene can be retained for later use. Photographs at first considered insignificant may later be critical in establishing liability or a valid defense. For example, photographs taken far from the actual point of origin may clearly

15

show a tripped circuit breaker, a switch left in an "on" position, an unplugged appliance, a closed sprinkler valve, or an entire area of alleged origin in pristine condition. Such photographs or films can, at a single showing, easily corroborate or refute the testimony of many witnesses. Complex scientific theories which may otherwise be misunderstood can be amply illustrated and demonstrated to a jury through the use of a single photograph.

Photographs and other visual aids should document the fire case throughout the relevant time frame. There are many sources of photographs which depict either the prefire condition of the building or the fire scene. A search for media photographs, news films, and the work of amateur photographers sometimes can disclose a significant and helpful piece of photographic evidence. The most useful photographic evidence, however, is that which the attorney and the experienced fire investigator obtain at the fire scene.

The second objective of photographic evidence in a fire case is to create a picture display in order to illustrate important testimony for the jury. Here the main rule is to move from the general to the specific. The problem with many fire scene photographs is that they are too narrowly focused on the precise point of origin or the instrumentality which is the suspected cause of the fire. Photographs which are too narrowly focused may cheat the jury out of a view of the broad area in which the suspected cause is located. In so doing, the jury may be unable to visualize or confirm the existence of significant burn patterns or other important evidence in the area described by witnesses. Accordingly, photographs of the fire scene should begin with exterior views of the structure from all directions in order to give the jury a general view of the premises. Thereafter, the interior of the building should also be photographed from all directions. When a critical area of significant burn patterns or other critical evidence is photographed, a broad view should first be taken; thereafter, narrower views can focus in on the precise instrumentality or point of origin.

In following this sequence, a logical presentation can be made to the jury illustrating the overall view of the premises, the areas of specific concern, the precise point of origin within the area, and the specific instrumentality of causation. Where physical evidence is to be removed from the scene of a fire for further examination, the physical evidence should be photographed in place before it is disturbed, while it is being removed, and after removal. A detailed log should be prepared indicating the date, time and direction of each photograph as well as the identity of the photographer. After the photographs are developed, they should be collated in a logical presentation and should be accompanied by narrative descriptions of the areas shown. Several of the most highly relevant photographs can later be enlarged and mounted for use at trial.

It is suggested that photographs be in color rather than black and white, simply because color photographs better illustrate the sharp distinctions in definition and condition which often become disputed issues in fire litigation cases.

Because today's jurors are familiar with television, consideration should be given to using videotape equipment to show the scene of the fire, relevant areas of origin, burn patterns, and the first discovery and removal of an instrumentality which is the suspected cause of the fire. The narrated videotape should include the date, time and camera direction, so there will be no dispute as to authenticity at a later time. Objective narration of a video presentation done on the day of the investigation can be both a valuable illustrative and testimonial piece of evidence.

In addition to photos taken by the fire investigator, other sources of photographs and film should be investigated. Local or state fire marshals may have had their own photographer or borrowed a police photographic unit to assist in their investigative effort by documenting their findings with photographs.

In fact, the fire marshal's file may contain the earliest photographs of the fire scene in its undisturbed conditions, since a fire marshal or arson investigator will often be called to the scene of the fire for investigative purposes before the fire is completely extinguished. Also, in many circumstances the fire scene itself is kept off limits to nonofficial personnel during the time that the fire marshal conducts an investigation. Therefore, official fire marshal photographs are often the best available source of immediate postfire conditions.

Media photographs and newsreels are a second source of potentially valuable photographic evidence. Media photographers and camera people try mightily to be on the scene to show a fire in progress. The short film clip seen on the evening news may represent only a small percentage of the actual footage taken and it is wise to secure the entire film footage from the appropriate party. In cases where the issues concern the intensity and spread of the fire, this type of evidence can prove invaluable.

Television stations have varying retention and disclosure policies concerning film footage. Depending on the station policy, obtaining newsreel footage may be a simple matter of request or may require the issuance of a subpoena. Experience shows that the latter route is more frequently followed by the news media; however, a viewing of the tape, as contrasted to its production, is often freely permitted upon request.

Efforts should also be made to tap other sources of photographic evidence. Amateur photographers, former owners or others often possess preloss photographs.

Finally, aerial photography may be appropriate. In large fires involving a series of buildings or an unusually large plant, the analysis of cause and origin may be greatly assisted through the use of this technique. An aerial photograph may be the most precise way to illustrate overall fire patterns and the spread of fire from one structure to another. The same techniques used in presenting other fire scene photographs apply to aerial photography.

The ultimate objective of a fire investigation is the determination of cause and origin. Photographs play an important part in the investigative process and, ultimately, in a trial presentation. Accordingly, no amount of effort spent securing good photographs or video presentations is too much in a complicated fire case.

§ 1.4.5   Preservation of Physical Evidence

A thorough inspection at the scene of the fire should yield numerous pieces of physical evidence that will be of significance later at the trial. It is obvious, however, that all relevant physical evidence cannot be removed from the scene and preserved. Therefore, as previously noted, reliance must be placed on accurate photographs, videotape, and the testimony of witnesses in describing burn patterns and physical conditions. Certain kinds of physical evidence, however, are critical. They can and should be carefully removed from the scene for examination and later evaluation by a scientific expert and the jury.

Physical remains of an instrumentality which is the suspected cause of a fire should be kept for later use at trial. This might include wiring, fuse boxes, circuit breakers, appliances, fixtures or machinery. Before such items are removed from the scene of a fire they should be photographed in their original position. It is also important to immediately photograph the area surrounding the evidence in order to show its relationship to other physical objects. The evidence then should be photographed as it is being removed from the scene of the fire in order to document its condition before and after removal.

In many cases the physical evidence itself will be seriously damaged from the fire. Great care should be taken to ensure that all damaged pieces of the instrumentality are retained. An important component of machinery or equipment which is later found missing may result in a deficiency in proofs and give rise to other arguments as to the adequacy or completeness of scientific examinations of the evidence.

Preservation of physical evidence should not be limited to the suspected cause of the fire. Other objects, located in the same general area, which could have caused the fire, such as appliances, electric line cords, wall outlets, or other potential sources of ignition, should also be photographed and retained. This applies even where careful examination of all potential ignition sources in the area of origin has ruled out other potential causes. Such items should not be discarded simply because they have been ruled out as a potential cause of the fire. It is important that they be retained in order to corroborate the accuracy of the cause and origin conclusion and to document the thoroughness of the examination. If such items are not saved, a jury could very well be persuaded that these alternative sources of ignition were as likely to have caused the fire as the instrumentality which was saved. Preservation of these other objects allows the attorney to refute such arguments as well as present the jury with physical evidence to reinforce the opinions reached.

It is not unusual in locating physical evidence to find that a particular appliance or piece of equipment suspected of causing the fire is almost totally destroyed by the fire itself. This situation, however, does not present insurmountable obstacles to the attorney. Sometimes, the premises may contain additional identical fixtures or pieces of equipment in good condition. In such circumstances, this equipment should be retained to facilitate comparisons between the severely damaged evidence and its twin. For example, in stores or commercial buildings, ballasts in fluorescent fixtures may all have been installed at the same time. Therefore, numerous fixtures installed in series may be identical in age, make, and model.

A similar situation may prevail in residential structures. In large apartment complexes, for example, individual apartments are often outfitted with identical equipment or appliances. If an identical model is available on the premises, it should be retained. If not, it is a good idea to purchase an identical appliance or piece of equipment elsewhere. Appliances and equipment are continually improved and old models are discontinued. Therefore, it is prudent to acquire an identical model at the time of loss rather than become involved in a long search years later for an item no longer available. This procedure sometimes yields a bonus, for you may find specific warranties, warnings and operating instructions with the new purchase. These warnings and instructions may have been improved to give more adequate instructions or warnings of potential hazards.

In addition to equipment or appliances, there are some cases which require that actual parts of the structure itself be removed and retained. This is often true in situations that involve chemical reactions or explosions. Here chemical residues, which can be collected and later analyzed, permeate components of the structure. Through the use of gas chromatography and mass spectroscopy, such residues left in wood, on porous surfaces, in tank bottoms and on other surfaces can be analyzed and identified. In turn, this analysis may lead to the development of a viable theory of causation or defense. It is important in removing chemical samples to place the material in a clean, nonreactive container so that an accurate chemical analysis can be performed at a later date. Finally, as with other physical evidence, the materials should be photographed before they are removed from the site.

Arson cases are a second area where actual portions of the structure will constitute physical evidence. In such instances, there may be multiple points of origin where accelerants were used to quickly propagate and intensify the fire and its spread.

There are many types of accelerants. The most commonplace are flammable liquids in the petroleum family, such as gasoline, kerosene, paint thinners or other solvents. As previously noted, burning flammable liquids create distinct irregular burn patterns. When such patterns are identified in a cause and origin investigation they should be properly photographed and samples should be taken within the area of irregular burning. As is true with other chemicals, the fact that the floor or wall surface is seriously burned does not preclude the later identity of the chemical properties of the flammable liquid poured in the area before the fire. The residues of such flammable liquids often exist in burned wood, floor tiles, wallboards, rugs, and even in concrete. The usual precautions should be taken in photographing this evidence and placing it into clean nonreactive containers.

Building materials themselves are a third factor where part of the structure is itself evidence. In cases where the cause and origin investigation suggests that the fire traveled at an unusually rapid speed, unaided by weather conditions, and where materials used in the building include insulation, ceiling tiles or highly flammable contents, samples of such materials should be taken from the building. Many building materials have been rated or tested by independent agencies for fire characteristics or flame spread. Ratings are also available for other components such as rugs, wallpaper, and fabrics. Some of these may have been advertised as "fire retardant," "fire resistant," or as conforming to certain standards regarding flame spread and fire propagation. The terms "fire retardant," "fire resistant" and the like have defined meanings under specified flame spread tests. A good discussion of these distinctions is contained in reports and proceedings concerning polyisocyanurate (polyurethane) foam prepared by the Federal Trade Commission during investigations of advertising by the foam plastics industry. Certain materials tested under laboratory conditions often perform quite differently in actual fires. By retaining samples, one can properly identify the materials and, thereafter, compare their actual characteristics and behavior under actual fire conditions to those advertised or warranted. Information on product design, testing and improvements and investigation of defects can be obtained from the Consumer Product Safety Commission in Washington, D.C. Also, numerous trade associations in specific industries can provide detailed product information on improvements in warnings and safety developments. Accordingly, the retention of samples of the materials themselves leaves open the option of intelligently evaluating potential product liability claims involving the use or advertisement of these materials.

§1.5   Use of Scientific Experts

Of necessity, a cause and origin investigation in a fire case involves specialized investigative techniques in analyzing the physical fire scene. Analysis of burn patterns and debris, depth of char and path of the fire has its basis in logic. The clear foundation of the investigative method is derived from established principles of science.

The behavior and causes of fire have been the subject of many scientific tests and studies. While a general overall investigation can pinpoint the origin of a fire, and even explain its cause and spread, the assistance of scientific experts is essential to a complete understanding and explanation of the cause and origin of fire. In a fire case, proof of theories of liability and defense involves application of the basic laws of physics, electricity, metallurgy, chemistry, materials science, structural engineering, mechanical engineering and even nuclear physics. Chapter 3 treats in detail the role of scientific experts in the preparation and trial of a fire case. In evaluating the results of a fire investigation, however, it is sufficient to say, at this point, that it is extremely important to engage the proper expert or experts to assist in the overall explanation of cause and origin. This serves both to educate the lawyer during trial preparation and benefit the jury at trial.

Many cases require the involvement of several scientific disciplines to properly explain a theory of causation or defense. This is because fires involve the ignition of various kinds of fuels and materials. In addition, a single case may involve multiple parties with multiple theories of causation, all of which require scientific proofs. For example, an identified electrical source of ignition may require evaluation by an electrical engineer to explain complicated circuitry or the reason for failure. Combustion of the materials immediately adjacent to this suspected source of ignition may require the testimony of a materials scientist on the ignition temperatures and ability of certain materials to sustain combustion. In the same case, a nonoperational sprinkler system and the absence of water pressure may involve the testimony of a fire protection engineer and a hydraulics engineer. Finally, an organic chemist may be called on to evaluate the burning characteristics of building materials or the toxicity of fumes which may be causally related to injuries sustained during the fire or fire fighting operations.

Other commonly used types of experts can include metallurgists to evaluate fire damage to metal, pathologists to examine tissue, architects and construction engineers to evaluate conformity to building codes, meteorologists to evaluate weather, and numerous others.

In the preparation stages of the case it is obviously important to determine which disciplines will be needed in order to arrange critical pieces of circumstantial proof into a cohesive and explainable sequence of events for the jury. Scientific experts may also be utilized in early stages of the investigation in order to eliminate or confirm basic scientific phenomena so that an intelligent decision can be made as to the proper direction and scope of further investigation.

In addition to scientific experts themselves, there are numerous fire science-related articles and literature which serve to explain the basic terms of fire science and the use and application of experts. NFPA codes contain extensive bibliographies of authoritative articles on many areas of interest to fire litigators. These codes should be utilized to their fullest extent as education and preparation tools. Education of the attorney is of the utmost importance to effective communication with his or her expert. If the expert's language and terminology is not understood by counsel, the expert's advice and opinions can be misapplied, resulting in confusion and the appearance of unpreparedness.

§1.6 Other Useful Sources of Information

In one place, it is virtually impossible to outline every source of useful information relevant to a particular fire case. Because fire cases arise in numerous contexts, each is unique in defining the boundaries of relevance. Some of the following sources of information, however, have proved generally relevant in a number of cases.

Weather data is sometimes significant in evaluating the conditions under which a fire occurs. Certain weather conditions may preclude a theory of spontaneous combustion or ignition; wind direction and velocity may explain the path or intensity of the fire; temperature conditions may explain difficulties in fire fighting or inadequate water pressure. Fire department records will often record relevant weather data in official reports. The United States Weather Service's National Weather Data Bank is another source of useful atmospheric data which can be obtained at a nominal charge. A simple letter to the local weather service offices will obtain the weather reports.

Public records can also be extremely relevant to a fire case. These are of several types. When blueprints or drawings of relevant buildings are unavailable, a search should be made of the local building inspector's records. Relevant information concerning the initial construction, subsequent modifications, improvements, or changes in the building are also usually found there. These records can also include specific plans for sprinkler systems and fire suppression equipment, as well as listings of building materials and a history of conformance to local codes and ordinances. Reconstruction of exact dimensions, room locations, piping diagrams, gas meters, electric service, and other details of construction can be facilitated with a simple check at the building inspector's office.

Fire code enforcement records and licensing can also be of significance. Many large cities have departments which are specifically concerned with the issuance of licenses for the handling of flammable or explosive materials. The same or similar department may also conduct firesafety inspections and record notices issued to property owners mandating the correction or elimination of hazardous conditions. An examination of such records often provides the foundation for theories of liability or defense pertaining to the creation or maintenance of fire hazards or violations of municipal, city or state fire codes.

In addition to those records maintained by the investigative unit, there are numerous other fire department records which can be obtained by simple letter request. Each responding fire company generally is required to maintain records detailing their work. These records, while not containing the results of an investigation, often provide recorded times of alarms, arrivals and other pertinent data which can be pieced together to develop an official sequence of events which may later be useful at trial. Since the fire department is usually under an official duty to record this type of information, and maintain the records in the ordinary course of business, their conclusions as to recorded times may achieve a greater degree of reliability than the recollections of witnesses.

In cases involving gas and electric utilities, public and private utility companies are a source of meticulous records concerning gas installations, repair and maintenance calls, responses to reports of leaks and other such information. Utilities generally investigate cases involving gas explosions, gas leaks, power failures and similar occurrences. Reports of these investigations are often required by public utility commissions or similar administrative agencies who regulate the utilities. Such records are a valuable source of basic investigative material involving these types of incidents. Where reporting to a public utility commission or similar agency is not required, traditional forms of discovery must be used to secure utility investigations.

Finally, in cases involving specific products, the Consumer Products Safety Commission is a good source of information. The Commission maintains indexes and records of complaints involving fires and failures in numerous types of consumer products. This information is generally available by request through the Freedom of Information Act (FOIA). Such requests are generally processed promptly. With the exception of the prohibition against disclosure of proprietary or privileged information, these reports and data can be a valuable source of basic information concerning fire-related product failures.

CHAPTER 2

NEGLIGENCE

§2.1 Introduction

Issues of liability and defense in fire-related cases necessarily invoke the common-law principles of negligence familiar to most attorneys. The four basic elements of a common-law negligence action, (the duty of reasonable care, a breach, proximate cause, and damages) as they apply to fire causation and damages, must be present. Accordingly, a sound working knowledge of general negligence principles in the applicable jurisdiction is an indispensable foundation for developing negligence theories of liability and defense in a fire case.

A careful consideration of the results of the cause and origin investigation is central to constructing a theory of negligence in the fire case. The relationship of the plaintiff or defendant to the fire source and the nature of each party's conduct as it relates to cause, origin and spread of the fire are crucial facts that will dictate the logical choice of available negligence theories: contract, simple negligence, strict liability or a combination of all three. For example, a cause and origin investigation which discloses that an overheated furnace was the cause of a fire will require an analysis of potential products liability theories revolving around the furnace design, controls, thermostats or valves, or theories of negligence involving the actions of an installer or service technician who worked on the unit sometime prior to the fire.

Some negligence theories will be more evident from the facts of an investigation than others. A critical analysis of structures, equipment, occupancies and activities, as well as analyses of materials and substances, is needed to develop negligence theories of liability and defenses. Obviously, every conceivable theory of fire-related negligence cannot be discussed in a handbook format. However, those theories which arise frequently and are commonly used and developed will be discussed. These theories have general application and can be tailored to fit many particular factual circumstances.

§2.2   *Basic Sources of Relevant Law —*
Statutes, Codes, Regulations

Firesafety is a topic which has been studied in depth by numerous governmental and private agencies. As a result, there are many firesafety codes extant in the form of statutes and ordinances, as well as private industry codes. All can be effectively used for reference and evidentiary purposes. In general, the codes reflect the myriad circumstances in which fires have occurred in the past. These circumstances have been studied and evaluated and, as a result, code provisions have been developed. A firesafety code, therefore, is a likely source of rules and guidelines which may deal with the precise occupancy, system or product relevant to the particular fire in question.

In applying a common-law negligence theory, the lawyer in a fire case should make a preliminary search of relevant firesafety statutes, building codes or private codes in the particular jurisdiction in question. After brief research, it will be apparent that numerous statutes, regulations and industry safety standards will be relevant to the preparation of the case (and to the development of proofs during a cause and origin investigation).

From a plaintiff's standpoint, violation of a statute or ordinance designed to protect the particular class of injured individuals may constitute negligence per se in some jurisdictions; in others, it may at least be considered by a jury as evidence of negligence. Similarly, an examination of relevant statutory or code criteria is also important from a defense standpoint because conduct in compliance with the statute or relevant private safety codes will be persuasive evidence that the defendant has met the relevant and accepted standard of care.

Although the evidentiary value of statutes and private codes is treated further in Chapter 3 on fire evidence, a brief statement at this point is necessary in order that proper direction be given. It is not unusual to find that national safety codes promulgated by private agencies have been adopted either in whole or in part as the official fire or safety code of a municipality or other governmental body. The largest group of authoritative codes and standards prepared by a private agency that have been so adopted are sections of the National

Fire Codes® published by the National Fire Protection Association (NFPA). As with publicly developed fire codes, the NFPA Codes provide extremely broad, general firesafety standards as well as narrowly drawn, highly specialized and technical applications. The most frequently cited codes which have been adopted in many jurisdictions are the NFPA National Electrical Code® and the Building Officials and Code Administrators' (BOCA) basic fire prevention code. The National Electrical Code provides numerous standards pertaining to the installation and maintenance of virtually every conceivable kind of electrical installation from relatively simple residential wiring to unusual commercial and industrial applications. The BOCA basic fire prevention code contains numerous building standards related to firesafety.

Once a relevant statute or private industry code is located, its potential legal effect should be analyzed. The approach taken by the *Restatement (Second) of Torts* illustrates the analysis applied by many courts. Section 286 of the *Restatement (Second) of Torts* provides:

§286. When Standard of Conduct Defined by Legislation or Regulation or Regulation Will Be Adopted

The court may adopt as the standard of conduct of a reasonable man the requirements of a legislative enactment or an administrative regulation whose purpose is found to be exclusively or in part
  (a) to protect a class of persons which includes the one whose interest is invaded, and
  (b) to protect the particular interest which is invaded, and
  (c) to protect that interest against the kind of harm which has resulted, and
  (d) to protect that interest against the particular hazard from which the harm results.

In evaluating the role of statutes or private codes it is important to note that fire codes often provide very broad as well as extremely narrow guidelines. These phenomena cut both ways. On the one hand, narrowly drawn code provisions have enhanced significance when the particular facts fall within the narrow code provision. For example, certain classes of occupancies are classified as to hazard and, therefore, trigger specific requirements as to sprinklers, fire protection equipment or safe practices. On the other hand, broad general fire protection provisions designed to promote general firesafety can have general application to many fact situations where conduct or conditions might be considered unsafe from a fire standpoint.

For example, the Philadelphia Fire Code (section 5-3102) provides general restrictions concerning the collection of combustible or flammable waste or litter tending to create a fire hazard. In addition, this provision of the code provides reasonable regulations for security of vacant buildings in order to prevent entry by unauthorized persons (and presumably fires caused by vandalism). This type of general fire code provision is useful in analyzing fact patterns or occupancies which are not specifically regulated under a code. Furthermore, this type of code provision is relevant to the common-law duties of care owed by a possessor of land where dangerous or artificial conditions cause injury to third persons outside the land. (*See, e.g., Restatement (Second) of Torts* §364.)

Case law appears to be divided as to the admissibility of private codes. Many courts have held that, with proper foundation testimony, private safety codes are properly admissible where the code is shown to represent a consensus approved by a significant segment of a particular industry. On the other hand, other courts have taken the position that such codes represent solely the opinions of their promulgators and, accordingly, the codes represent hearsay and are not admissible in evidence. The merits of these divergent positions, the methodology of introducing such codes into evidence and the significance, relevancy and weight to be given to such codes will be considered in Chapter 3 concerning fire evidence (§3.2). It is sufficient, at this point, to note that the codes are relevant to the preparation and evaluation of a proper negligence case and, even if not found admissible, may provide highly relevant areas of cross-examination or expert testimony which will bear on the ultimate liability issues in the case.

In summary, it is important to analyze the facts of a particular case under common-law negligence principles. Thereafter, when general theories of negligence are developed, a search should be made of available codes and standards to determine if the conduct has been specifically regulated by statute, private code or other type of regulation. Such an examination is extremely helpful in fulfilling the elements of the negligence cause of action, and in refining the theory through the later use of experts.

§2.3 The Negligence Time Frame

Most claims for fire injury or damage are derived specifically from liability-producing conduct related to the cause of a fire. The negligent conduct which gives rise to liability for fire damage may occur immediately before the fire or may have occurred weeks, months or even years earlier. In other cases, the defendant's conduct may have no relationship to the actual cause of the fire, and yet, that conduct, after the start of the fire, may have been responsible for most or all of the damage. Accordingly, the precise relationship between the acts and omissions of the tortfeasor and the damage sustained must be closely analyzed to determine if the actionable conduct: (1) preceded the fire by the creation of hazardous conditions; (2) actually caused the fire itself; or (3) after the fire began, enhanced its intensity, scope, or spread. Analysis of liability-producing conduct in these relevant time frames is discussed in the following sections.

§2.4 Common Negligence Theories

§2.4.1 Creation or Maintenance of Dangerous Artificial Conditions by Possessors of Land

Liability for fire injuries or damages is frequently found in acts or omissions which create a fire hazard despite the defendant's freedom from any responsibility for the actual start of the fire. Section 364 of the *Restatement (Second) of Torts* is a concise encapsulation of this theory. It imposes liability upon a landowner who unreasonably creates, or allows to be created, artificial conditions or structures which pose an unreasonable risk of physical harm to persons outside the land. This section, while used frequently in personal injury litigation, has relevant application in fire litigation. The section provides:

§364 Creation or Maintenance of Dangerous Artificial Conditions

A possessor of land is subject to liability to others outside of the land for physical harm caused by a structure or other artificial condition on the land, which the possessor realizes or should realize will involve an unreasonable risk of such harm, if,

    (a) the possessor has created the condition, or

    (b) the condition is created by a third person with the possessor's consent or acquiescence while the land is in his possession, or

    (c) the condition is created by a third person without the possessor's consent or acquiescence, but reasonable care is not taken to make the condition safe after the possessor knows or should know of it.

Section 365, a companion section, states an additional rule which imposes liability for failure to remedy conditions that make a building, originally safe, dangerous to persons outside the land. Section 365 states:

§365. Dangerous Disrepair

A possessor of land is subject to liability to others outside of the land for physical harm caused by the disrepair of a structure or other artificial condition thereon, if the exercise of reasonable care by the possessor or by any person to whom he entrusts the maintenance and repair thereof

    (a) would have disclosed the disrepair and the unreasonable risk involved therein, and

    (b) would have made it reasonably safe by repair or otherwise.

Facts producing liability are more obvious in those cases where the condition created involves a recognized high degree of risk, such as the unattended storage of flammables or highly combustible materials.

Less obvious are those cases in which the basic materials present may not, in and of themselves, pose a high risk of harm — but where a subtantial risk of harm arises due to the quantity present, storage methods, or to a combination of commodities. The principles of the above sections can apply to deteriorating industrial areas or areas where absentee owners neglect or disregard the condition of buildings to the point where they become attractive nuisances to children, vagrants, and vandals. Storage conditions of warehoused materials can be evaluated under these rules even where specific storage regulations may not apply, but where the quantities or configurations of storage present a hazard to adjoining buildings in the event of fire.

Sections 364 and 365 of the *Restatement* are also particularly applicable to cases imposing liability for fire spread resulting from the poor condition of the landowner's property. *(See, e.g., Ford v. Jeffries, infra.)*

§2.4.2   Liability for Spread of Fire Due to Negligent Conduct of Possessor of Land

In the negligence context, fire spread refers to some act or omission which directly enhances the intensity, scope or spread of fire. This is distinguished from fire spread cases predicated on theories of products liability involving flammability of materials or failure of suppression systems. In the context of negligence, liability for fire spread often falls into two categories: (1) where the fire originated by negligent conduct and then spread; or (2) where the fire was purposefully and lawfully set, but then spread out of control. It is important to note that in the first instance liability for the spread of fire may be imposed whether it was started negligently or nonnegligently. Liability here arises from failure to use reasonable care to prevent the fire from spreading to other property (upon discovery of the fire). In the second instance, liability may only be imposed where the landowner was negligent in starting or controlling a fire.

The general rule applied by the courts is that an owner of property on which a fire originates (because of the owner's negligence) or that the owner's employee acting within the scope of employment may be held liable for damage resulting from the spread of fire to other property. This result may vary in those jurisdictions accepting the defenses of contributory negligence by the plaintiff or the operation of an intervening cause. However, only those intervening conditions which are unusual or extraordinary occurrences beyond the foreseeability of the defendant will relieve the defendant from liability. A high wind or other natural phenomenom generally falls outside this rule.

In examining what types of negligent conduct will impose liability for the spread of the fire, courts have imposed liability in a variety of circumstances. Two of the most common occur when a property owner fails to secure a building against known dangers of fire presented by vagrants or where there is a failure to provide people to watch or guard a building against known or foreseeable dangers of intrusion. For example, in *Aetna Insurance Company v. 3 Oaks Wrecking and Lumber Co.,*[1] where a fire started by vagrants in a partially dismantled building spread to two nearby buildings, the court held that a demolition contractor had a duty to secure the building against known dangers of vagrants frequenting the property looking for a place to sleep and cook food.

Similarly, in *Ford v. Jeffries*[2] an action was brought by the owner of a neighboring house which was damaged when fire spread from the defendant's property. The trial evidence showed that the property where the fire originated was left in a general state of disrepair which created an unreasonable risk of fire to the plaintiff's adjoining property. The court expressly applied *Restatement (Second) of Torts* §365, as well as statutory fire code authority recognizing that structures in a "state of disrepair" or in a "dilapidated condition" may both constitute a fire menace or hazard to nearby property and require removal or repair of the structure. The open invitation to vagrants or vandals and the ensuing likelihood of fire was sufficient to impose liability on the landowner when, in fact, unknown vandals started the fire which ultimately spread, causing damage to the adjoining property.

---

[1] 382 N.E.2d 283 (Ill.App. 1978).

[2] 379 A.2d 111 (Pa. 1977). *See generally, Annot., Liability of One on Whose Property Accidental Fire Originates for Damage from Spread Thereof,* 18 A.L.R.2d 1081 (1951).

§2.4.3   Liability for Spread of Fire Purposefully and Lawfully Lit by Possessor of Land

Although the early English rule imposed almost strict liability for the spread of fire set for a lawful purpose, the general rule today is less certain. In order to impose liability upon an owner of property who sets a fire for a lawful purpose, it must be shown that the owner was negligent in starting it or controlling it. Ordinary care is the measure of diligence required to free one from liability for spread of fire set for a lawful purpose. For example, in *Criscola v. Guglielmelli*,[3] the defendant lawfully burned trash in a trash receptacle on premises specifically licensed for that purpose. The defendant, however, left the fire believing it was out but did not put dirt or water on it. Later that evening, a fire in the trash spread to an adjoining property. The defendant was held liable for negligent failure to ensure that the fire was out.

In *Douglas v. Nielsen*,[4] a town was held liable for damage caused by negligent burning of refuse where the dump was located at an elevation which exposed burning trash to high winds. The town was held to have breached its duty to take adequate precautions to control and contain the burning area and to prevent the fire from spreading out of control.

Care should be taken to determine the extent and application of contributory negligence in spread cases of this nature. For example, in one reported case, *Andrews v. East Texas Theaters, Inc.*,[5] the plaintiff was found to be contributorily negligent in maintaining scrap lumber, firewood, weeds, grass and other waste material from a sawmill on the plaintiff's property which in itself constituted a fire hazard.[6]

---

[3] 308 P.2d 239 (Wash. 1957).

[4] 409 P.2d 240 (Wyo. 1965).

[5] 289 S.W.2d 781 (Tex.Civ.App. 1956).

[6] *See generally, Annot., Liability for Spread of Fire Purposely and Lawfully Kindled,* 24 A.L.R.2d 241 (1952).

§2.4.4   Explosion Cases

There are a great number of potentially explosive substances which, if not properly handled, may cause explosions and ensuing fires. Natural gas, highly explosive dusts, and numerous hazardous chemicals are only a few examples. As in other types of negligence cases, the factual investigation should identify the substance involved in the explosion so that it can be traced to its source. Thereafter, a proper assessment of theories of liability can be made.

Natural gas is a very common explosive substance and provides a convenient model for the application of negligence theory in explosion cases. Its transmission, handling, delivery and sale are highly dangerous activities and subject to a high degree of regulation. Consequently, utilities, pipeline companies, common carriers, service agencies and others handling natural gas have been held to a very high degree of care. Gas explosions inside occupied buildings often raise questions both as to the source of gas leaks and involvement of gas company-owned equipment or owner-serviced equipment. In many instances, gas line leaks occur far from the point of ignition and explosion. Gas often follows openings in the ground and the path of underground pipes before finding its way into a building and a source of ignition. Therefore, a complete analysis of the surrounding area of a gas explosion is often necessary to identify the responsibility for the leak. An investigation into gas explosions will often support potential theories of liability concerning installation, servicing or repair of the gas equipment, the failure to discover leaks in existing equipment during routine maintenance or inspection, inadequate odorization of the gas preventing discovery of the leak, improper delivery, excessive pressure, or the impurities in the gas itself.

Again, because gas transmission is highly dangerous it should be apparent that an analysis of both public and private regulations is necessary to the proper preparation and determination of liability in a gas case. An examination should be made of relevant NFPA and ANSI (American National Standards Institute) standards, ICC (Interstate Commerce Commission) regulations and local public utility laws concerning the installation and delivery of gas and the maintenance of gas pipelines and equipment.

Generally, in gas cases there is no difference between the general rules imposing liability for negligence and those applicable to fire cases. However, the often dire consequences of the escape of gas support the rule that a seller or distributor of natural gas must use a degree of care commensurate with its highly dangerous character. In this regard a number of courts have held the conduct of suppliers of gas to be negligent where: (1) a gas supplier's employee delivered gas and failed to inspect the system which supplied the gas; (2) the seller of gas and installer of a gas line system failed to check for open valves in a customer's household system or sold nonodorized gas; and (3) a propane gas delivery truck was delivered to a garage for repairs without adequate or proper draining and venting of all valves and hoses, and without a warning that the truck should not be placed in the path of a heating system blower.

In addition to theories of common-law negligence, gas explosions and related fire cases generally can involve theories of warranty and strict liability, especially in view of the fact that gas is a highly dangerous product and has highly dangerous propensities. (See §4.3.2.) Numerous cases have arisen involving manufacturers of either the gas itself or of gas transmission equipment or appliances that have been found defective. Finally, in gas-related cases there is a greater tendency to permit use of the doctrine of res ispa loquitor to establish liability where the gas supplier made delivery, installation or repair which was circumstantially associated with a fire or explosion.

As in all negligence cases, facts pointing to contributory negligence or to the assumption of the risk should be thoroughly investigated. These defenses involve some of the following considerations: (1) a consumer misplacement or faulty installation of the consumer's own equipment or appliances; (2) negligent operation or management of equipment in the control of the consumer which involves the handling of gas; (3) disregard of communicated notices or warnings concerning leaking or defective equipment; and (4) disregard of objective manifestations of leaking gas, such as gas odor or unusual flames.

The rules of law governing liability and the standard of care in gas combustion cases are equally applicable to other highly explosive products and activities connected with such products. For instance, liability may be imposed for negligent storage, transportation, delivery, loading or unloading of petroleum products.[7] The same standard of care has been applied in dealing with high explosives such as dynamite, gunpowder, and nitroglycerin. However, in cases where the plaintiff is injured on the premises of one storing explosives, the practitioner should also be aware of judicial recognition of varying standards of care dependent upon the plaintiff's status as an invitee, licensee, or trespasser.[8]

The choice of remedy may be crucial for the plaintiff's attorney and if the defenses of contributory negligence or governmental immunity are anticipated, it may be advisable to plead nuisance or strict liability, or both, rather than to limit the action to the theory of negligence. The plaintiff's attorney should also be aware that persons other than the storer or transporter of the explosives or petroleum products may be potential defendants. These persons include contractors, manufacturers, retailers, wholesalers, and municipalities. As in every case of negligence, one factor which may have considerable impact is if the allegedly negligent defendant violated a statutue or ordinance. Such a violation may constitute negligence per se or at least make a jury finding of negligence more likely.

§2.4.5   Negligent Performance of Services

A large body of fire litigation arises from negligence of contractors and others in rendering services. Section 323 of the *Restatement (Second) of Torts* provides:

§323. Negligent Performance of Undertaking to Render Services.
One who undertakes, gratuitously or for consideration, to render services to another which he should recognize as necessary for the

---

[7] *See generally, Annot., Liability in Connection with Fire or Explosion Incident to Bulk Storage, Transportation, Delivery, Loading or Unloading of Petroleum Products,* 32 A.L.R.3d 1169 (1970).

[8] *See, e.g., Carlysle v. Aetna Insurance Co.,* 248 So.2d 64 (La.App. 1971); *Bailey v. Bakers Air Force Gas Corp.,* 376 N.Y.S.2d 212 (App.Div. 1975). *See generally, Annot., Liability of One Selling or Distributing Liquid or Bottled Fuel Gas for Personal Injury, Death or Property Damage,* 41 A.L.R.3d 782 (1972); *Annot., Liability in Connection with Fire or Explosion or Explosives While Being Stored or Transported,* 35 A.L.R.3d 1177 (1971).

protection of the other's person or things, is subject to liability to the other for physical harm resulting from his failure to exercise reasonable care to perform his undertaking, if:

    (a)   his failure to exercise such care increases the risk of harm, or

    (b)   the harm is suffered because of the other's reliance upon the undertaking.

Section 323 mixes elements of contract and warranty theory with negligence theory based upon the "undertaking" language of the opening sentence of that section. Its application is extremely broad and can be utilized to measure the liability of any variety of individuals or companies which have a relationship to the performance of services, including highly skilled and experienced service technicians rendering a service for a fee, municipal inspectors' failure to discover fire code violations, or a security or guard service.

As with all cases dealing with the creation of a fire hazard, the negligent acts or omissions of the wrongdoer may be deemed to occur well in advance of the fire itself. The following subsections show examples of situations involving negligent performance of services but in no way are intended to be exhaustive. Cases dealing with on-site faulty installations or negligent repairs which are immediately apparent as a fire cause are excluded in the belief that the practicing lawyer will recognize such situations.

§2.4.6   Negligent Performance of Services: Excavators

Excavating, both in the street and on private premises, poses the often difficult problem of avoiding various underground pipes, mains, and electric lines. Various specific duties have been imposed upon an excavator or an excavator's employer in this context. Among these are: (1) the duty to avoid striking a pipeline; (2) the duty to make reasonable inquiry to locate lines which might be encountered; (3) the duty to ensure that the street or alley right-of-way is clear of obstruction before attempting to perform construction work; and (4) the duty to notify the utility company that one of its pipelines has been struck or damaged. With respect to a litigator's investigation of events occurring considerably prior to the time of the fire, it has been found that the liability of an excavator who struck a gas main while digging a sewer trench six months before the explosion and fire was a jury question.[9]

The status of the excavator will determine the potential liability. Two questions arise in this context: Can the employer be held liable under the doctrine of respondeat superior for the actions of an employee; and, of more difficult resolution, is the excavator an independent contractor. In the second instance, the rule is that the employer is generally not liable for the tortious acts of an independent contractor. This situation commonly arises in cases involving damage from an explosion following actual excavation by an independent contractor. The governing factor in the relationship is the employer's right to control the manner in which the work is to be done by the independent contractor. However, results will vary according to the particular jurisdiction.

The variety of potential defendants is well illustrated in *Brown v. Wisconsin Natural Gas Co.*[10] In that case, an explosion occurred as a result of a leak in a gas line damaged by an excavator employed by a telephone company for the purpose of laying telephone cables. The court apportioned negligence among the telephone company, gas company, contractor and subcontractor based on the following breach of duties: (1) the gas company which had knowledge of the excavation failed to inspect or supervise work of others digging near its pipes; (2) the telephone company failed to adequately reveal gas lines to its excavator, to inspect the work, and to report damage to the gas company; (3) the contractor failed to supervise an inexperienced subcontractor; and (4) both contractor and subcontractor failed to locate gas lines, to inspect lines for damage, and to ensure damage was repaired before refilling the trench.

§2.4.7 Negligent Performance of Services: Inspections

Liability arising out of the negligent performance of services often involves inspections where there is either a failure to recognize a

---

[9] *Kalmn, Inc. v. Empiregas Corp.*, 406 So.2d 276 (La.App. 1981).

[10] 208 N.W.2d 769 (Wis. 1973).

potential hazard or to recommend taking precautions to guard against it. In the fire context, a number of cases have arisen both out of the failure of a municipal or state employee to discover a fire code violation and the failure of a gas company to inspect the condition of its mains or lines and failure to supervise the work of contractors digging near gas mains.

Traditionally, municipal governments and other governmental agencies, when performing official functions, have enjoyed some form of governmental immunity from liability for harms attendant to negligent performance of official duties. In recent years, however, many jurisdictions have abolished or diminished governmental immunity with the enactment of tort claims statutes which permit suits against governmental bodies for negligence. Under these statutes, negligent fire code inspections as well as failures to enforce fire codes may give rise to municipal liability. An illustrative case involving these principles is *Adams v. State*.[11] In *Adams*, a hotel fire occurred causing serious personal injuries and a number of deaths. Suit was brought by the injured hotel patrons and the representatives of the estates of deceased patrons against the State of Alaska and other defendants. By statute, the Alaska Department of Public Safety had jurisdiction to inspect hotels and other buildings and had enforcement authority to seek correction of discovered fire code violations. The case revealed that state fire marshals were aware that ongoing construction at the hotel posed serious fire hazards to patrons and employees. The court also found that four marshals had failed to enforce code provisions giving them the authority to abate the existing hazards despite their knowledge of an inoperative fire alarm. It was held that, by statute, the state assumed the duty to enforce the safety codes. It breached that duty in the manner in which fire code inspections were undertaken. Accordingly, this failure to take affirmative steps to abate the fire hazards was found to impose liability on the state arising out of the negligence of its inspectors. Although liability was predicated upon a failure to take affirmative action to abate discovered violations, it should be noted that the court further indicated that liability would also attach to failure to discover existing violations as well.

With respect to inspections of gas transmission lines or equipment, the courts have generally recognized a duty by gas companies to make

[11] 555 P.2d 235 (Alas. 1976).

reasonable inspections of their equipment so as to avoid leaks and explosions. A determination that gas leaks could have been discovered and repaired with reasonable inspections may impose liability on a gas company. The extent to which courts will apply this principle is illustrated in *Karle v. National Fuel Gas Distribution Corp.*[12] In this case, a gas company was found liable for an explosion and ensuing fire which resulted from leaks in a pipe which had been installed 23 years prior to the explosion. The original installation did not provide proper corrosion protection to the pipe, giving rise to the need for additional inspections. Because the pipe lead had leaked on several other occasions, the gas company was found to have breached its duty to adequately inspect on a regular basis and thereby avert the explosion.

The duty to inspect imposes liability in most any factual circumstances where services or repairs are made to equipment or machinery. The services of oil burner service technicians, alarm repair technicians, sprinkler service companies, electrical inspection agencies, building inspectors, construction superintendents and others can be scrutinized under an inspection theory where the facts suggest that a more comprehensive inspection should have disclosed dangerous conditions which caused or contributed to fire damage. The attorney is well advised to consider all the parties or individuals who may have had an impact on the condition of such equipment and machinery.

§2.4.8   Duties of Warehousers to Protect Goods Against Fire

1. Source of the Warehouser's Duty. The source of a warehouser's duty to protect against the loss of goods by fire while stored in its warehouse emanates from two sources. The first source is the common-law duty which imposes upon the warehouser the obligation to take those precautions which a person of ordinary, reasonable prudence in the same business would take.[13] The warehouser's duty at

---

[12] 448 F.Supp. 753 (W.D.Pa. 1978).

[13] 8 Am.Jur.2d *Bailments* §213 (1980) (It should be noted that the standard of care may vary depending upon if the storage of goods is gratuitous or for the benefit of

common law falls under the "reasonable man" standard employed in judging negligence actions generally.[14] However, the law of bailments, in some respects, alters this duty because it creates certain presumptions against the warehouser and alters the typical burden of production of evidence.

Under general legal principles, a bailment is created when the bailor (owner of the goods) delivers them to a bailee (the warehouser) for some purpose.[15] In the absence of an express contract between the parties, such as a warehouse receipt, an implied contract is created which imposes upon the bailee the duty to redeliver the goods to the bailor in the same condition in which the goods were delivered.[16]

Although the warehouser's liability is created by judging its conduct in light of the "reasonable man" standard, the law of bailments creates a presumption that the bailee (warehouser) is negligent when it cannot redeliver the goods in the condition in which the bailor delivered them.

Unlike the typical burden of production of evidence, the bailee must then come forward with evidence to show that the bailee was *not* negligent.[17] In other words, from this point forward, the warehouser must show that its conduct did not cause the loss or destruction of the goods.

The second source of the warehouser's duty is found in the Uniform Commercial Code (UCC). Article 7 of the UCC (Documents of Title), contains several sections which govern the duty of a warehouser. Article 7 also contains various provisions concerning presentation of claims and the applicable burden of proof.

Section 7-204(1) sets forth the duty of a warehouser.

---

both or either the bailor and bailee.) *See generally, Annot., Sufficiency of Warehousman's Precautions to Protect Goods Against Fire*, 42 A.L.R.3d 908 (1972). *See also*, W. Prosser, *Law of Torts* §32 at 149 (4th ed. 1971).

[14] *Restatement (Second) of Torts* §282 & §283 (1965).

[15] *See generally*, 8 Am.Jur.2d *Bailments* §54 (1980); 8 C.J.S. *Bailments* §15 (1962).

[16] 8 Am.Jur.2d *Bailments* §154 & §156 (1980); 8 C.J.S. *Bailments* §37 (1962).

[17] *Annot., Presumption and Burden of Proof Where Subject of Bailment is Destroyed or Damaged by Fire*, 44 A.L.R.3d 171 (1972); 8 Am.Jur.2d *Bailments* §322 (1980); 8 C.J.S. *Bailments* §50 (1962).

A warehouseman is liable for damages for loss of or injury to the goods caused by his failure to exercise such care in regard to them as a reasonably careful man would exercise under like circumstances but unless otherwise agreed he is not liable for damages which could not have been avoided by the exercise of such care.[18]

Although the UCC imposes upon the warehouser the same duty as the common law, §7-403(1)(b) suggests that the burden of establishing its negligence should be upon the person claiming the goods, or damages for their destruction.[19] In this respect, the UCC does not permit the bailor "the presumption of negligence" available under the common law. From a practical standpoint, the bailor must prove negligence under the UCC, whereas under the common law, the bailee must prove he or she was not negligent.

Section 7 of the UCC also permits the parties to establish "reasonable provisions as to the time and manner of presenting claims and instituting actions based on the bailment...."[20] These provisions must be expressly contained in the warehouse receipt.[21] A warehouser's liability can also be established by a failure to include required terms in the warehouse receipt, but only where such omission is causal of the injury or damage.[22]

2. *The Warehouser's Breach of Duty.* Under the common law, negligence is established by proof of a party's duty and its breach of that duty.[23] In warehouse cases, a breach is established by showing

---

[18] Subsection 4 of section 7-204 permits reference to state statutes which may impose a different degree of responsibility upon the warehouser. Therefore, it is necessary to review the particular jurisdiction's version of the UCC to determine if there are variances with subsection 1 of section 7-204.

[19] Section 7-403(1)(b) contains the following bracketed language "[the burden of establishing negligence in such cases is on the person entitled under the document]." Again, this variance requires that the particular jurisdiction's UCC be consulted to determine upon which party the burden to establish negligence rests.

[20] UCC §7-204(3).

[21] *Id.* Section 7-204(2) sanctions limitations of liability in warehouse receipts and storage agreements.

[22] UCC §7-202(2).

[23] *Restatement (Second) of Torts* §282 (1965); W. Prosser, *Law of Torts* §30 at 143 (4th ed. 1971).

that the actions or inaction of the warehouser failed to comport with the "reasonable man" standard.[24]

The acts or omissions of a warehouser which may establish its negligence are limited only by the imagination. However, the following list suggests the principal areas of concern which need to be addressed in determining if a viable claim exists against a warehouser.

1. Was it negligent for the warehouser to accept the goods, given their characteristics and the condition of the warehouse? (For example, it may be negligent for a warehouser to store automobile tires or other highly flammable materials in an unsprinklered warehouse.)

2. Did the warehouser perform adequate housekeeping? (For example, was the warehouse free from debris, flammable materials or other items capable of combustion?)

3. Did the warehouser provide adequate fire suppression systems? (For example, was the warehouse sprinklered; did it contain operable fire extinguishers; were there sufficient fire extinguishers at the warehouse; was a fire alarm system in place; did the warehouse contain a smoke detection system?)

4. Did the warehouser maintain adequate security to protect against vandals or other intruders who might enter the premises to set a fire? (For example, did the warehouse have adequate security guards or maintain an adequate burglar alarm system?)

5. Was the warehouse structurally deficient? (For example, was the warehouse insulated with flammable cellular plastic materials or other fire-spreading building materials?)

6. Was the property in the warehouse properly stored? (For example, were the goods stored so that a fire would not spread or so that a fire could be extinguished?)

7. Did the warehouser take precautions against the start of a fire? (For example, did the warehouser caution its employees against careless smoking or should the warehouser have prohibited smoking in the warehouse?)

8. Did the warehouser misrepresent that the warehouse was fireproof? (For example, did the warehouser advertise to the public

---

[24] *Id.*

that the warehouse was fireproof, or were such oral representations made to any particular bailor?)

9.   Were the warehouser's employees properly instructed as to the steps to be taken in the event of a fire? (For example, were the employees instructed on the use of fire extinguishers, fire hoses, or other procedures such as blocking off air supply in the event of a fire?)

Although not strictly an issue of the warehouser's liability, the following is important to consider in the warehouse context.

10.   Should the bailors not have stored their goods in this particular warehouse? (For example, should a manufacturer of goods store them in a warehouse which it knows to be inadequate?)

The above list of potential breaches of the warehouser's duty is not meant to be all inclusive. Certainly, in the fire context, the points addressed are those most likely to serve as bases of a warehouser's liability to a bailor.

3.   Local, State, and Federal Regulations. Local or state building codes, various federal regulations, and industry standards, customs and usages may also form the basis of a warehouser's liability. The standard of liability imposed by the use of codes or regulations is known as negligence per se. Under the concept of negligence per se, the claimant must establish both that the warehouser violated the particular code or statutue and that its violation was a cause of the injury or damage.[25]

Even if it cannot be shown that the warehouser is guilty of negligence per se as to the cause or origin of the fire, the doctrine can be useful in establishing an aggravation of damages beyond normal measures. For example, it may be determined that the fire started as a result of vandalism. However, if it can also be shown that the warehouser's improper storage of flammable materials, improper housekeeping, or other code or statute violations permitted the fire to spread in an unusual manner, these acts or omissions will become a substantial factor in the determination of the amount or character of the damages sustained. For example, it may be that the fire was actually started by vandals but that the warehouser's improper

---

[25] *Restatement (Second) of Torts* §286-§288C (1965); W. Prosser, *Law of Torts* §36 at 190 (4th ed. 1971).

storage of flammable materials caused the fire to become a conflagration. Although the warehouser would not be responsible for the cause of the fire, it would, under general negligence principles, or the principles of negligence per se, be responsible for the aggravation of the damages.[26]

§2.4.9   Liability of an Innkeeper to a Guest

1.   Source of the Innkeeper's Duty. An innkeeper may be held liable to any guest for a fire occurring upon the innkeeper's premises under two different theories. The first of these theories is the common-law negligence standard which imposes upon the innkeeper the duty to take reasonable precautions for the safety of guests.[27] This is, of course, similar to that duty imposed upon warehousers. The second source of the innkeeper's duty is local codes or statutes which require notices to the guests of the precautions they should take in the event of a fire, or which impose upon the innkeeper certain duties to procure and maintain firesafety devices or equipment.[28]

2.   Breach of the Duty. The innkeeper is not a guarantor of the guests' safety; however, the innkeeper is responsible for taking reasonable measures to provide for guests' safety. These may include:

1.   The design and construction of these premises as they relate to the spread of the fire or the ability of the structure to retard fire;

2.   The duty to eradicate fire hazards such as the accumulation of debris, proper maintenance of kitchen equipment, and the storage of solvents and other flammable materials;[29]

3.   The posting and lighting of exits;

4.   Instructions concerning leaving the premises and directions on how to leave the premises in the event of a fire;

[26] *See, e.g., Ford v. Jeffries,* 379 A.2d 1211 (Pa. 1977); *Restatement (Second) of Torts* §354 (1965).

[27] *Annot., Liability of Innkeeper to Guest for Injury Due to Fire,* 60 A.L.R.3d 1217 (1974).

[28] Building or maintenance codes may require innkeepers to maintain sprinklered premises, smoke detectors or other fire fighting equipment, for example.

[29] W. Prosser, *Law of Torts* §36 at 200-02 (4th ed. 1971).

5. The guests' access to and the availability of exits to leave the premises;

6. The installation and proper maintenance of fire detection equipment such as smoke and fire alarms;

7. The installation and maintenance of fire suppression equipment such as fire hoses, fire extinguishers, and sprinkler systems;

8. The failure to warn of a fire upon the premises.

3. Codes, Regulations, and Statutes. The violation of a building or maintenance code, regulation, or statute can establish negligence per se. As with the application of negligence per se to other entities (see §2.4.8), the violation of the statute must have a causal connection to the fire or to the aggravation of the damages.[30]

4. Property Damage. In addition to seeking recovery for personal injury sustained in a hotel fire, hotel guests may also seek recovery for damage to their property destroyed in a fire. State statutes limiting the innkeeper's liability for loss or damage to the guests' property come into play here. Many jurisdictions have such statutes which limit an innkeeper's liability for loss or damage to the guests' personal property to a posted, often minor, amount. Full liability may attach, however, if the guests' property is deposited in the hotel's safe; if the innkeeper fails to post a notice which sets out the limitation of liability; or if the innkeeper fails to advise the guests of the option of delivering their goods to the hotel's safe.[31] These statutes either expressly govern the innkeeper's liability or limit the amount of damages available to the hotel guests.[32]

§2.4.10   Duty of Owner or Occupant to Professional Fire Fighter

In typical circumstances, a fire fighter who enters upon the premises of another to fight a fire is considered a licensee to whom the owner of the premises owes a duty, limited to protecting the fire

---

[30] This list is not meant to be exhaustive. The cases are collected at 60 A.L.R.3d 1217, note 27, *supra.*

[31] 40 Am.Jur.2d *Hotels* §153 (1968); *Annot., Statutory Limitations Upon Innkeeper's Liability as Applicable Where Guest's Property is Lost or Damaged Through Innkeeper's Negligence,* 37 A.L.R.3d 1276 (1971).

[32] *Id.*

fighter from wanton or willful acts and injuries.[33] This rule has been applied despite an "invitation" from the property owner to the fire fighters to respond to a fire at the owner's premises.[34] Concomitantly, the property owner has no duty to use reasonable care to make the premises safe for the fire fighter. However, the property owner does have a duty to warn of hidden dangers on the premises when the owner is aware of the danger and has an opportunity to warn the fire fighters.

In those jurisdictions which give fire fighters the status of an invitee, it is the duty of the property owner to exercise reasonable care. In other words, an owner who negligently creates a fire hazard or causes a fire may be held liable to a fire fighter who is injured.[35]

Although this falls below the general standard of reasonable care, it is applied nevertheless under the policy that fire fighters enter upon the premises in the discharge of their duties. Such duties, of course, include facing the hazards associated with fires. In some respects, this doctrine amounts to an "assumption of risk" by the fire fighters.[36]

Whether the violation of a code, ordinance or statute relating to fire or fire hazards may be used to impose liability upon a landowner has been the subject of considerable discussion; the courts are equally divided on this point.[37] Those jurisdictions which impose liability upon the owner reason that the fire fighter has the right to assume that the landowner has complied with all applicable laws. Those courts which take the opposite approach and refuse to use negligence per se as a basis of liability, reason that the statutes, codes and regulations were not enacted for the benefit of protecting fire fighters, but rather, for the benefit of preventing fires. Under this rationale, the doctrine of negligence per se does not apply because fire fighters are not within the scope of those the laws seek to protect.[38]

---

[33] *Annot., Liability of Owner or Occupant of Premises to Fireman Coming Thereon in Discharge of His Duty*, 11 A.L.R.4th 597 (1982). *See also, Annot., Modern Status of Rules Conditioning Landowner's Liability Upon Status of Injured Party as Invitee, Licensee, or Trespasser*, 32 A.L.R.3d 508 (1970).

[34] 11 A.L.R.4th 597, note 33, *supra*.

[35] *Id.*

[36] *Id.*

[37] *Id.*

[38] *Id.*

However, the duty of the landowner to fire fighters has also been extended in other ways. For example, the courts have imposed a duty upon landowners to provide safe access onto their premises for fire fighters.[39] Finally, some courts have imposed liability upon landowners when they are negligent in the creation of risks for the fire fighter after the fire is in progress.[40]

§2.4.11   Duties of Landlords and Tenants

There are three basic sources of the duty which a landlord owes to a tenant and the tenant to the landlord regarding either's liability for damages caused by fire to the landlord's property. These three sources are: the common law; statutes, codes and regulations; and any contracts between the parties, i.e., a lease.

1.   Duty of the Landlord Under Common Law. At common law, the rule of caveat emptor (let the buyer beware) controlled, exculpating the landlord from liability.[41] Recently, however, this rule has been eroded by a number of exceptions, such that in some jurisdictions, the exceptions have become the major premise. This is especially true in the context of residential property.[42] One of the first such exceptions imposed liability on the landlord for latent defects upon the landlord's premises.[43] Under this exception it must typically be shown that the landlord is in a better position to discover the latent defect, and in some cases, that the landlord actually is aware of the

---

[39] *Id.*

[40] *Id.*

[41] *Annot., Modern Status of Landlord's Tort Liability for Injury or Death of Tenant or Third Person Caused by Dangerous Condition of Premises,* 64 A.L.R.3d 339 (1975).

[42] *See, e.g., Green v. Superior Court of San Francisco,* 517 P.2d 1168 (Cal. 1973). Cf., *Webster v. Heim,* 399 N.E.2d 690 (Ill.App. 1980). *Cappaert v. Junker,* 413 So.2d 378 (Miss. 1982). (Landlord's obligation to use reasonable care in maintaining the premises applies only to common areas and not areas leased by individual lessees.)

[43] *Restatement (Second) of Torts* §358 (1965); W. Prosser, *Law of Torts* §63 at 401-02 (4th ed. 1971).

latent defect.[44] For example, the landlord could be held liable for faulty wiring in the walls. The second exception concerns the landlord's knowledge of a tenant's proposed use of the premise.[45] For example, the landlord could be held liable for leasing the premises to a welder, if the premises were not adequately protected to guard against the spread of fire. A third exception has imposed liability upon the landlord for damages or injuries arising in the common areas of the premises maintained by the landlord.[46] A fourth exception imposes liability upon the landlord when the landlord retains physical control of the premises.[47]

2. *Duty of the Tenant Under Common Law.* Under common-law negligence principles, tenants have been found liable where the tenant has control of the premises,[48] where the tenant maintains a nuisance upon the premises,[49] or where the tenant fails to keep the premises in good repair where the tenant has a duty to do so either by contract[50] or where the tenant undertakes to perform that duty absent a legal duty to do so.[51]

3. *Duty Arising from Codes, Statutes, and Regulations.* A landlord or tenant may be found to be negligent per se based upon violations of codes, statutes, or regulations.[52] In the fire context, a frequent basis of liability may be either the landlord's or tenant's failure to maintain building appliances which may give rise to a fire, such as heating units, electrical wiring, and gas or electrically run

---

[44] *Annot., Modern Status of Rule Requiring Actual Knowledge of Latent Defect in Leased Premises as Prerequisite to Landlord's Liability to Tenants Injured Thereby,* 88 A.L.R.2d 586 (1963).

[45] *Annot., What Constitutes a "Public" Use Affecting Landlord's Liability to Tenant's Invitees for Defects in Leased Premises,* 17 A.L.R.3d 872 (1967).

[46] 64 A.L.R.3d 339, note 41, *supra.*

[47] *Id.*

[48] *Annot., Tenant's Obligation Under Lease as Basis of Tort Liability to Third Person,* 44 A.L.R.3d 943 (1972).

[49] *Id.*

[50] *Id.*

[51] *Id. See also, Restatement (Second) of Torts* §323 & §324 (1965).

[52] See note 13, *supra,* §2.3.4.

appliances. Liability may be imposed for inadequate maintenance or for faulty repairs.[53]

4. Duty Based on a Contract (the lease). The most common source of the landlord's or tenant's duties, however, is found in the contract or lease between the parties. There are several common provisions which govern the landlord-tenant relationship in this context.

The first of these are indemnification clauses. Many standard-form lease agreements contain provisions wherein the tenant agrees to indemnify the landlord for the landlord's own negligence.[54] These indemnification provisions, however, have come under strict scrutiny by the courts and the following rules of construction have been applied. First, the courts will strictly construe any indemnification clause, and require that the parties set out with specificity the action or conduct to which indemnification is to be applied. For example, it has often been found that the term "negligence" must be utilized in the document.[55] In addition, the courts often impose a requirement that in viewing the lease as a whole, it must evidence an intent by the parties for one to indemnify the other for the other's own negligence.[56]

Second, a lease may also contain exculpatory provisions by which the landlord, typically, exculpates the landlord from liability to the tenant. While such exculpatory provisions are similar to the indemnification provisions discussed above, the exculpatory clause does not require the tenant to be responsible to third parties for the landlord's negligence because exculpatory clauses control only the rights and liabilities of the parties inter se.[57]

The courts also strictly scrutinize exculpatory provisions.[58] Some states have enacted statutes which prohibit the landlord from requiring the tenant to indemnify the landlord or which prohibit the

---

[53] 64 A.L.R.3d 339, note 41, *supra.*

[54] *Annot., Tenant's Agreement to Indemnify Landlord Against All Claims as Including Losses Resulting from Landlord's Negligence,* 4 A.L.R.4th 798 (1981).

[55] *Id.*

[56] *Id.*

[57] *Id.*

[58] *Id.*

landlord from exculpating the landlord from liability in a residential lease agreement.[59]

A third typical provision found in lease agreements is the "fire clause." Pursuant to the fire clause, either the landlord or tenant specifically assumes the responsibility for damages due to fire.[60] These clauses have generally been upheld as a bargain between private parties and concomitantly, the courts have ruled that they do not violate public policy.[61] Some courts, however, in an effort to alleviate a tenant of the responsibility for a fire, have determined that the clause is inapplicable when the fire results from the landlord's negligence,[62] or willful conduct.[63]

Finally, many leases contain "benefit of insurance" clauses wherein either the landlord or tenant is under a duty to secure insurance or where one party agrees to secure insurance for the benefit of the other. The presence of such a clause in a lease agreement often affects the rights of the parties when they attempt to maintain suits against each other for fire damage.[64] A growing number of jurisdictions have prohibited such suits when the lease contains the benefit of insurance provision.[65]

§2.4.12   Careless Smoking — Employment Relationship

Investigations in a great number of fire cases reveal that careless smoking is a major cause of accidental fires. A negligence theory of

---

[59] *Id.*

[60] *Annot., Validity, Construction, and Effect of Provision of Lease Exempting Landlord or Tenant from Liability on Account of Fire*, 15 A.L.R.3d 786 (1967).

[61] *Id.*

[62] *Id.*

[63] *Id.*

[64] *See generally, General Mills, Inc. v. Goldman*, 184 F.2d 359 (8th Cir. 1950), *cert. denied*, 340 U.S. 947 (1951); *Alaska Insurance Co. v. R.C.A. Alaska Communications*, 623 P.2d 1216 (Alas. 1981); *Gordon v. J. C. Penney Co.*, 86 Cal.Rptr. 604 (Ct.App. 1970); *Woodruff v. Wilson Oil Co.*, 382 N.E.2d 1009 (Ind.App. 1978); and *Sears, Roebuck & Co. v. Poling*, 81 N.W.2d 462 (Iowa 1957).

[65] *Id.*

liability can be developed against a potential defendant, where the circumstantial proof in a cause and origin investigation points to careless smoking. In examining the facts, particular attention should be paid to the employment relationship where careless smoking is indicated in the context of performance of business duties.

The imposition or nonimposition of vicarious liability on employers arising out of careless smoking by their employees presents a number of difficult issues. The courts are less than uniform in their holdings on this point. Some have applied the doctrine of respondeat superior where an employee smokes during the performance of duties for the employer and a fire has resulted. Even though the employee may have been satisfying his or her own personal needs by smoking, liability is found because the employee's negligence occurs during the performance of required duties.

On the other hand, a number of jurisdictions have held that the act of smoking while on the job is not within an employee's scope of employment. The underlying rationale employed by these courts has been that the servant's conduct in smoking on the job constitutes turning away from the master's work and engaging in a purely personal matter, thus negating application of the respondeat superior doctrine.

In a number of other cases, the courts have found that no liability should be imposed upon the master in circumstances where the servant temporarily abandoned the master's work in order to smoke, or where the fire occurred at a time when the employee was off duty or "loaned" and was under the supervision and control of a new employer.[66]

In those cases where smoking takes place under a warehousing or bailment situation, a number of courts have found that careless smoking will impose liability on the warehouser under an enterprise theory, even under circumstances where the employee's conduct may be considered a temporary abandonment of his or her duties at the time of the negligent act. A good example of this rationale is *George v.*

---

[66] *See generally, Annot., Master's Liability for Injury to or Death of Person, or Damage to Property, Resulting from Fire Allegedly Caused by Servant's Smoking,* 20 A.L.R.3d 893 (1968).

Bekins Van & Storage Co.[67] There, the court found that a bailee's enterprise required the presence of employees in the warehouse and, as such, the enterprise of being a custodian of goods involved the inherent risk that smoking by an employee would set fire to the bailed goods. The court reasoned that the bailee's employees, while on the job, act as custodians of the goods and any conduct on their part which creates an unreasonable risk of damage to the goods renders them negligent as custodians. Therefore, careless smoking by a custodian was considered to be a risk arising out of the employment, and circumstantial proof of an employee's careless smoking on the job was a basis for imposing liability on the warehouser.

Another circumstance similar to the bailment situation is where an employee's smoking on the job creates an unreasonable risk of fire because of the nature of the employer's business. For instance, where the employer's business involves the production or handling of highly flammable materials requiring special precautions against smoking, smoking by an employee would amount to performance of the work in a negligent manner for which the employer may be held responsible.

In the employment situation, the employer may seek to avoid liability on the grounds that the employer exercised reasonable care to prevent fires by having extinguishing equipment available and by developing rules and regulations prohibiting smoking. Nevertheless, liability has been imposed where it can be shown that the employer, although having developed reasonable nonsmoking regulations, failed to enforce those regulations and thereby authorized or ratified careless smoking on the premises.[68]

The analysis required to impose liability on an employer for an employee's careless smoking will depend necessarily upon the definitions applied in determining the employee's course and scope of employment as defined in a particular jurisdiction.

---

[67] 205 P.2d 1037 (Cal. 1949).

[68] *See generally*, 20 A.L.R.3d 893, note 66, *supra*.

CHAPTER 3

BASIC FIRE EVIDENCE

§3.1 *Introduction*

Prior sections of this handbook discussed the fire cause and origin investigation, development of facts concerning the spread of the fire, and legal theories upon which liability or defenses are predicated. This chapter discusses the use of evidence in the presentation of a fire liability case.

Evidentiary considerations in fire cases are not materially different from those which exist in other types of negligence or product liability cases. However, one major difference is that proof in fire cases is derived more heavily from two factors: the interrelationship of circumstantial evidence and expert opinions; and the interpretation of private fire codes. This chapter on basic fire evidence does not cover all types of evidence that can be used in a fire case, given that that topic varies with the particular facts of any individual case. Thus, the chapter highlights the use of circumstantial proofs, expert testimony and fire codes as they are relevant to the presentation of proofs in a fire case and as they fit into the overall objectives of proving cause and origin.

It is important to understand at the outset the interrelationships among circumstantial evidence, expert testimony, and testimony concerning fire codes. First, all further the general objectives of proof of cause, origin, and spread. Second, and more specifically, the presentation of evidence in a fire case will typically involve explanation of circumstantial facts by an expert.

§3.2 *Circumstantial Evidence*

Fire cases require the use of circumstantial proofs on the issues of cause and origin. The reasons are simple. First, the actual start of a fire often occurs under circumstances where no direct eyewitness testimony is available to document the occurrence itself. Second, in many cases, clear, direct physical evidence of the cause of the fire may be damaged or destroyed in the fire. Accordingly, proofs must be developed by piecing together numerous isolated facts and circumstances which, when considered together, will support reasonable inferences on the ultimate conclusions of cause and origin. As a practical matter, the evidentiary issues in fire cases do not generally revolve around the admissibility of the circumstantial facts but, rather, the relevancy and sufficiency of those facts offered to establish cause and origin and other issues concerning the fire itself.

One practical hurdle to overcome is the old notion commonly shared by many laypeople (and even some lawyers) that circumstantial evidence is, in some way, inadequate or less probative than direct evidence. In fire cases, the fallacy of this assertion is borne out by practical experience. Experience has shown that a credible chain of circumstantial proofs developed in the fire investigation can result in a very logical and persuasive evidentiary picture. Furthermore, when properly constructed, such a chain cannot be effectively cross-examined. For example, photographs of burn patterns and other relevant areas, coupled with eyewitness testimony concerning early events in the course of the fire, can be highly persuasive before a jury when tied together by expert testimony. The status of circumstantial evidence is borne out in the law; circumstantial proofs are of equal value and should be given equal weight to other evidence offered in a case. *See, Lukon v. Pennsylvania R.R.*, 131 F.2d 327, 329 (3d Cir. 1941); 286 F.2d 306, 312 (9th Cir. 1960); *United States v. Turner*, 528 F.2d 143 (9th Cir. 1975), *cert. denied*, 423 U.S. 996 (1975); Wigmore, *Wigmore on Evidence* §26 (Tillers, rev. ed., 1983).

Most courts have recognized that proof of liability in fire cases is made difficult by the nature of the case itself. As a result, the courts have been quite uniform in recognizing that proof of liability in fire cases can be sustained solely upon the presentation of adequate circumstantial evidence. *See, e.g., Metzer v. Weller,* 130 Pa.Super. 573 (1934); *Miller & Dobrin Furniture Co. v. Camden Fire Ins. Co.,* 55 N.J.Super. 205, 150 A.2d 276 (1959); *Duffy v. National Janitorial Services, Inc.,* 429 Pa. 334, 240 A.2d 532 (1968); *Connelly Containers, Inc. v. Pennsylvania Railroad,* 22 Pa.Super. 4, 292 A.2d 437 (1972).

The *Connelly Containers* case provides a good illustration of how properly developed circumstantial evidence, combined with expert opinion, can be used effectively. In that case, the plaintiff's theory of liability was that the fire was caused by a hot welding stub which had been negligently dropped into a railroad car filled with bags of corn starch. The circumstantial facts were numerous. First, six days before the fire, the boxcar had undergone repairs at a railroad service facility 100 miles away from where the boxcar burst into flames. The repairs in question involved welds made on an extension of the metal running board at one end of the boxcar. After the fire was extinguished, a 2 ½-inch welding rod stub was found inside the same end of the boxcar where the repairs had been made. In addition, the stub was found directly beneath a large heat scar on the inside of the boxcar. Burn patterns within the boxcar itself pointed to the area where the welding rod stub was found. The boxcar had sat undisturbed on the plaintiff's railroad siding for a period of two days prior to the outbreak of the fire. Eyewitnesses placed the fire at its earliest stage in the boxcar.

These circumstantial proofs were supplemented by expert testimony indicating that the slow burning and insulating characteristics of corn starch made it likely that a fire could smolder within the confines of the boxcar for days before bursting into flames. These circumstantial facts were found sufficient to take the issue of the railroad's negligence to a jury.

Circumstantial evidence tending to support each phase of the cause and origin investigation, if presented in a logical and orderly fashion, sustain a burden of proof on the issues of liability in a fire case.

As noted, the issue most often presented in a fire case is not the admissibility of the circumstantial proofs, but their sufficiency in establishing a prima facie case of liability or defense.

The "preponderance of the evidence" test is the usual burden of persuasion in civil cases. A number of courts have considered whether circumstantial proofs meet this test. Generally, if the inferences to be drawn from the circumstantial proofs reasonably support the conclusion which the jury is asked to reach, the proofs will be held sufficient to meet this burden of persuasion. An illustration of this rule is the Pennsylvania case of *Smith v. Bell Telephone Co.*, 397 Pa. 134, 153 A.2d 477, 480 (1959), where the court stated:

> Therefore, where a party who has the burden of proof relies upon circumstantial evidence and inferences reasonably deducible therefrom, such evidence, in order to prevail, must be adequate to establish the conclusion sought and must so preponderate in favor of that conclusion as to outweigh in the mind of the fact finder any other evidence and reasonable inferences therefrom which are inconsistent therewith. The rule has been applied in substance in many cases.

An excellent statement of the rule was also set forth by a New Jersey court in *Miller & Dobrin Furniture Co. v. Camden Fire Ins. Co.*, 55 N.J.Super. 205, 150 A.2d 276 (L.Div 1959), (citations omitted) where the court, in an arson and fraud case, analyzed the sufficiency of circumstantial evidence as follows:

> But the issue, after all, as in other civil cases, is whether the burden of proof has been sustained. "Reasonable probability" is the standard of persuasion, that is to say, evidence in quality sufficient to generate belief that the tendered hypothesis is in all human likelihood the fact. Does the evidence reasonably give rise to a circumstantial inference of the requisite casual relation? Circumstantial or presumptive evidence, as the basis for deductive reasoning in the determination of civil issues, is defined as "a mere preponderance of probabilities and, therefore, a sufficient basis for a decision." It need not have the attribute of certainty, but it must be a presumption well-founded in reason and logic; mere guess or conjecture is not a substitute for legal proof. The determinative inquiry is whether the evidence demonstrates the offered hypothesis as a rational inference, that is to say, a presumption grounded in a preponderance of the probabilities according to the common experience of mankind. The accepted standard of persuasion is that the determination be probably based on truth. A bare quantitative preponderance is not enough. The evidence must be such in quality as to lead a reasonably cautious mind to the given conclusion. The measure of the weight of evidence is "the feeling of probability which it engenders."

These case quotations clearly reveal that circumstantial evidence which engenders the "feeling of probability" will often be the deciding evidence in a fire case. Knowing the likelihood that direct evidence may be unavailable, the trial lawyer in a fire case should build circumstantial pictures which, when enhanced by proper expert testimony, will engender the feeling of probability, making out a prima facie case for jury consideration.

§3.3   Expert Testimony

§3.3.1   Function of Experts in Fire Cases

Expert testimony is essential in a fire case. Depending on the case, the expert may be the cause and origin investigator or other scientific expert. As previously noted, (§1.4) a thorough cause and origin investigation should develop all the essential circumstantial facts which will later be used at trial to support the theory of liability or defense on the issues of cause, origin, and spread. It is the job of the expert to explain the facts to the jury.

The cause and origin expert plays several important roles at trial. Initially, the cause and origin expert who examined the scene of the fire will introduce essential facts — developed from fire-scene photographs and physical evidence obtained from the scene, and offer a narrative description of the fire damage. When presented in factual testimony from the cause and origin expert, the comprehensive scope of a cause and origin investigation lays the proper foundation for the conclusions on cause and origin which the expert will later draw in expressing an opinion to the jury. Because fire cases often present difficult evidentiary issues, the cause and origin investigator/expert also plays an essential educational role in explaining to the jury the logic and methodology of the proper investigative techniques used in determining cause and origin, and ties together the circumstantial facts which lead to his or her conclusion as to the cause and origin, or spread of the fire. Usually, the cause and origin investigator will also be the first witness who will shed light on the ultimate conclusions which the jury will be asked to reach.

Second, the testimony of the cause and origin expert is also the essential predicate to the presentation of other scientific expert testimony. The expert's testimony establishes the reliability of foundation facts upon which scientific or other testimony is based. For example, if properly trained, the cause and origin expert will be able to establish the chain of custody of debris samples or instrumentalities removed from the scene which later may be the subject of scientific or laboratory examination and more narrowly focused testimony by other experts. In addition to affirmative proofs, the cause and origin expert may be required to present evidence that negates certain theories of cause and origin. Proof that eliminates one or more of these sources may prove critical in establishing negligence or the absence of negligence. Although the expert may not be able to state to a reasonable degree of scientific certainty that any particular instrumentality caused the fire, the expert may be able to eliminate all other causes which circumstantially infers that the cause not eliminated was the cause (substantial factor) of the fire. *See* discussion of res ipsa loquitor in §3.3.6, *infra*. For example, the circumstantial evidence, as developed, may suggest several ignition sources for the fire.

After presenting expert testimony on cause and origin, scientific experts may then be used to confirm the cause and origin opinion using general principles of fire science or by specific reference to chemicals or electrical equipment found in the area of origin.

In many cases, it may also be necessary to offer scientific testimony to explain the physics of combustion; the characteristics of fire behavior under certain controlled conditions; the combustion characteristics of materials or liquids found at the scene; the functioning of electric circuitry; the design of suppression equipment; or the meaning and application of codes and standards. Experience in litigating fire cases teaches that in all but the simplest of factual cases, not only will a cause and origin expert be needed, but also an additional scientific expert will be necessary in order to present convincing proofs to the jury.

In addition to the investigative and testimonial function, experts in fire cases serve an educational function for counsel. This, of course, is especially true where the trial attorney lacks familiarity with fire investigation, practices, personnel, equipment, or scientific principles. Experts can assist counsel to: (1) identify areas for investigation and research; (2) formulate theories of liability or defense; (3) describe theoretical weaknesses and strengths in an opponent's case; (4) assist in preparing cross-examination questions; (5) perform experiments; and (6) coordinate the overall factual and scientific presentation to be made to the jury.

In order to properly present factual and scientific testimony before a jury, trial counsel must be conversant with the areas of expertise and the scientific principles which will be advanced by experts on both sides of the case. Knowledgeable experts can recommend background reading material to counsel for educational purposes and identify authoritative treatises which may be useful in cross-examination of opposing experts.

§3.3.2   Selection of the Expert Witness

Because expert testimony is essential in most fire cases, selecting the proper experts for investigation, consultation and trial is an extremely important task of trial counsel. Obviously, the selection of experts is always a balancing process that involves weighing many factors which may be ultimately involved in the trial of the case. Experts in fire cases can be utilized for many other purposes aside from trial testimony and fire investigation. Accordingly, counsel should have in mind the general purposes for which each expert will be retained.

An initial determination should be made with respect to the function of each expert. First, most jurisdictions allow broader discovery with regard to persons who are expected to be called as trial experts than with those retained soley for consultation. *Compare, Fed.R.Civ.P.* 26(b)(4)(A) (Trial Experts) with *Fed.R.Civ.P.* 26(b)(4)(B) (Non-Trial Experts). In cases with highly theoretical or technical issues, counsel may wish to consider engaging a consulting expert solely for the purpose of providing ideas, scientific background or information concerning the development of theories or for an analysis of opposing theories. In such circumstances, the main criterion for selecting the expert will be substantive skill rather than the numerous other factors which must be considered for a testifying witness. A determination should be made at the outset if the expert will be engaged to express opinions in a broad general area such as general principles of fire investigation, of physics or chemistry, or in a narrower and more specific field. Consideration should be given to the expert's availability to consult with counsel during the course of the litigation as discovery proceeds.

Substantive expertise is of primary importance with respect to fire cause and origin investigators. This expertise may be evidenced by academic credentials, degrees or seminar attendance. However, the most important factor is practical experience. This is derived from conducting investigations, making observations at fire scenes, interpreting circumstantial facts, and documenting those facts. In this regard, it is also important to examine the proposed expert witness' prior testimony in judicial forums, depositions, or administrative proceedings. Such examinations will reveal whether the expert's testimony in similar areas is consistent or in conflict with the case at hand, and if in conflict, if that result can be rationalized by the circumstances of the other cases.

Other professional achievements of the fire investigation expert should be analyzed. Written reports or published articles concerning investigative techniques or procedures, participation in panel presentations, and service on firesafety committees are all relevant here and should be examined. Additionally, an examination should be made to determine if the expert has received honors for professional achievement. All of these background facts will be necessary to determine not only the general qualifications of the expert witness, but whether the expert will be suitable to give testimony in a particular case.

For credibility, the local fire marshal or fire investigator, serving in an official capacity and possessing the necessary expertise, should be utilized to the fullest extent permissible. Juries generally give great credence to the independent official investigation of a well-qualified fire marshal. Furthermore, if a well-qualified local fire marshal's opinion is consistent with that of an independent expert engaged by either side of the case, this can often serve to tip the scales where competing, independent experts tend to cancel out each other's testimony.

Finally, there are additional criteria to consider when experts actually testify at trial. These include speaking ability, communication skills, objectivity, and respect for the court. These should all be critically evaluated in addition to the substantive expertise of the expert. An ability to assemble facts and to tie them together into a logical and persuasive description of the most probable cause of a fire is a critical asset for a testifying fire expert.

Needless to say, there are numerous other general criteria which should be analyzed in engaging any expert witness, whether related to fire cases or not. In some circumstances, experts engage in self-puffery, advertising degrees and experience which have not actually been obtained, or professing expertise in areas where such expertise truly does not exist.

The limits to the use of the expert at trial should also be carefully evaluated. First, it is important to have the expert limit testimony to those areas in which he or she is truly an expert. Although a fire cause and origin investigator may be extremely well qualified to express opinions as to the cause and origin of a fire through examination of the fire scene, he or she may be unqualified to express opinions concerning electrical phenomena, chemical reactions, product assembly, product design, and other areas. Because credibility is an all-important issue in jury cases, it is important in this regard to understand the inherent limitations of a fire cause and origin expert. Often it is necessary to supplement the expert's testimony with qualified testimony of other specialized experts. Such individuals can fill in scientific gaps which may be lacking in the fire investigator's background. There will always be thorough cross-examination concerning the extent of the investigation itself and there is no need to create additional problems by attempting to have a cause and origin expert testify beyond the area of the individual's true expertise.

In summary, the most important general considerations in selecting an expert will be his or her substantive expertise, communication skills, availability, objectivity, honesty, and sincerity. These general criteria must be evaluated with any expert.

§3.3.3 Establishing Competence and Qualifications of Experts

Most courts have recognized that opinion evidence as to the cause and origin of a fire is admissible when given by a witness who is properly qualified to give opinions on matters not within the common knowledge or expertise of ordinary people. In general, to be competent to testify, fire experts must be qualified by learning, training or experience in the relevant fire field. Accordingly, a person who is experienced in determining the cause and origin of fires can be found competent to testify and give an opinion on that subject. *See, Annot., Evidence as to Cause or Origin of Fire,* 88 A.L.R.2d 230(1978).

Generally, the question of who is competent to testify as an expert in a fire case depends upon the particular type of fire in question. However, it is important to note that a witness who is found to have had special training, knowledge or experience in determining cause and origin will usually be found competent to testify as an expert whether his or her job description be fire investigator, fire fighter, police officer, arson investigator, or similar titles. Furthermore, experience has shown that many highly qualified fire investigators do not gain their experience solely in the classroom. Many qualified fire investigators have obtained their expertise in the field through many years of fighting fires as members of fire departments. Attendance at special programs, courses and seminars which are particularly tailored to fire science and fire investigation on a practical level is also relevant. Many jurisdictions require local licensing, formal education, or continuing education for official or private fire investigators. Finally, many fire investigators, in addition to practical hands-on experience, are members of professional organizations and societies which conduct practical seminars and other educational programs to update and refine expertise at different investigative levels.

The question of how much expertise or experience and training is required before a fire scene investigator will be qualified to testify as an expert is generally within the discretion of the trial court, and depends upon the evidence and information offered to qualify the witness. There are no definitive standards concerning minimum education, training, or experience required to qualify a cause and origin investigator to testify as an expert. In considering a potential cause and origin expert, however, it is prudent to look for witnesses who either are or were formerly members of official fire investigative agencies. Such individuals often are designated as fire marshals or arson investigators by major city fire departments. As individuals involved in fire investigation on a daily basis these persons are normally required to attend courses and seminars on basic fire science, fire behavior, identification of electrical faults and causes, behavior of liquids and gases, identification of multiple origins, identification of liquid burn patterns, and other relevant subjects. This background training, coupled with firsthand experience in fire investigation, will generally serve to qualify a cause and origin expert to offer opinions concerning the investigation and conclusions as to cause and origin.

In those cases where credentials are at issue, most courts will permit an individual to testify as an expert, leaving the issue of the witness' credibility or expertise to the jury.

Of course, not all fires take place in major cities. Often small towns do not have the budgets or programs in fire investigative training that are offered in large cities or in state programs. The qualifications of fire department personnel in small local or volunteer fire companies can become an issue where official investigators serve only in a part-time capacity and have little or no actual experience in investigation. When doubt exists as to the expert's qualifications, or background of fire personnel, prudent counsel will engage a supplementary fire expert with demonstrated experience. In circumstances where a court may not permit an experienced fire fighter to testify as to his or her opinion on cause and origin in the specific case, it should be remembered that trained fire fighters *may* be permitted to testify to factual conclusions based on experience in observing numerous fires. For example, the type, color and density of smoke, the presence or absence of certain types of fumes such as gasoline or other petroleum-based products, as well as the areas of dense flames are all within the knowledge and expertise of the average fire fighter. The fire fighter who has no special training, however, relative to determining the cause and origin of a fire would not be permitted to testify as to the fire's cause and origin unless he or she saw the fire start. This type of testimony will often be as useful and persuasive as that directly related to the fire at issue.

§3.3.4 Foundation Evidence

In order for any expert to express opinions in a fire case, there must exist an adequate factual foundation for the opinions. No matter how properly qualified, an expert will not be permitted to guess or speculate but must give opinions limited to facts actually observed, or based upon facts which are or will be introduced into the record at trial. The terminology of Rule 703 of the Federal Rules of Evidence essentially describes the facts or data upon which an expert may express an opinion as "those perceived by or made known to him at or before the hearing." *Fed.R.Evid.* 703. The Federal Rule is more liberal than rules in some jurisdictions in that an expert in federal court may express opinions based upon facts or data of a kind reasonably relied upon by experts in a particular field, even if they may not necessarily be otherwise admissible.

The proper foundation for an opinion to be expressed by the cause and origin fire investigator involves several steps. The first is a description of prefire conditions. This should be developed through the use of witnesses, photographs or other data. Secondly, a description should be made of the investigative efforts of the fire cause and origin investigator from the beginning through the conclusion of the investigation. Here an effective method is to have the investigator go step-by-step through each procedure of the on-site inspection of the fire scene. During this testimony, the investigator can describe the general condition and configuration of the building and document his or her observations with photographs or physical evidence obtained at the fire scene. Thereafter, a full description of the expert's observations; description of burn patterns; identification and examination of heat or ignition-producing sources; materials; fire load and other matters can be completely and comprehensively described. As each relevant area is discussed from a fire-science perspective, its relationship and significance to the ultimate conclusion can be explained to the jury. These explanations should include an evaluation of each potential ignition source and its assignment or elimination as the potential cause of the fire. Each visit to the fire scene should be explained, and detailed testimony should include the circumstances under which photographs were taken and under which physical evidence was removed from the scene.

Documentation of foundation evidence in fire cases must be done carefully and systematically. In detailing what is shown in the photographs, the fire investigator should identify each photograph as designating a specific area in the building, describe the scope of each photo, and delineate its significance in denoting burn patterns or other indicia which can be used to determine cause and origin under recognized principles of fire investigation. Each item of physical evidence ultimately removed from the fire scene should be documented through the testimony of the fire investigator by indicating the date, time and locations from which physical evidence or samples were taken. Thereafter, testimony can be offered to show the chain of custody of the physical evidence throughout the time it was removed from the scene, examined by other experts and ultimately presented in the courtroom. It is not sufficient merely to produce the evidence in court. In order to be admissible, it must be shown that the object removed remained in substantially the same condition as it was at the time of the fire or immediately thereafter. An adequate "chain of custody" is a predicate for the testimony of additional experts who have examined the ignition instrumentality or performed laboratory analyses of samples of debris removed from the scene.

Foundation evidence establishing the need for other experts must be offered by the on-scene fire investigator. His or her testimony will establish the need for other experts such as electrical engineers, chemical engineers, metallurgists, materials scientists or structural engineers to complete the cause and origin picture. For example, the testimony of a fire cause and origin expert may lay the factual foundations of the burn patterns and debris analysis in pinpointing origin of the fire to a particular area where electrical wires or an electrical apparatus was found. Other foundation evidence documented through photographs may show short circuiting or beading on electrical wires in the same area. The establishment of these foundation facts, however, will be the predicate to further examination and testimony by an electrical engineer whose testimony may be confined to such issues as: the size of the wires; the reason for the short circuit or overheating; whether the wire was heated and suffered damage as the result of external heat or fire or whether the damage occurred as the result of an internal short circuit or defect in the wiring itself. In a like fashion, areas similarly isolated as points of origin may disclose the existence of chemicals or other substances which might be subject to spontaneous combustion or may suggest the existence of an incendiary fire, depending upon their chemical characteristics.

The absence of complete and adequate foundation evidence can be fatal as it can serve to exclude the logical and persuasive expert opinion evidence upon which many fire cases are based. An illustration of this is the case of *LaClaire v. Silverline Manufacturing Co., Inc.*, 393 N.E.2d 867 (Mass. 1979). In *LaClaire*, a worker sustained fatal injuries after an explosion and fire at the employer's factory. The employer was a defendant in the action but was found not liable based upon workers' compensation principles. The defendant was a manufacturer of aluminum powder which was used at the factory. Plaintiff's theory of liability maintained that a spark resulting from static electricity ignited aluminum dust in suspension in the air located between decedent and the floor. Testimony from a chemical expert established: first, that in a dry condition, aluminum powder dispersed in the air could create an explosive atmosphere; and second, that pouring aluminum powder from paper bags could produce an electrostatic charge sufficient to cause the powder to explode.

The plaintiff's proffered expert testimony from the state fire marshal that the explosion and fire was caused by ignition of the aluminum dust suspended in the air, was precluded by the trial judge on the ground that adequate foundation evidence had not been introduced to establish the existence of aluminum powder in suspension prior to the explosion. Unfortunately, the expert was merely asked to express opinions as to whether, under the circumstances, a dust explosion could occur in a mixture of aluminum dust and air. At that time the record was barren of any evidence, establishing that a cloud of aluminum had existed prior to the explosion. At a later point in the trial, testimony from a coworker of the plaintiff established the necessary predicate facts in the existance of the aluminum "cloud." A further attempt to recall the expert to express opinions was not pursued by counsel. In affirming a judgment in favor of the defendant, the court offered the following conclusions concerning the foundation testimony necessary to sustain an expert's opinion:

> While it can be observed that LaClaire's co-worker, Richard Smith, later testified to viewing aluminum dust prior to the explosion, the plaintiff's attorney never sought to use this statement as a foundation to

re-examine [the expert] regarding his opinion as to the accident's causation. Hence, when the evidence closed, the record was barren of any proof supporting the proposition that [defendant manufacturer's] powder was a cause of the explosion.

The problem illustrated by the *LaClaire* case is one which often occurs where the expert is asked to express an opinion based upon a hypothetical question. Prior to asking the hypothetical question, counsel must insure that adequate foundation evidence exists from which the expert can draw the required conclusion. This problem can also occur in cases where the expert has, in fact, performed the factual investigation personally. Accordingly, a careful review of all of the circumstantial facts is necessary in order to provide the necessary foundation for the opinion. Absent adequate foundation, an expert's opinion will be reduced to sheer guess or conjecture and, accordingly, will be excluded by the court. *See, e.g., Hahn v. Eastern Illinois Office Equipment Co.*, 355 N.E. 2d 336 (Ill. App. 1976).

Considering the prevalence of circumstantial proofs in a fire case, the relevance and significance of each circumstantial fact should be adequately and completely evaluated and presented into evidence in order to lay the proper foundation for the expression of expert opinions both by fire cause and origin investigators and by scientific experts in a narrower field.

§3.3.5 Effective Expert Opinion

Given the proper testimony and adequate foundation evidence, most courts permit expert testimony as to the cause and origin of a fire. The Federal Rules of Evidence, in Rule 704, authorize testimony in the form of an opinion, or an inference to be drawn therefrom, even where that opinion embraces the ultimate issue to be decided by the trier of fact. During the course of a fire liability case, expert opinions may be necessary on multiple issues, all of which lead to the ultimate conclusion or "ultimate issue" to be decided in the case.

Effective expert opinion testimony in a fire case is that which not only correctly identifies the area of origin and most probable cause of the fire, but also effectively rules out each and every alternative potential cause in that area. This, of course, is the essence of fire investigative techniques of cause and origin. In fact, the methodology of this investigative technique often depends almost entirely upon the analysis of circumstantial evidence and elimination of possible causes to reach a conclusion that given conduct or a specific instrumentality was the most probable cause of the fire.

The technical portions of this handbook (Chapters 6-11) discuss the elements necessary to sustain combustion, i.e., fuel, oxygen and heat, which comprise the fire triangle. An expert's analysis of circumstantial facts and resultant opinions will generally relate these elements of combustion to the particular facts found at the scene of the fire. In turn, the fire scene investigation will require a meticulous examination of ignition sources present in the area of origin for the purpose of eliminating them or including them as potential causes of the fire. This methodology will then be incorporated as a part of the cause and origin expert opinion.

The case of *Gichner v. Antonio Troiano Tile & Marble Co.*, 410 F.2d 238 (D.C. Cir. 1969), is an excellent illustration of such cause and origin expert opinion testimony based upon an analysis of circumstantial facts and elimination of other potential causes.

In *Gichner*, the plaintiff landlords owned a warehouse which they leased to the corporate defendant. On the day in question, defendants' employees and others entered the warehouse at 3:00 a.m. after a drinking spree and left the warehouse within an hour. A fire was reported at the warehouse at 5:35 a.m. One of the plaintiff's theories of liability was that the fire was caused by the careless smoking of the defendant tenant's employees. The principal expert witness at trial was the District of Columbia Fire Department investigator. After eliminating all other possible causes of the fire (such as faulty electrical equipment, heating equipment, etc.), he concluded that the fire was caused by careless smoking. This conclusion was reached although there was no direct physical evidence of smoking, and in the face of denials by defendant's employees that they were smoking in the warehouse prior to the time of the fire.

The trial judge, sitting without a jury, granted judgment for the defendants on the ground that the evidence, while adequate for fire department purposes of determining cause and origin, was insufficient to justify inferring that careless smoking was a cause of the fire. The trial court reached this conclusion even though the investigator had eliminated every other possible cause in reaching a conclusion. On appeal, the D.C. Court of Appeals reversed, finding that the testimony of the fire investigator was sufficient to establish a jury question on the issue of careless smoking.

The fire investigator who testified was a lieutenant employed by the fire department for 26 years. He had vast experience in examining fire scenes to determine cause and origin. Within an hour of the first alarm he had entered the warehouse and begun his investigation into the cause of the fire. A detailed examination of char patterns on the rafter beams pinpointed the area of origin of the fire inside a door located in the defendant's part of the warehouse. After examination of all available sources of ignition in the area, the fire investigator concluded by the process of elimination that careless smoking was the cause of the fire. The building and its contents did not include electrical equipment, heating equipment, spark-producing devices, or other ignition-producing devices. The testimony also eliminated vandalism as a cause of the fire; there was no evidence of forceable entry into the building. Based upon this circumstantial evaluation, the court of appeals, in expressing its opinion on the sufficiency of the expert opinion testimony and investigation of the fire investigator, stated as follows:

> Further, we think that where a fire investigator identifies the cause of a fire in terms of probability, as opposed to mere possibilities, by eliminating all potential causes of the fire but one, that testimony is not only relevant, but in some circumstances may be a basis for decision.
>
> Bearing on the credibility of such testimony is the thoroughness with which the expert identifies all the potential causes and the soundness of his reasoning in eliminating each one.
>
> We recognize that the matter of prevention, detection and control of fires is a major governmental concern. The men trained in this science and exposed to experience to the problem are called upon to execute their jobs with exactitude. We also recognize that when smoking causes a fire it is highly likely that the offending cigarette butt or ash will dissolve in smoke in the ensuing conflagration. Thus direct evidence of smoking will be hard for the fire investigator to find, and they will have to form their opinions by way of exclusion of other possibilities.
>
> Often the plaintiff in a fire case, at least where smoking is the cause,

can hardly be expected to do more than to present expert testimony of this sort as to the fire's cause, and support that testimony by other, direct evidence of smoking on the premises shortly before a fire began. His burden might well be impossible if his experts also had to discover some direct evidence of a cigarette butt or ash; and placing such a burden upon him would be unnecessarily harsh in view of the reliability of expert testimony even when arrived at by a process of elimination.

410 F.2d at 247.

A second excellent illustration of the weight given the expertise of a fire investigator is *Cullen v. Archibald Plumbing & Heating Co.*, 555 S.W.2d 676 (Mo. App. 1977). In this case, the investigator's experience, evaluation of circumstantial evidence and inferences drawn therefrom were most persuasive. At issue was the cause of an explosion allegedly due to the negligence of a painting contractor in mishandling lacquer and lacquer thinner during construction of an apartment complex. Expert testimony of three experienced fire fighters was offered as to the cause of the explosion. The sufficiency of the fire fighter's expert testimony was directly at issue. In affirming judgment in favor of the plaintiff, the court of appeals made the following statement concerning the testimony of the fire experts:

> [The defendant] contends that the testimony of Clark, Yackie and Hyder [the three testifying firemen] as to the cause of the explosion and fire should have been excluded because they did not have sufficient facts or other competent evidence to support their opinions. Each of the witnesses were and have been long time firemen (27 to 28 ½ years) with the St. Joseph, Missouri Fire Department. Each had extensive university training, self-study, and on-the-job training, designed to determine the cause and origin of fires. Clark testified that when he arrived at the scene, the north wall on the west side of the building was blown out with the force hitting the bottom part of the wall. There was lacquer odor in all parts of the building, and he knew that lacquer thinner and vapors, which are heavier than air and which settle to the lowest point, would explode very violently when properly concentrated with air and ignited. He did not smell natural gas which will rise in the air.... At the time he was there, the windows of the building were closed. With respect to causation, Clark testified: "This explosion in

the area of the east side apartments was a low-type explosion, because it blew the bottom part of the wall out. Now a gas explosion could possibly blow a wall from the top and lay it out this way, but this wall was kicked out from the bottom. So it had to be vapor fumes that were heavier than air and settled to the lowest point .... Lacquer thinner fumes will settle at the lowest point. Q. Did you find evidence of such an explosion in the northwest corner of the apartment you fought the fire in? A. Yes, judging from the way the wall was blown out into the courtyard."

Yackie's opinion was of similar import, except that he added, "Flammable liquids are almost always followed by a heavy fire. A natural gas explosion 99 out of 100 times will blow itself out." The fire did not blow itself out, but continued as a heavy fire. He eliminated electricity because it was off, and natural gas because he saw no damage to the gas pipe, no smell of gas, or a fire from leaking gas.

Hyder directed the fire fighting activities. He smelled a paint or lacquer odor — very strong vapors — on both sides of the building, and all the windows were closed. The explosion was localized into one area of the building, which was not compatible with explosions he had investigated which occurred as the result of an accumulation of natural gas. Based upon this personal observation at the scene, his knowledge he obtained in fighting the fire, his training, background and experience, his opinion was that the explosion was a result of ignition of a mixture of air and vapors from Class I flammable liquids of which lacquer thinner is one.

Obviously each of these ranking firemen gave their expert opinions based upon personal observation, knowledge and long experience, and the court did not err in receiving their testimony. (citations omitted).

The *Cullen* case clearly highlights the worth of fire marshals' testimony based on knowledge and first-hand experience. It should also be noted in this case that the factual testimony and opinion served as a foundation for the testimony of other experts, specifically chemists, who testified as to the physical properties of lacquer and lacquer thinners. The testimony of the experienced fire fighters, however, was obviously critical to the theory of liability and served as a predicate for the testimony of other experts. The integration of other experts' testimony with fire investigative technique and testimony provided an ample and clear picture from which a jury could evaluate and draw reasonable conclusions as to the cause of the fire.

§3.3.6   Applicability of the Doctrine of Res Ipsa Loquitur

One further topic must be considered to complete the discussion of expert testimony in fire cases. This is the doctrine of res ipsa loquitur. This is essentially an evidentiary doctrine, based on exclusive control of the casual agent, that establishes a presumption of negligence by the controlling party. The following elements must be met:

1.   The event is of a kind which ordinarily does not occur in the absence of negligence;

2.   Other responsible causes, including the conduct of the plaintiff and third persons, are sufficiently eliminated by the evidence;

3.   The indicated negligence is within the scope of the defendant's duty to the plaintiff; and

4.   The defendant was in exclusive control of the instrumentality which caused the fire.[1]

In fire cases, the use of res ipsa is often warranted when the exact cause of the fire cannot be determined, but all other nonnegligent causes can be eliminated.[2] The doctrine of res ipsa has been applied under the following factual circumstances: smoking;[3] blow torches and welding equipment;[4] the use of flammable materials;[5] gas leaks;[6] storage of flammable materials;[7] appliances and machinery;[8] heating equipment;[9] and electrical wiring.[10]

---

[1] *Restatement (Second) or Torts* §328D (1965).

[2] *Annot., Res Ipsa Loquitur as to Cause of or Liability for Real-Property Fires*, 21 A.L.R.4th 929 (1983)

[3] *Olswanger v. Funk*, 470 S.W.2d 13 (Tenn.App. 1970).

[4] *Taylor v. ROA Motors, Inc.*, 152 S.E.2d 631 (Ga.App. 1966).

[5] *Megee v. Reed*, 482 S.W.2d 832 (Ark. 1972); *Arrington v. Hearin Tank Lines, Inc.*, 80 So.2d 167 (La.App. 1955).

[6] *Wichita City Lines, Inc. v. Packett*, 295 S.W.2d 894 (Tex. 1956).

[7] *Standard Oil Co. v. Midgett*, 116 F.2d 562 (4th Cir. 1941).

[8] *Maryland Casualty Co. v. Saunders*, 394 So.2d 1256 (La. 1981); *Granata v. Schaefer's Bake Shop, Inc.*, 232 A.2d 513 (Conn. Cir. 1967).

[9] *Roselip v. Raisch*, 166 P.2d 340 (Cal.App. 1966).

[10] *Evans v. VanKleek*, 314 N.W.2d 486 (Mich.App. 1981).

The key to the application of res ipsa loquitur may depend upon the testimony of an expert that it was probable that the fire started by the defendant's negligence, including the expert's elimination of all other potential fire sources.[11] It is not sufficient, however, if the expert testifies that it was possible, rather than probable, that the plaintiff's negligence caused the fire.[12]

If res ipsa loquitur is applied, a presumption arises that the fire occurred as a result of the defendant's negligence. However, this is a rebuttable presumption and the defendant may introduce evidence to counter the presumption.[13] For example, the defendant may introduce expert testimony which conflicts with that of the plaintiff. The defendant may also present evidence that it did not have exclusive control of the fire-causing instrumentality, i.e., a landlord may produce evidence that tenants had access to certain areas of a building where the fire occurred. This would negate the inference of control, and could preclude the use of the doctrine of res ipsa loquitur.

---

[11] *See, e.g., Lanza v. Poretti*, 537 F.Supp. 777 (E.D.Pa. 1982), *aff'd, 696* F.2d 983 (3d Cir. 1983); *Gichner v. Antonio Troiano Tile & Marble Co.*, 410 F.2d 238 (D.C.Cir. 1969).

[12] *Lanza, supra*, note 11. *See also*, Federal Rule of Evidence 705; *Breider v. Sears, Roebuck & Co.*, 722 F.2d 1134 (3d Cir. 1983).

[13] *See generally*, 21 A.L.R.4th 929, note 2, *supra*.

CHAPTER 4

PRODUCTS LIABILITY

§4.1 Introduction

In the fire liability context, product defect cases fall into several categories. First, and most obvious, are those cases where the product is the cause or ignition of a fire. Manufacturing or design defect cases in this area usually involve an analysis of energy-producing appliances or equipment which can produce ignition. Wiring defects, improper wiring connections, improper sizing of wires, the absence of or malfunction of necessary fusing devices, and the inadequate selection of materials or improper quality control testing are common concerns in this realm.

A second category of manufacturing defect cases involves failure in the operation of fire suppression systems and alarm equipment specifically designed to operate in fire conditions. In these cases, the product failure is not directly related to the cause of the fire but may be causally related to the extent of property damage or personal injury.

For example, many buildings are required to have properly operating sprinkler systems or other types of suppression devices. These kinds of equipment utilize mechanical and electronic devices to trigger a response and to relay alarms. There are also specific types of suppression systems designed to protect commercial cooking equipment which are also operated by mechanical spring mechanisms and triggering devices. If not properly manufactured and designed, these can result in failure and fire spread.

A third area of liability under product liability theory involves the design of fire suppression equipment itself. For example, many pieces of electrical equipment should be equipped with fail-safe devices (such as thermal fuses or limiters) or other types of shut-off devices designed to prevent overheating. In a given case involving sprinkler design and fire suppression, location and coverage may be as important as proper functioning.

Finally, an often-litigated area involves the absence or adequacy of instructions and warnings associated with hazardous products. These are commonly known as "failure to warn" cases. For example, if not properly ventilated or dispersed, flammable vapors can create serious fire hazards when subject to sources of ignition such as sparks from electric light switches or heater pilot lights. Here, the adequacy of instructions and warnings are central to numerous product defect rulings.

In addition to strict liability, fire cases often involve alternative or overlapping theories under negligence, warranty and contract law. Although modern procedure allows pleading under all of these theories, the pleadings must still be tailored to meet the specific requirements of each theory as dictated by the facts and the law in the particular jurisdiction in which the case is brought.

§4.2   General Theories of Products Liability

§4.2.1   Strict Liability in Tort (*Restatement (Second) of Torts* §402A)

In 1965, the American Law Institute adopted the doctrine of strict products liability for inclusion in the *Restatement (Second) of Torts*. Although it is the general goal of the Institute to "restate" existing law, §402A represented a departure from existing law at the time of its adoption. Since then, however, §402A or variations of the theme of strict liability have been adopted in a majority of jurisdictions. Entitled "Special Liability of Seller of Product for Physical Harm to User or Consumer," section 402A provides:

> (1)   One who sells any product in a defective condition unreasonably dangerous to the user or consumer or to his property is subject to liability for physical harm thereby caused to the ultimate user or consumer, or to his property, if
> > (a)   the seller is engaged in the business of selling such a product, and

(b)   it is expected to and does reach the consumer without substantial change in the condition in which it is sold.
(2)   The rule stated in subsection (1) applies although
    (a)   the seller has exercised all possible care in the preparation and sale of his product, and
    (b)   the user or consumer has not bought the product from or entered into any contractual relation with the seller.

Based upon §402A of the *Restatement*, it is necessary for the claimant to establish five elements in order to recover: (1) that the product was in a defective condition; (2) that the product was unreasonably dangerous[1]; (3) that physical harm was caused to persons or property[2]; (4) that the defendant was in the business of selling the product; and (5) that the product reached the consumer without substantial change from the condition it was in when it left the hands of the seller. The *Restatement* imposes strict liability upon the product seller by stating in subsection (2) of §402A that the exercise of reasonable care by the seller will not exculpate the seller from liability. Finally, §402A is distinguished from warranty law by subsection (2) which states that privity of contract is not required between the seller and the injured party.

Although §402A represents the basic outline of strict products liability theory, a number of variations have arisen based upon this general scheme. In some jurisdictions, a "risk/utility" analysis is

---

[1] The unreasonably dangerous requirement has been the source of continued debate. Some jurisdictions have abolished this element of a §402A claim, on the basis that it injects negligence principles into the strict liability concept. *See, e.g., Greenman v. Yuba Power Products, Inc.*, 377 P.2d 897 (Cal. 1963); *Berkebile v. Brantley Helicopter Corp.*, 337 A.2d 893 (Pa. 1975); *See generally*, J. Beasley, *Products Liability and the Unreasonably Dangerous Requirement*, Chap. 6, "'Green' Jurisdictions 'Unreasonably Dangerous' Inapplicable)" (1981).

[2] There has also been disagreement among the courts as to whether physical damage to the product itself is compensable under the strict liability doctrine. Some courts classify this type of damage as "economic loss," while other jurisdictions permit recovery on the basis that the type of damage sustained as a result of a product defect, which is hazardous, should not govern the application of strict liability principles. *See generally*, C. Fallon, *Physical Injury and Economic Loss—The Fine Line of Distinction Made Clearer*, 27 Vill. L. Rev. 483 (1982).

utilized whereby the risk posed by the defect is balanced against the utility of the product.[3] The factors to be evaluated in a strict products liability action under the "risk/utility" analysis have been identified as follows:

(1) The usefulness and desirability of the product—its utility to the user and to the public as a whole.

(2) The safety aspects of the product—the likelihood that it will cause injury, and the probable seriousness of the injury.

(3) The availability of a substitute product which would meet the same need and not be as unsafe.

(4) The manufacturer's ability to eliminate the unsafe character of the product without impairing its usefulness or making it too expensive to maintain its utility.

(5) The user's ability to avoid danger by the exercise of care in the use of the product.

(6) The user's anticipated awareness of the dangers inherent in the product and their avoidability, because of general public knowledge of the obvious condition of the product, or of the existence of suitable warnings or instructions.

(7) The feasibility, on the part of the manufacturer, of spreading the loss by setting the price of the product or carrying liability insurance.[4]

Some jurisdictions have adopted this "risk/utility" analysis to govern the imposition of strict liability.[5] Typically this analysis is used in design defect cases.

In several jurisdictions, the "dangerousness" requirement of §402A has been substantially weakened. The California Supreme Court's decision in *Greenman v. Yuba Power Products*, 59 Cal.2d 57, 377

---

[3] *See generally,* J. Wade, *Strict Liability of Manufacturers,* 19 Sw. L.J. 17 (1965).

[4] J. Wade, *On the Nature of Strict Tort Liability for Products,* 44 Miss. L.J. 825, 837-38 (1973).

[5] *See, e.g., Roach v. Kononen,* 525 P.2d 125 (Or. 1974); *Ford Motor Co. v. Matthews,* 291 So.2d 169 (Miss. 1974); *Seattle-First National Bank v. Tabert,* 542 P.2d 774 (Wash. 1975). *See generally,* J. Beasley, *supra* note 1 Chap. 7, "Wade and Negligence Per Se Jurisdictions (Defining Unreasonably Dangerous in Other Than Consumer Expectation Terms in Design Defect Cases)."

P.2d 897, 27 Cal.Rptr. 697 (1962), is a good example of this trend. Pursuant to the *Greenman* formulation, it is not necessary to prove that the product was unreasonably dangerous, but merely that the product contained a defect.[6] The courts adopting the *Greenman* formulation have done so upon the basis that the unreasonably dangerous requirement injects into the strict products liability context the concept of negligence, and reason that negligence has no place in the law of strict products liability because the seller's exercise of reasonable care is not at issue.[7] Variations on the *Greenman* formulation include a prohibition from instructing the jury on the unreasonably dangerous requirement, leaving the court to make this determination in the first instance. *Berkebile v. Brantly Helicopter Corp.*, 462 Pa.83, 337 A.2d 893 (1975).[8]

Other jurisdictions have denoted strict products liability as negligence per se.[9] The Wisconsin Supreme Court, in *Dippel v. Sciano*, 155 N.W.2d 55 (Wis. 1967), adopted the view that strict products liability is negligence per se in order to permit application of Wisconsin's comparative negligence and comparative contribution theories.[10] Alabama, in *Caserell v. Altec*, 335 So. 2d 128 (Ala. 1976), also adopted a similar negligence per se theory.[11]

Under all of these strict products liability variations, however, it is necessary to prove that the product was defective. From a factual standpoint, defects are typically classified as: (1) manufacturing defects; (2) design defects; (3) lack of warning or instruction defects; and (4) inadequate warning or instruction defects.

A manufacturing defect is present when the particular product was not manufactured in conformity with the seller's manufacture of the same product. In other words, the defect is not generic to the entire

---

[6] *Greenman v. Yuba Power Products, Inc., supra* note 1.

[7] *Id.*

[8] *Berkebile v. Brantly Helicopter Co., supra* note 1.

[9] J. Beasley, Chap. 8, *supra* note 1.

[10] *Dippel v. Sciano*, 155 N.W.2d at 64-65.

[11] The Alabama Supreme Court denominates its negligence per se doctrine as "extended manufacturer's liability." Under this doctrine, a manufacturer who places in the stream of commerce an unreasonably dangerous product is negligent as a matter of law. *See generally*, J. Beasley, *supra* note 1 at 264-65.

line of products, but rather is specific to the item at issue.[12] For example, a product may contain a thermostat that normally shuts off the device at a certain temperature. If the particular device contained a thermostat which failed to shut off at the predetermined temperature, the product would contain a manufacturing defect.

Design defects are typically those endemic to an entire line of products. It may be that the seller or manufacturer failed to incorporate sufficient safeguards to minimize or prevent accidents or injury into its product. For example, it may be alleged that the particular equipment should have been manufactured with a fail-safe device or a fire suppression system,[13] or that the product was defectively designed because it permitted the rapid spread of fire.[14]

As noted above, there are two aspects to "defective" warnings. In the first, a product without warnings concerning its use or attendant dangers may be defective if warnings are necessary to safe use.[15] Second, a product defect may also be established with proof that warnings given were inadequate for safe use.[16] For example, products which emit harmful fumes should contain both a warning concerning the harmful fumes and instruction concerning proper ventilation. The absence of either could establish a product defect.

In both inadequate warning and failure to warn cases, some jurisdictions require the plaintiff to prove that adequate warnings would have been heeded or followed.[17] Testimony that the user did not review or follow even an inadequate warning may result in a determination adverse to the user on the issue of liability.[18] The user's

---

[12] See generally, J. Beasley, supra note 1 at 69-72.

[13] See, e.g., Pennsylvania Glass Sand Corp. v. Caterpillar Tractor Co., 652 F.2d 1165 (3d Cir. 1981).

[14] See, e.g., Cloud v. Kit Manufacturing Co., 563 P.2d 248 (Alas. 1977).

[15] See generally, Annot., Failure to Warn as Basis of Liability Under Doctrine of Strict Liability in Tort, 53 A.L.R.3d 239 (1973).

[16] Id.

[17] Restatement (Second) of Torts §402A, comment (j); Technical Chemical Co. v. Jacobs, 480 S.W.2d 602 (Tex. 1972). See generally, J. Beasley, supra note 1, Chap. 16, "Failure to Provide Adequate Warnings." Cf. Nissen Trampoline Co. v. Terre Haute First National Bank, 332 N.E.2d 820 (Ind.App. 1975), rev'd on procedural grounds, 358 N.E.2d 974 (Ind. 1976).

[18] J. Beasley, supra note 1, Chap. 16.

conduct in failing to heed warnings is relevant in determining if the product defect was a substantial factor in causing the user's injuries or damage.[19]

Another defect theory often used in fire cases is the "malfunction theory." This is typically applied where the product is relatively new and where other explanations for the cause of injury can be eliminated. In some respects, the malfunction theory is similar to the doctrine of res ipsa loquitur in the negligence context.[20] In other words, there is no explanation for the damage or injury other than a defect in the product.[21] For example, if it can be shown that a fire originated in a clothes dryer, and there is no other explanation for the fire other than the dryer's malfunction, the plaintiff, in some jurisdictions, may be able to meet the burden of proving a product defect.[22] However, the malfunction theory differs from res ipsa loquitur in that control of the instrumentality by the defendant need not be shown. This theory can be especially useful when a product is relatively new or has received infrequent use. For example, if an automobile engine catches on fire two days after its delivery to the customer, but a specific defect cannot be pinpointed, a defect may be presumed; this presumption is, however, a rebuttable one.

§4.2.2   Negligence

Negligence is another legal theory which can be used to establish a manufacturer, seller or supplier's liability for the manufacture or distribution of a defective product. The basic legal requirements of the negligence theory do not differ when it is applied to products liability actions. The *Restatement (Second) of Torts* §388 outlines, in detail, the application of negligence principles to product manufacturers, suppliers and sellers:

---

[19] *Id.*

[20] *See* §3.8, *Res Ipsa Loquitur, supra.*

[21] *See, e.g., Cornell Drilling Co. v. Ford Motor Co.*, 359 A.2d 822 (Pa.Super. 1976).

[22] *See, e.g., Cassisi v. Maytag Co.*, 396 So.2d 1140 (Fla.App.), *petition for review denied*, 408 So.2d 1094 (Fla. 1981).

One who supplies directly or through a third person a chattel for another to use is subject to liability to those whom the supplier should expect to use the chattel with the consent of the other or to be endangered by its probable use, for physical harm caused by the use of the chattel in the manner for which and by a person for whose use it is supplied, if the supplier

(a)   knows or has reason to know that the chattel is or is likely to be dangerous for the use for which it is supplied, and

(b)   has no reason to believe that those for whose use the chattel is supplied will realize its dangerous condition, and

(c)   fails to exercise reasonable care to inform them of its dangerous condition or of the facts which make it likely to be dangerous.

Under this *Restatement* section, liability is imposed if physical harm to a forseeable user arises from the use of the product.[23] There need be no privity of contract between the supplier and the claimant,[24] and it should be noted that the supplier is liable *only* when physical harm results.[25] Essentially, §388 sets up a negligent "failure to warn" theory.[26]

The *Restatement* imposes a duty to warn as to dangerous chattels (products),[27] and a duty to carefully manufacture.[28] These duties may be imposed even if the dangerous condition is discoverable by an inspection. Therefore, in essence, the manufacturer's duty to warn and carefully manufacture is nondelegable.[29] Furthermore, the seller of the product is subject to the same liability as the manufacturer.[30]

---

[23] *Restatement (Second) of Torts* §388, comment (e). *See also, Restatement (Second) of Torts* §389 & §390.

[24] *Restatement (Second) of Torts* §388.

[25] *See* note 2, *supra.*

[26] *Restatement (Second) of Torts* §388(c).

[27] *Restatement (Second) of Torts* §394.

[28] *Restatement (Second) of Torts* §395.

[29] *Restatement (Second) of Torts* §396. *Cf. Jonescue v. Jewel Home Shopping Service*, 306 N.E.2d 312 (Ill.App. 1973); *Shuput v. Heublein, Inc.*, 511 F.2d 1104 (10th Cir. 1975) (The user or consumer is not entitled to compensation if the danger is open, obvious or patent).

[30] *Restatement (Second) of Torts* §§399-401.

Unlike strict liability, the focus in a negligence action is the manufacturer's conduct. Common allegations in such actions include negligent manufacture, negligent design, negligent failure to warn, negligent testing, and negligent inspection. As with other types of negligence actions, such as forseeability in terms of the user, the use of the product and the resultant harm are necessary elements in establishing the negligence of the seller, manufacturer or supplier.[31]

§4.2.3 Breach of Warranty

In products liability, liability for breach of contract may attach under the law of warranties. A seller, supplier or manufacturer's liability for breach of warranty is governed by the Uniform Commercial Code (UCC), which has been adopted, in some form, in all fifty jurisdictions.[32] The UCC establishes two types of warranties: express warranties and implied warranties.

Section 2-313 governs the creation of express warranties and provides:

(1)   Express warranties by the seller are created as follows:
  (a)   any affirmation of fact or promise made by the seller to the buyer which relates to the goods and becomes part of the basis of the bargain creates an express warranty that the goods shall conform to the affirmation or promise.[33]
  (b)   any description of the goods which is made part of the basis of the bargain creates an express warranty that the goods shall conform to the description.
  (c)   any sample or model which is made part of the basis of the bargain creates an express warranty that the whole of the goods shall conform to the sample or model.[34]

---

[31] *See generally*, J. Beasley, *supra* note 1 at 505-10.

[32] The Uniform Commercial Code has been adopted, in some form, in all fifty jurisdictions.

[33] *See generally, Annot., What Constitutes "Affirmation of Fact" Giving Rise to Express Warranty Under UCC §2-313(1)(a)*, 94 A.L.R.3d 729 (1979).

[34] In order to establish liability under subsection (b) of §2-313(1), the buyer must rely upon the sample or model and it must form the basis of the bargain. Official comments to §2-313.

(2)   It is not necessary to the creation of an express warranty that the seller use formal words such as "warrant" or "guarantee" or that he have a specific intention to make a warranty, but an affirmation merely of the value of the goods or a statement purporting to be merely the seller's opinion or commendation of the goods does not create a warranty.

Two implied warranties arise under the UCC; they are the implied warranty of merchantability[35] and the implied warranty of fitness for a particular purpose.[36] The implied warranty of merchantability creates the assumption that so long as the seller is a merchant of goods of the kind purchased, the goods will be fit for the ordinary purpose for which such goods are used.[37]

The implied warranty of fitness for a particular purpose is created when the seller knows of the use to which the goods are to be put and when the buyer relies on the seller's skill or judgment in purchasing the goods. The warranty of merchantability differs from the fitness for a particular purpose warranty in that the former anticipates customary use of the goods, while the latter anticipates special use by the buyer.[38] The warranty of fitness for a particular purpose is not created when the buyer furnishes the specifications or where the buyer insists on a particular model or brand. In these instances, there is no reliance by the buyer on the seller's skill or judgment in selecting or furnishing the goods.[39] In most jurisdictions, privity is not required in order to successfully assert a breach of warranty claim.

§4.2.4   Alternative Theories of Liability

While the most widely used and enforced theories of liability relating to the sale, supply or manufacture of products are strict

---

[35] UCC §2-314.

[36] UCC §2-315.

[37] UCC §2-314(2)(c). *See also*, official comments to §2-314.

[38] *See* official comments to §2-314.

[39] *Id.*

products liability, negligence and breach of warranty, a number of alternative theories of liability have been developed. Some of these are variations of standard tort concepts. Others are quite novel and have not gained wide acceptance. The five most common theories are discussed below.

Misrepresentation. Section 402B of the *Restatement (Second) of Torts* sets forth a cause of action for misrepresentation in the products liability context. Section 402B provides:

> One engaged in the business of selling chattels who, by advertising, labels, or otherwise, makes to the public a misrepresentation of a material fact concerning the character or quality of a chattel sold by him is subject to liability to physical harm to a consumer of the chattel caused by justifiable reliance upon the misrepresentation, even though
> (a)   it is not made fraudulently or negligently, and
> (b)   the consumer has not bought the chattel from or entered into any contractual relation with the seller.[40]

To make a successful claim under §402B, the claimant must satisfy four elements: that the supplier or seller made a misrepresentation to the public in advertising or labeling; that the misrepresentation was of a material fact regarding the character or quality of the product; that physical harm resulted; and that the consumer justifiably relied upon the misrepresentation. It is important to note that the existence of misrepresentation of a material fact is, in itself, sufficient to establish liability; fraud or negligence need not be proven.[41]

Market Share Liability. The concept of market share liability initially arose in the context of diethylstilbesterol (DES) cases. DES

---

[40] *See* comments and caveats to §402B.

[41] *See, e.g., Winkler v. American Safety Equipment Corp.*, 604 P.2d 693 (Colo.App. 1979), *rev'd on other grounds, American Safety Equipment Corp. v. Winkler*, 640 P.2d 216 (Colo. 1982); *Klages v. General Ordnance Equipment Corp.*, 367 A.2d 304 (Pa.Super. 1976); *Rowland v. Cessna Aircraft Co.*, _____ S.E.2d _____ (Tenn.App. 1984); *Nugent v. Utica Cutlery Co.*, 636 S.W.2d 805 (Tex.App. 1982). *See generally*, J. Sales, *The Innocent Misrepresentation Doctrine: Strict Tort Liability Under Section 402B*, 16 Hous. L.Rev. 239 (1979).

was a fertility drug widely used in the United States. In DES cases, it is frequently impossible for the plaintiffs, who claim increased risk of cancer due to their mothers' ingestion of DES, to identify the particular manufacturer. To overcome this proof problem, the California Supreme Court, in *Sindell v. Abbott Laboratories, Inc.*, 26 Cal.3d 588 (1980), adopted the market share theory of liability.[42] Under the *Sindell* approach, each DES manufacturing defendant must share in any verdict in proportion to its share of the DES market.[43] Although the market share theory, enterprise liability or similar theories[44] may seem appropriate only in drug or latent disease cases, this approach has been used in the fire context.

In the case of *In Re Beverly Hills Fire Litigation*, 695 F.2d 207 (6th Cir. 1982), cert. denied ————U.S.————, 1035 S.Ct.2090, 77L.Ed.2d 300 (1983), a class of plaintiffs, consisting of the legal representatives of persons killed and individuals injured in the Beverly Hills Supper Club fire in Southgate, Kentucky in 1977, sued a group of aluminum branch circuit wiring manufacturers. Their central claim was that the wiring, classified as "old technology," caused the fire. The plaintiffs based their claims on three theories of liability, including enterprise liability.[45] Although the court did not permit the plaintiffs to proceed on this theory, ruling that Kentucky did not recognize market share or enterprise liability, the court did permit the plaintiffs to go forward on a theory of concert of action.[46] This theory evidences the potential use of the market share or enterprise theories in a fire context, where generic products are involved or where specific product identity may be difficult. Concert of action is analyzed in the next section.

Concert of Action. The American Law Institute has recently adopted a section dealing with concert of action. This theory is outlined in §876 of the *Restatement (Second) of Torts:*

---

[42] Market share liability is often called enterprise liability.

[43] 26 Cal.2d at 611-12.

[44] *See, e.g., Collins v. Eli Lilly and Co.*, 342 N.W.2d 37 (Wis. 1984).

[45] 695 F.2d at 210 & n. 1.

[46] *See* section c., *infra.*

For harm resulting to a third person from the tortious conduct of another, one is subject to liability if he

    (a)   does a tortious act in concert with the other or pursuant to a common design with him, or

    (b)   knows that the other's conduct constitutes a breach of duty and gives substantial assistance or encouragement to the other to so conduct himself, or

    (c)   gives substantial assistance to the other in accomplishing a tortious result and his own conduct, separately considered, constitutes a breach of duty to the third person.

As noted in the previous section, the concert of action theory was sanctioned in the *Beverly Hills Fire Litigation*.[47] In that instance the court determined that the doctrine imposed joint liability against all persons who pursued the common plan or design to commit a tortious act and who actively participated in the plan or design.[48] The elements of this cause of action identified by the district court in the *Beverly Hills Fire Litigation* were: (1) causal relationship between the act and injury; (2) cooperation or concerted activities by the defendants; and (3) violation of a legal standard of care.[49] These elements substantially comport with the *Restatement (Second) of Torts* §876.[50]

Although the concert of action theory has seldom arisen in the reported products liability cases, the use of the theory could prove fruitful in actions against manufacturers, sellers or suppliers where agreements among groups of manufacturers, sellers or suppliers can be shown. In industries where the number of suppliers is limited, there may be extensive cooperation among them. The concert of action theory would seem to fit best under those circumstances.[51]

[47] *See* note 45, *supra*.

[48] *In Re Beverly Hills Fire Litigation*, No. 77-79, E.D.Ky., Nov. 14, 1979, slip. op. at 8. 695 F.2d at 210 n.1.

[49] Slip op. at 8-9.

[50] *See, e.g., Bichler v. Eli Lilly and Co.*, 436 N.E.2d 182 (N.Y. 1982).

[51] *Restatement (Second) of Torts* §433B. *See, e.g., Borel v. Fibreboard Paper Products Corp.*, 493 F.2d 1076 (5th Cir. 1973), *cert. denied*, 419 U.S. 869 (1974).

Alternative Liability. The doctrine of alternative liability first arose in the context of a hunting accident. In *Summers v. Tice*, 33 Cal.2d 80 (1948), plaintiff Summers was hunting with two companions who discharged their weapons at the same time. Summers was injured when struck twice. It was not ascertainable whether the shots which struck Summers were fired by one or both of the defendants.

The court found that both defendants brought about the situation which resulted in Summers' injuries. In doing so, the court shifted the burden of proof to the defendants to show that they had not caused the harm or to show what portion of the injuries was attributable to each of them. One of the bases upon which the court rested its decision was that redress of the injured party is a more equitable alternative to exculpation of parties whose conduct brings about the harm. Under this theory of liability, the defendants are considered joint tortfeasors if an apportionment of the harm is incapable of proof.

The alternative liability theory has also been applied in a products liability context in *Abel v. Eli Lilly & Co.*, 289 N.W.2d 20 (Mich.App. 1979). In *Abel*, a DES case, the court determined that all of the manufacturers of DES joined as defendants in the action could be held jointly liable unless a particular defendant could absolve itself of liability and shift that burden to the other defendants. The plaintiff's burden was to establish that each of the defendants had breached its duty of care in producing the product, and that the injuries to the plaintiff were caused by DES.

Another decision permitting the use of a similar theory is *Hall v. E.I. DuPont de Nemours & Co.*, 345 F. Supp. 353 (E.D.N.Y 1972). The plaintiff here alleged that a group of blasting cap manufacturers and their trade association cooperated and acted in concert in failing to label the blasting caps with a warning. The claimants, thirteen children injured by the caps, also alleged that the members and trade association jointly considered the labeling of the caps and collected data regarding accidents. Because the caps were destroyed upon detonation, the claimants were unable to identify the manufacturer of each offending cap.

In permitting the claimants to proceed against the manufacturers and their trade association, the court adopted a concert of action-enterprise liability theory similar to that adopted in *Summers v. Tice*. The court determined that joint control over the risk-producing activity could be found by proof of an explicit agreement or joint action in relation to the warnings. Similar behavior by the manufacturers could support an inference of cooperation sufficient to impose joint liability. Finally, industry-wide custom, adopted independently by each manufacturer, could also serve to support the inference of joint action. The court explicitly noted, however, that these theories should be limited to industries comprised of a small number of manufacturers.

Products Liability Statutes. In some jurisdictions, the law of products liability is governed by statute. For example, in Alabama,[52] Indiana,[53] Maine,[54] and South Carolina,[55] the duty owed by a product manufacturer is governed by statute. Several jurisdictions also govern by statute what entities may be held liable for product defects. Some of these jurisdictions are Arizona,[56] Colorado,[57] Connecticut,[58] Idaho,[59] Illinois,[60] Kentucky,[61] Minnesota,[62] Rhode Island,[63] Tennessee,[64] and Washington.[65]

For a number of years, legislation has been introduced in Congress to adopt a federal statute governing the law of products liability.[66] The most recent, and, perhaps, most publicized attempt to enact a federal products liability statute has been introduced in the 98th Congress by Senator Robert Kasten (R.Wis.). This proposal, if enacted, would preempt both state statutes and state common law (S.44)(Sec. 3).

The proposed act would require proof that the product was unreasonably dangerous (Sec. 4), a requirement which has been

---

[52] Ala. Code §6-5-501.

[53] Ind. Code Ann. §§34-4-20A-2.5 & 3.

[54] Me. Rev. Stat. Ann. tit. 4 §221.

[55] S.C. Code Ann. §15-73-10 (legislative adoptment of §402A).

[56] Ariz. Rev. Stat. Ann. §12-684.

[57] Colo. Rev. Stat. §13-21-402.

[58] Conn. Cen. Stat. Ann. §52-572m.

[59] Idaho Code §6-1402.

[60] Ill. Ann. Stat. ch 110 §2-261.

[61] Ky. Rev. Stat. §411.320.

[62] Minn. Stat. Ann. §544.41

[63] R.I. Gen. Laws §9-1-32.

[64] Tenn. Code Ann. §§29-28-105 & 06.

[65] Wash. Rev. Code §§7.72.030 & .040.

[66] *See, e.g.,* 70 *ABA Journal* 12 (Feb. 1984).

abandoned in some jurisdictions. (See §4.2.1.n.1, *supra*). This proposal also encompasses the risk/utility analysis in respect to design defects (Sec. 5).

Comparative responsibility would be applicable (Sec. 9) and misuse, alteration, contributory negligence and assumption of the risk would be available defenses (Sec.10). Capital goods would be subject to a twenty-five year statute of repose, running from the date of delivery (Sec. 12). This statute of repose would change the law in many jurisdictions where the statute of limitations runs from the date of injury and the claimant has from one to six years, depending upon local law, to bring suit regardless of the date of manufacture or sale.

In many jurisdictions, the proposed federal statute would have little effect on the outcome of products suits. In jurisdictions such as California and New Jersey, for example, where the courts have eliminated some aspects of plaintiffs' proofs such as the unreasonably dangerous requirement and have eliminated defenses such as "state of the art," the federal statute may have a significant impact.

§4.3   Defenses to Products Liability Actions

§4.3.1   Contributory/Comparative Negligence

Contributory[67] or comparative negligence[68] can be used in some jurisdictions as defenses in products liability actions. Conduct of the

---

[67] Prosser defines contributory negligence as follows:

Conduct on the part of the plaintiff, contributing as a legal cause to the harm he has suffered, which falls below the standard to which he is required to conform for his own protection....

Although the defendant has violated his duty, has been negligent, and would otherwise be liable, the plaintiff is denied recovery because his own conduct disentitles him to maintain the actions. In the eyes of the law, both parties are at fault; and the defense is one of the plaintiff's disability, rather than the defendant's innocence.

W. Prosser, *Law of Torts* §65 at 416-17 (4th ed. 1971) (footnote omitted).

[68] Under the doctrine of comparative negligence, the plaintiff's damages are reduced in proportion to the percentage of negligence attributable to the plaintiff.

plaintiff, which is a substantial factor in causing the plaintiff's injuries, may amount to contributory or comparative negligence.[69] For example, while the defendant's product may have started a fire, the plaintiff's failure to erect fire walls (when under a duty to do so) also may be a substantial factor in causing the damages. In contributory negligence jurisdictions, the plaintiff would not be entitled to recover; in comparative negligence jurisdictions, the plaintiff would be entitled to recover only a portion of his or her damages.[70]

In breach of warranty actions, on the other hand, the negligence of the parties is not at issue, as the focus is not on the defendant's conduct but rather, is on the product. The plaintiff's contributory negligence will not affect the defendant's liability,[71] or the amount of damages the plaintiff can recover[72] as is the case in negligence actions

---

(For example, if the plaintiff's damages are $1,000 and the plaintiff is 20 percent negligent, the plaintiff will be awarded $800 in damages.)

There are several types of comparative negligence: (1) pure, in which the plaintiff recovers even if the percentage of his negligence exceeds that of the defendant. (For example, if the plaintiff is 80 percent negligent and his damages are $1,000, the plaintiff will recover $200.) The second type of comparative negligence permits the plaintiff to recover so long as his negligence is not greater than that of the defendant. Under this circumstance, if the plaintiff is 50 percent negligent or less, the plaintiff recovers. (For example, if the plaintiff is 50 percent negligent, the plaintiff is entitled to recover one-half of the damages awarded.) The final type of comparative negligence, often called modified comparative negligence, permits the plaintiff to recover so long as the plaintiff's negligence is not equal to or greater than that of the defendant. Under these circumstances, if the plaintiff is 50 percent negligent, the plaintiff does not recover. In other words, the plaintiff's negligence must be 49 percent or less. (It should be noted that all of the above examples are based upon the assumption that the plaintiff is seeking to recover against only one defendant.) *See generally,* J. Beasley, *supra* note 1 at 547-48.

[69] The leading case applying comparative negligence to a strict liability action is *Daly v. General Motors Corp.,* 575 P.2d 1162 (Cal. 1978). J. Beasley, *supra* note 1 at 549-57. *See generally, Annot., Strict Liability: Applicability of Comparative Negligence Doctrine to Actions Based on Strict Liability in Tort,* 9 A.L.R. 4th 633 (1981).

[70] *See* Notes 67 & 68, *supra.*

[71] J. Beasley, *supra* note 1 at 43. *Cf.* UCC §2-316(3)(b). (Implied warranties cannot be the basis for recovery where the buyer has the opportunity before entering into the contract to examine the goods and where such an examination would have revealed the defects.)

[72] UCC §2-714-715.

and strict liability actions in those jurisdictions which apply the comparative negligence defense.

The application of the principles of contributory or comparative negligence to a strict products liability action has met with mixed results among the jurisdictions.[73] The most frequently advanced reason for the proposition that comparative negligence principles are inapplicable to strict products liability cases is the conceptual irrelevance of injecting negligence concepts into the "liability without fault" concept of strict products liability.[74] In other words, strict liability is akin to a no-fault concept, whereas comparative or contributory negligence is a doctrine based solely on fault.

The basic strict liability provisions in section 402A of the *Restatement (Second) of Torts* indicate that contributory negligence should not be a defense under that section.[75] Thus, the *Restatement* view espouses the fault/no-fault distinction between negligence and strict liability. Many of the courts holding that comparative negligence is inapplicable have applied this distinction.[76] In those jurisdictions which continue to employ the total bar of contributory negligence, most courts have refused to apply the doctrine to strict liability cases because of its harshness as a complete bar to the plaintiff's recovery.[77]

On the other hand, many jurisdictions have applied comparative negligence principles to strict liability actions. Wisconsin was one of the earliest jurisdictions to apply its comparative negligence statute to strict products liability cases in *Dippel v. Sciano*, 155 N.W.2d 55 (Wis. 1967). There, however, the court's application of comparative

---

[73] *See generally, Annot.,* 9 A.L.R.4th 633 *supra* note 68.

[74] *See. e.g., Kinard v. Coats Co.,* 553 P.2d 835 (Colo.App. 1976); *Seay v. Chrysler Corp.,* 609 P.2d 1382 (Wash. 1980) *superseded by statute, Klein v. R.D. Werner Co.,* 654 P.2d 94 (Wash. 1982). *See generally, Annot.,* 9 A.L.R.4th 633, 638-41, *supra* note 69.

[75] *Restatement (Second) of Torts,* §402A comment n.

[76] *Annot.,* 9 A.L.R.4th 633, 638-41, *supra* note 69.

[77] *Reese v. Chicago, Burlington & Quincy Railroad Co.,* 303 N.E.2d 382 (Ill. 1983) (prior to the adoption of the comparative negligence); *Azzarello v. Black Brothers Co.,* 391 A.2d 1020 (Pa. 1978) (prior to the adoption of comparative negligence); *Holt v. Stihl, Inc.,* 449 F.Supp. 693 (E.D.Tenn. 1977).

negligence was justified by its characterization of strict liability as negligence per se.[78] Most courts adopting comparative negligence to strict liability have based their reasoning on the equity of apportioning loss based upon causation.[79]

Finally, some courts have blended the doctrine of assumption of risk into comparative negligence and have applied assumption of the risk in a comparative context.[80] Essentially, these courts have opted for a system of comparative causation.[81]

§4.3.2 Assumption of the Risk

While the courts have been somewhat reluctant to apply the bar of contributory negligence or the partial bar of comparative negligence to strict products liability actions, this reluctance has not been exhibited in their application of the doctrine of assumption of the risk as a bar to recovery.[82]

The *Restatement (Second) of Torts* contains several sections which outline the assumption of the risk defense. In §496C, an implied assumption of risk is defined as follows:

> (1) Except as stated in subsection (2), a plaintiff who fully understands a risk of harm to himself or his things caused by the defendant's conduct or by the condition of the defendant's land or chattels, and who nevertheless voluntarily chooses to enter or remain, or to permit his things to enter or remain within the area of that risk, under circumstances that manifest his willingness to accept it, is not entitled to recover for harm within that risk.

[78] 155 N.W.2d at 64-65. *See also, Sun Valley Airlines, Inc. v. Avco-Lycoming Corp.*, 411 F.Supp. 598 (D.C. Idaho 1976).

[79] *Annot.*, 9 A.L.R.4th 633, 642-47, *supra* note 69.

[80] *Baccelleri v. Hyster Co.*, 597 P.2d 351 (Or. 1979); *Busch v. Busch Construction, Inc.*, 262 N.W.2d 377 (Minn. 1977).

[81] *Murray v. Fairbanks Morse*, 610 F.2d 149 (3d Cir. 1979); *Thibault v. Sears, Roebuck & Co.*, 395 A.2d 843 (N.H. 1978).

[82] *See generally, Annot., Products Liability: Contributory Negligence or Assumption of Risk as Defense Under Doctrine of Strict Liability in Tort*, 46 A.L.R. 3d 240 (1972).

(2) The rule stated in subsection (1) does not apply in any situation in which an express agreement to accept the risk would be invalid as contrary to public policy.

From this definition alone, the reason for applying this doctrine to strict products liability actions is evident. Subsection (1) of §496C clearly states that the risk of harm may be presented by not only the defendant's conduct but by the condition of the defendant's products. In this respect, assumption of the risk would be applicable to both negligence and strict products liability actions.

Unlike simple negligence, contributory negligence or comparative negligence, which all employ the "reasonable man" standard, the doctrine of assumption of the risk is grounded in subjective terms. Section 496D of the *Restatement (Second) of Torts* sets out this subjective standard:

> Except where he expressly so agrees, a plaintiff does not assume a risk of harm arising from the defendant's conduct unless he then knows of the existence of the risk and appreciates its unreasonable character.[83]

This subjective standard takes several forms in its application to the plaintiff. The plaintiff's subjective knowledge of the risks inherent in a particular product may also bar his or her recovery.[84] An assumption of the risk may be found where the plaintiff continues to use the "defective product" despite knowledge of previous injury or malfunction; this constitutes evidence of the plaintiff's subjective appreciation of the risk.[85] Misuse of the product may also amount to an assumption of the risk.[86] Finally, the plaintiff may be found to have assumed the risk if the danger presented is open and obvious.[87] For

[83] *See also, Restatement (Second) of Torts* §496D, comments c, d & e.

[84] *Wojciechowski v. Long Airdox Div. of Marmon Group, Inc.*, 488 F.2d 1111 (3d Cir. 1973) (prior malfunction); *Hagenbuch v. Snap-on-Tools Corp.*, 339 F.Supp. 676 (D.N.H. 1972) (prior injury).

[85] *See, e.g., Benjamin v. Deffett Rentals, Inc.*, 419 N.E.2d 883 (Ohio 1981), *superseded by statute, Anderson v. Ceccardi*, 451 N.E.2d 780 (Ohio 1983).

[86] *See, e.g., Genteman v. Saunders Archery Co.*, 355 N.E.2d 647 (Ill. 1976).

[87] *Turcotte v. Ford Motor Co.*, 494 F.2d 173 (1st Cir. 1974); *Micaleff v. Miehle Co.*, 348 N.E.2d 571 (N.Y. 1976). *See generally, Annot., Products Liability: Duty of*

example, a welder would be a likely candidate against whom to apply this doctrine. The welder should be aware of the possibility of stray sparks causing a fire and that fire suppression equipment should be kept immediately at hand in the event a fire is started.

In some jurisdictions, however, assumption of the risk is not a complete bar to recovery, but is to be considered in application of comparative negligence principles. Here recovery may be reduced but not completely barred.[88] In this context, assumption of the risk is abolished as a separate defense and is assumed within comparative negligence principles.[89]

§4.3.3   Product Alteration

Section 402A of the *Restatement (Second) of Torts* requires that the product reach the consumer without substantial change in its condition in order to hold the seller liable.[90] The reason for the requirement is that an alteration may be an intervening or superseding cause of the plaintiff's injuries.[91] Generally, the alteration to the product must be substantial.[92]

In this context some courts have imposed a foreseeability requirement; if the alteration was not foreseeable, the plaintiff's recovery is

---

Manufacturer to Equip Product with Safety Device to Protect Against Patent or Obvious Danger, 95 A.L.R. 3d 1066 (1979).

[88] *South v. A.B. Chance Co.*, 635 P.2d 728 (Wash. 1981); *Baccelleri v. Hyster Co.*, 597 p.2d 351 (Or. 1979); *Butaud v. Suburban Marine & Sporting Goods, Inc.*, 555 P.2d 42 (Alas. 1976).

[89] *See, e.g., Dippel v. Sciano*, 155 N.W.2d 55 (Wis. 1967).

[90] *Restatement (Second) of Torts*, §402A(1)(b). *See also*, comment p to §402A.

[91] *See generally, Annot., Products Liability: Alteration of Product After it Leaves Hands of Manufacturer or Seller as Affecting Liability for Product-Harm*, 41 A.L.R. 3d 1251 (1972); *Annot., Products Liability: Proof, Under Strict Tort Liability Doctrine, That Defect was Present When Product Left Hands of Defendant*, 54 A.L.R. 3d 1079 (1973).

[92] *Id. See also, Speyer, Inc. v. Humble Oil & Refining Co.*, 403 F.2d 766 (3d Cir. 1968), *cert. denied*, 394 U.S. 1015 (1969); *State Stove Manufacturing v. Hodges*, 189 So.2d 113 (Miss. 1966), *cert. denied*, 386 U.S. 912 (1967); *Martinez v. Clark Equipment Co.*, 382 So.2d 878 (Fla.App. 1980); *Cox v. General Motors Corp.*, 514 S.W.2d 197 (Ky. 1974).

barred.[93] For example, a fire suppression system may be removed from a piece of equipment for various reasons; such a substantial alteration may preclude the manufacturer's liability as the product is, in essence, no longer the same product manufactured by the defendant. The manufacturer is liable, however, when it is determined that the alteration undertaken was reasonably foreseeable.[94]

Special attention should be paid to those products which may be considered used or second-hand at the time of their purchase. Used products have typically been purchased from someone other than the original distributor or manufacturer. The type or amount of use or abuse to which a previous purchaser has subjected the product, possible changes made by previous users, and repairs to the product are important considerations in the used product context.[95]

§4.3.4   Contribution and Indemnification

Contribution or indemnification may be available to a products liability defendant as a means of distributing its loss. Contribution distributes a loss among joint tortfeasors by requiring them to pay a pro rata or proportionate share of the loss. Indemnity, on the other hand, shifts the entire burden of the loss from one party to another.[96]

---

[93] *Helene Curtis Industries, Inc. v. Pruitt*, 385 F.2d 841 (5th Cir. 1967), *cert. denied*, 391 U.S. 913 (1968); *Smith v. Hobart Manufacturing Co.*, 302 F.2d 570 (3d Cir. 1962); *Talley v. City Tank Corp.*, 279 S.E.2d 264 (Ga.App. 1981).

[94] *Steinmetz v. Bradbury Corp.*, 618 F.2d 21 (8th Cir. 1980); *Kennedy v. Custom Ice Equipment Co.*, 246 S.E.2d 176 (S.C. 1978); *D'Antona v. Hampton Grinding Wheel Co.*, 310 A.2d 307 (Pa.Super. 1973).

[95] *See generally, Manufacturers Liability for Defects in Used Products*, 22 FOR THE DEFENSE 10 (Dec. 1980); *Annot., Strict Liability in Tort: Liability of Seller of Used Product*, 53 A.L.R.3d 337 (1973).

[96] *See generally, Annot., Comment Note — Contribution or Indemnity Between Joint Tortfeasors on Basis of Relative Fault*, 53 A.L.R.3d 184 (1973); *Annot., Products Liability: Right of Manufacturer or Seller to Contribution or Indemnity from User or Product Causing Injury or Damage to Third Person, and Vice Versa*, 28 A.L.R.3d 943 (1969); 18 Am.Jur.2d *Contribution* §7 (1965); 41 Am.Jur.2d *Indemnity* §3 (1968).

Contribution. At common law, no contribution was permitted between joint tortfeasors.[97] This common-law rule, however, has been changed by statute in many jurisdictions. Many jurisdictions have adopted variations of the Uniform Contribution Among Joint Tortfeasors Act proposed in 1955 by the Commissioners on Uniform State Laws.[98] Under the uniform statute, contribution is permitted among joint tortfeasors, whether judgment has been recovered against all or some of them.[99] Joint tortfeasors are defined as two or more persons jointly or severally liable in tort for the same injury to persons or property. The right of contribution does not arise, however, until a joint tortfeasor has, by payment, "discharged the common liability or has paid more than his pro rata share" of that liability.[100]

The Uniform Act also changes the common law in respect to the release of a joint tortfeasor. Under the common-law rule, the release of one joint tortfeasor was the release of all.[101] Under the Uniform Act, the discharge and release of one joint tortfeasor reduces the claim against the other tortfeasors either "in the amount of the consideration paid...or in any amount or proportion by which the release provides that the total claim shall be reduced if greater than the consideration paid."[102] The Uniform Act, however, does permit a release to act as a general release discharging all of the tortfeasors, but the release must specifically so provide.[103]

The respective liability of each joint tortfeasor may be on a pro rata or proportionate basis.[104] Under the former approach, two joint tortfeasors would be liable each for one-half of the damages, three tort feasors would be liable each for one-third of the damages, and so forth. Under the proportionate approach, each of the joint tortfeasors would be liable for its respective percentage of responsibility, similar

---

[97] W. Prosser, *Law of Torts* §50 at 305 (4th ed. 1971).

[98] *Id.* at 307 n. 63.

[99] Uniform Contribution Among Joint Tortfeasors Act §2 & §4.

[100] *Id.* at §4.

[101] *See* note 97, *supra.*

[102] Uniform Contribution Among Joint Tortfeasors Act §6.

[103] *Id.*

[104] *Id.*

to a comparative negligence approach. The proportionate share approach is also known as comparative contribution.[105]

Liability for damages sustained in a warehouse fire is a good illustration of the theory of contribution. Here joint liability may be imposed against the warehouser for permitting careless smoking by employees, against the manufacturer of the products stored for selecting an inadequate warehouse, and against a building materials manufacturer for failing to warn of the product's flammability.

Indemnification. Indemnity, unlike contribution, is not a proportionate sharing of the loss, but rather requires that one party bear the burden of the entire loss.[106] Indemnity, again unlike contribution, does not arise by operation of law absent special circumstances.[107] Indemnity, in the absence of an agreement providing for indemnification, is the exception rather than the rule. However, there are situations where the law shifts the burden of the loss to only one of the parties.

The most common situation in which indemnity is allowed between joint tortfeasors is where one of the tortfeasors is guilty of active negligence and one guilty of passive negligence. In this case the actively negligent tortfeasor will owe indemnification to the passively negligent tortfeasor.[108] One joint tortfeasor may also be permitted indemnification where he or she is only constructively, vicariously or derivatively liable.[109]

Absent these special circumstances, the most common situation in which indemnity is permitted is where the parties have an indemnifi-

---

[105] *Bielski v. Schulze,* 114 N.W.2d 105 (Wis. 1962); *Rogers v. Spady,* 371 A.2d 285 (N.J.Super. 1977); *Sitzes v. Anchor Motor Freight, Inc.,* 289 S.E.2d 679 (W.Va. 1982).

[106] *See generally, Annot.,* 53 A.L.R.3d 184, *supra* note 96.

[107] *Nelson v. Quimby Island Reclamation Dist. Facilities Corp.,* 491 F.Supp. 1364 (N.D.Cal. 1980); *Odom v. Monogram Industries,* 555 F.Supp. 378 (S.D.Tex. 1983).

[108] *Colt Industries Operating Co. v. Coleman,* 272 S.E.2d 251 (Ga. 1980); *Lewis v. Amchem Products, Inc.,* 510 S.W.2d 46 (Mo.App. 1974); *Universal Underwriters Insurance Co. v. Security Industries, Inc.,* 391 F.Supp. 326 (W.D.Wash. 1974).

[109] *South Austin Drive-In Theatre v. Thomison,* 421 S.W.2d 933 (Tex.Civ.App. 1967); *Burns v. Pennsylvania Rubber & Supply Co.,* 189 N.E.2d 645 (Ohio App. 1961); *Schipper v. Lockheed Aircraft Corp.,* 278 F.Supp. 743 (S.D.N.Y. 1968).

cation agreement. Quite often, vendors of a particular product will have an agreement with the manufacturer that the manufacturer will indemnify the vendor in the event the vendor is held liable; the same may be true with component part manufacturers.[110] These indemnity contracts are usually construed according to the rules governing any other type of contract,[111] but there are jurisdictions which apply strict construction principles to indemnification agreements.[112]

§4.3.5   Disclaimer of Warranties

Section 2-316 of the UCC provides that warranties may be excluded in writing or by the buyer's inspection of the goods.[113] In order to exclude the implied warranties of merchantability and fitness for a particular purpose, the exclusion must be in writing, and must be conspicuous. As to the implied warranty of merchantability, the language of the exclusion must mention "merchantability."[114] Implied warranties may also be excluded by expressions such as "as is," "with all faults" or by other language which brings to the buyer's attention the exclusion of the warranties.[115] When a buyer is afforded an examination or inspection of the goods, there are no implied warranties with respect to those defects which an examination would or should have revealed.[116] In the absence of a warranty, the plaintiff

---

[110] For example, the comprehensive general liability insurance policy may be subject to a vendor's endorsement which provides that the vendor is an insured under the manufacturer's policy with respect to the distribution of the manufacturer's products. *See, e.g., St. Louis - San Francisco Railway Co. v. Armco Steel Corp.*, 490 F.2d 367 (8th Cir.), *cert. denied*, 417 U.S. 969 (1974); *W.T. Grant Co. v. United States Fidelity & Guaranty Co.*, 421 A.2d 357 (Pa.Super. 1980); *Sears, Roebuck & Co. v. Reliance Insurance Co.*, 654 F.2d 494 (7th Cir. 1981).

[111] *Anthony v. Louisiana & A.R. Co.*, 316 F.2d 858 (5th Cir.), *cert. denied*, 375 U.S. 830 (1963); *Chicago, Milwaukee, St. Paul & Peoria Railroad Co. v. Famous Brands, Inc.*, 324 F.2d 137 (8th Cir. 1963).

[112] *Cate v. United States*, 249 F.Supp. 414 (S.D.Ala. 1966); *Ging v. Parker-Hunter, Inc.*, 544 F.Supp. 49 (W.D.Pa. 1982).

[113] UCC §2-316(2).

[114] *Id.*

[115] UCC §2-316(3)(a).

[116] UCC §2-316(3)(b).

would be required to prove that the defendant was negligent or that the product was defective, i.e., unreasonably dangerous. Cf. §4.2.1 *supra*.

§4.3.6   Limitation of Remedies and Damages

Sections 2-718 and 2-719 of the UCC permit the seller to modify or limit the remedies available to the buyer. Damages may be limited to a liquidated amount which is reasonable in light of the anticipated harm.[117] The buyer's remedy may also be limited to repair or replacement of the goods, repayment of the purchase price, or repair and replacement of the defective parts.[118] Repair or replacement may be the sole and exclusive remedy provided to the buyer.[119] The presence of such limitations is an important consideration in commencing actions against product manufacturers, and in developing theories of liability against other potential defendants.

Consequential damages may also be limited or excluded. Consequential damages are defined as:

> (a)   any loss resulting from general or particular requirements and needs of which the seller at the time of contracting had reason to know and which could not reasonably be prevented by cover or otherwise; and
>
> (b)   injury to person or property proximately resulting from any breach of warranty.[120]

The limitation or exclusion is valid unless the exclusion or limitation is unconscionable.[121] The limitation of consequential damages for personal injury related to consumer goods is unconscionable; such a limitation is not unconscionable per se when the loss is commercial.[122]

---

[117] UCC §2-718(1).

[118] UCC §2-719(1)(a)

[119] UCC §2-719(1)(b).

[120] UCC §2-715(2).

[121] UCC §2-719(3).

[122] *Id.*

The application of these warranty remedy limitations to negligence and strict products liability actions has been the subject of considerable debate. The argument for the inapplicability of such limitations to negligence and strict liability actions is that the exculpatory or limitation provisions of a written warranty usually do not specifically make the limitation provisions applicable to negligence or strict liability claims.[123] In addition, many courts have been reluctant to apply the UCC to strict liability actions, acknowledging that breach of warranty and strict liability claims are distinct and based upon different policy considerations.[124] Therefore, even if a warranty limits the damages available, plaintiff may recover its entire loss if the claim in negligence or strict liability is sustained. The warranty limitation will not affect the negligence or strict liability cause of action or the damages available under the latter two theories.

§4.3.7   Collateral Estoppel

The doctrine of collateral estoppel, or issue preclusion, prevents one from relitigating an issue which one has had a full and fair opportunity to litigate on a prior occasion. The basic elements of collateral estoppel are: (1) the issue decided in the prior adjudication is identical with the one presented for which collateral estoppel is sought; (2) there must have been a final judgment on the merits; (3) the party against whom it is asserted must have been a party or in privity with a party in the prior adjudication; and (4) the party against whom it is asserted must have had a full and fair opportunity to litigate the issue in the prior action.[125] In those jurisdictions which do not require mutuality, i.e., identity of parties,[126] the doctrine of collateral estoppel might be used to establish a product defect by its

---

[123] *See, e.g., Neville Chemical Co. v. Union Carbide Corp.*, 422 F.2d 1205 (3d Cir.), *cert. denied*, 400 U.S. 826 (1970).

[124] *Pennsylvania Glass Sand Corp. v. Caterpillar Tractor Co.*, 652 F.2d 1165 (3d Cir. 1981).

[125] *See, e.g., Parklane Hosiery Co. v. Shore*, 439 U.S. 322 (1979).

[126] *See, e.g., Bernhard v. Bank of America National Trust & Savings Association*, 122 P.2d 892 (Cal. 1942).

offensive use or the absence of a product defect by a defensive use.[127]

Where there exists a possibility of inconsistent judgments in similar actions, the application of the doctrine of collateral estoppel may be refused.[128] At least one decision has permitted the use of offensive collateral estoppel to establish the prior existence of a warning defect.[129] While a product manufacturer may be accorded a complete defense by the use of collateral estoppel (that the product was not defective), the offensive use of collateral estoppel by a plaintiff merely lessens the plaintiff's proof. It will not establish the defendant's liability because the plaintiff still bears the burden to prove that the defect established by collateral estoppel was the cause of the plaintiff's injuries.[130] The doctrine will most likely be applied in instances where litigation relating to product defects has been minimal, or where judgments for or against a product manufacturer have been consistent.

§4.3.8   State-of-the-Art Defense

The state-of-the-art defense attempts to establish that it was scientifically impossible for defendant manufacturers to test for or know of the defect complained of by the plaintiff.[131] For the manufacturer, the defense is a means to establish that there were no effective means for the manufacturer to discover and to remedy the risk. Hence, the manufacturer relies upon the state-of-the-art defense to argue that liability should not be imposed for the alleged design or warning defect.

This view has been judicially adopted in a number of jurisdictions and incorporated into a number of state statutes relating to products

---

[127] *See generally,* J. Beasley, *supra* note 1 at 702-04.

[128] *Hardy v. Johns-Manville Sales Corp.,* 681 F.2d 334 (5th Cir. 1982).

[129] *Ezagui v. Dow Chemical Corp.,* 598 F.2d 727 (2d Cir. 1979), *relying on Parke-Davis & Co. v. Stromsodt,* 411 F.2d 1390 (8th Cir. 1969); *See generally, Hardy: Offensive Collateral Estoppel Repulsed,* 26 FOR THE DEFENSE 10 (Jan. 1984); Update on *Hardy v. Johns-Manville,* 26 FOR THE DEFENSE 16 (Jan. 1984).

[130] *Restatement (Second) of Torts* §402A(1).

[131] *See generally, Abrogating the Defense of Foreseeability:* Beshada *Transforms Strict Liability In Absolute Liability,* 25 FOR THE DEFENSE 20 (Mar. 1983).

liability.[132] The claimant's common counterargument to the state-of-the-art defense is that it injects negligence principles into strict liability actions. Most descriptive of this position is a New Jersey decision, *Beshada v. Johns-Manville Products Corp.*, 447 A.2d 539 (N.J. 1982). Here the court ruled that the defense was purely a negligence defense which sought to explain away the defendant's culpability, and that the defense could not be used in a strict liability action. The court further reasoned that the reasonableness of the defendant's conduct was not at issue under strict liability principles.[133]

A plaintiff has several sources for analyzing a state-of-the-art defense. Most important among these are the defendant's research and development files and general patent files. For instance, if a manufacturer claims it could not have equipped its product with a fire supression system due to the lack of developed technology, the existence of a patent for such a system or recommendations for a system can rebut the defense. This defense can also be established by the introduction of patents which incorporate safety devices into the product. This can be invaluable to the defense in proving that its product conformed to the "state of the art" in cases where the plaintiff's employer, for example, may have removed the safety device for any number of reasons, such as cost of use or greater ease of operation.

§4.4   Burden of Proof

§4.4.1   Proving the Defect

The existence of a product defect may be shown in a variety of ways. Very often, the manner of proof selected will depend upon the

---

[132] *Heritage v. Pioneer Brokerage & Sales, Inc.*, 604 P.2d 1059 (Alas. 1979); *Boatland of Houston, Inc. v. Bailey*, 609 S.W.2d 743 (Tex. 1980); *Kerns v. Engelke*, 390 N.E.2d 859 (Ill. 1979). Ariz.Rev.Stat.Ann. §12-683; Ind.Code §31-1-1.5-4(4); Neb.Rev.Stat. §25-21; N.H.Rev.Stat.Ann. §507-D:4.

[133] 447 A.2d at 546. *See also,* 25 FOR THE DEFENSE 20, *supra* note 131.

availability of the allegedly defective product,[134] as well as the availability of expert testimony.

The several ways in which a product defect may be established are as follows:

(1)   Direct evidence, i.e., the product, may be examined by an expert to determine a specific defect.[135] This underscores the importance of preserving and photographing the physical evidence present at a fire scene.

(2)   Other possible causes of the injury may be negated through expert evidence.[136] For example, the expert may be able to negate arson, vandalism, spontaneous combustion, and natural gas or electrical sources, leaving only the product as a source of ignition in the area of origin.

(3)   Circumstantial evidence, primarily the testimony of experts, may form the basis of the expert's opinion that the product was defective.[137] The expert may be able to testify that the area where the product was located, and the product itself, was the area most severely burned and in the area of the fire's origin.

(4)   The plaintiff may rely upon the recency of manufacture of the product, its unusual performance and the conditions and circumstances present.[138] Under this approach, the burden of proof may shift to the manufacturer, as the party with superior knowledge of the product,[139] to rebut the inference of a defect. For example, it is unusual for a product to catch on fire one or two days after its purchase or delivery.

These methods of proof further support the necessity for a thorough

---

[134] This is a special concern in fire cases where the fire often consumes the product or mars it beyond identification.

[135] *Caprara v. Chrysler Corp.*, 417 N.E.2d 545 (N.Y. 1981); *Cassisi v. Maytag Co.*, 396 So.2d 1140 (Fla.App.), *petition for review denied*, 408 So.2d 1094 (Fla. 1981).

[136] *Scanlon v. General Motors Corp.*, 326 A.2d 673 (N.J. 1974); *Senco Products, Inc. v. Riley*, 434 N.E.2d 561 (Ind.App. 1982).

[137] *Farmer v. International Harvester Co.*, 553 P.2d 1306 (Idaho 1976); *Fenner v. General Motors Corp.*, 657 F.2d 647 (5th Cir. 1981).

[138] *Bollmeier v. Ford Motor Co.*, 265 N.E.2d 212 (Ill.App. 1970); *Cornell Drilling Co. v. Ford Motor Co.*, 359 A.2d 822 (Pa. Super. 1976).

[139] *Embs v. Pepsi-Cola Bottling Co.*, 528 S.W.2d 703 (Ky. 1975); *Campbell v. General Motors Corp.*, 649 P.2d 224 (Cal. 1982).

and adequate cause, origin and spread investigation. The extent of the investigation may dictate the method and types of proof available.

§4.4.2   Identification of the Product Manufacturer

Fire cases often present a serious challenge in identifying the manufacturer of the product. Very often, either the product is consumed in the fire or is marred to the extent that a ready identification is not possible.

One of the methods which may be utilized to identify the manufacturer is to trace the chain of purchase or acquisition of the product. For example, the identity of the immediate seller may be known. In turn, the seller may be able to identify either the wholesaler or the manufacturer. In the former instance, the wholesaler may then be the direct link with the manufacturer.[140] If the product is an unusual one, there may be only a small group of manufacturers, easing the process of indentifying the specific manufacturer. Patent documents provide another method by which the manufacturer may be determined.

Under the newer theories of strict liability, such as enterprise, alternative, or market share liability (4.2.4), it may be sufficient for the claimant to identify the manufacturers of a particular product and attempt to make claim against them as a group.[141]

§4.5   Evidence Issues in Products Liability

There are several important evidence issues that, although of general application, present special problems in product liability litigation. These are discussed in turn below.

---

[140] See generally, Annot., Products Liability: Necessity and Sufficiency of Identification of Defendant as Manufacturer or Seller of Product Alleged to Have Caused Injury, 51 A.L.R.3d 1344 (1973).

[141] In Re Beverly Hills Fire Litigation, 695 F.2d 207 (6th Cir. 1982).

§4.5.1   Subsequent Remedial Measures

Remedial measures taken by the manufacturer to correct product defects are often central to products liability. Federal Rule of Evidence 407 addresses the admissibility of subsequent remedial measures:

> When after an event, measures are taken which, if taken previously, would have made the event less likely to occur, evidence of the subsequent measures is not admissible to prove negligence or culpable conduct in connection with the event. This rule does not require the exclusion of evidence of subsequent measures when offered for another purpose, such as proving ownership, control, or feasibility of precautionary measures, if controverted or for impeachment.

The policy behind this rule is to encourage manufacturers to correct defects without fear of the correction being deemed an admission of liability.

Some courts have refused to apply the prohibition against admission of subsequent remedial measures on the basis that a strict liability action focuses not upon the negligence or culpable conduct of the manufacturer, but rather, upon the alleged product defect.[142] Under this approach, evidence of design, warning or manufacturing changes occurring after purchase, or even after the incident, is admissible.

Other courts have applied Rule 407 to strict liability actions upon the basis that the rule does not explicitly omit application to strict liability cases. Courts also apply the rule because the policy underlying it, encouragement of remedial measures, does not differ under either a negligence or strict liability theory.[143] The courts reason that the policy of encouraging correction of defects is equally applicable to both types of actions.

---

[142] *Robbins v. Farmers Union Grain Terminal Association*, 552 F.2d 788 (7th Cir. 1977); *Farner v. Paccar, Inc.*, 562 F.2d 518 (8th Cir. 1977); *Foster v. Ford Motor Co.*, 616 F.2d 1304 (5th Cir. 1980). *See generally*, Defense Research Institute *Products Liability Trial Notebook* (1982).

[143] *Cann v. Ford Motor Co.*, 658 F.2d 54 (2d Cir. 1981), *cert. denied*, 456 U.S. 960 (1982); *Werner v. Upjohn Co.*, 628 F.2d 848 (4th Cir. 1980), *cert. denied*, 449 U.S. 1080 (1981); *Knight v. Otis Elevator Co.*, 596 F.2d 84 (3d Cir. 1979).

Even in those jurisdictions where the rule is applied, and the evidence is inadmissible, the feasibility exception to the rule may permit introduction of subsequent remedial measures in the strict liability context. The feasibility exception is used where the manufacturer argues that the correction to the product was not possible or feasible at the time of manufacture. For example, if the manufacturer contends that the alternative design theory proffered by the claimant was not feasible, the subsequent changes adopted by the manufacturer may be admissible to show that this was not the case and that the manufacturer did have the knowledge and technology to correct the defect at the time of manufacture.[144] Evidence of subsequent remedial measures may also be admitted for the purpose of impeachment, typically of an expert witness, who testifies as to the feasibility of alternative designs or as to the necessity of a warning.[145]

§4.5.2   Evidence of Prior or Subsequent Accidents

The admission of evidence concerning the presence of or the absence of prior or subsequent accidents is governed by Federal Rule of Evidence 404(b):

> Evidence of other crimes, wrongs, or acts is not admissible to prove the character of a person in order to show that he acted in conformity therewith. It may, however, be admissible for other purposes, such as proof of motive, opportunity, intent, preparation, plan, knowledge, identity, or absence of mistake or accident.

Evidence of prior or subsequent accidents involving the same product is generally admissible.[146] Such evidence has been admitted to

---

[144] *Bauman v. Volkswagenwerk Aktiengesellschaft*, 621 F.2d 230 (6th Cir. 1980); *Davis v. Fox River Tractor Co.*, 518 F.2d 481 (10th Cir. 1975).

[145] *Kenney v. Southeastern Transportation Authority*, 581 F.2d 351 (3d Cir. 1978), *cert. denied*, 439 U.S. 1073 (1979); *Dollar v. Long Manufacturing Co.*, 561 F.2d 613 (5th Cir. 1977), *cert. denied*, 435 U.S. 996 (1978).

[146] *Wojciechowski v. Long-Airdox Division of Marmon Group Co.*, 488 F.2d 1111 (3d Cir. 1973); *Hoppe v. Midwest Conveyor Co.*, 485 F.2d 1196 (8th Cir. 1973). *See generally, Annot, Admissibility of Evidence of Absence of Other Accidents or Injuries*

prove that the manufacturer had notice of the defect, to establish the danger or defect, or to show that the defect caused the accident.[147] The absence of prior or subsequent accidents may be admitted to prove lack of a defect, the lack of notice of the defect, or that the alleged defect was not the cause of the accident. More generally such evidence may show that the particular product functioned over a period of time without mishap.[148]

The admissibility of such evidence may also be governed by general relevance rules,[149] or evidentiary rules which require that the probative value of evidence outweighs the danger of undue prejudice, confusion, undue delay or cumulative evidence.[150]

§4.5.3   Evidence of Product Recalls

The admissibility into evidence of a manufacturer's letter recalling a product for repair or replacement is intertwined with the subsequent remedial measure rule. This is because the recall, in most instances, may relate to the manufacturer's culpability.[151]

Because a recall notice relates to problems with a product line rather than a particular product, the courts have been reluctant to admit recall letters to prove the existence of the specific defect or to prove the defect was the cause of an accident.[152] The recall letter may be admissible, however, to show that a defect existed at the time the product was relinquished by the manufacturer.[153] As with evidence of

---

From a Customary Practice or Method Asserted to be Negligent, 42 A.L.R.2d 1055 (1955); *Annot., Admissibility of Evidence of Absence of Other Accidents or Injuries at Place Where Injury or Damage Occurred,* 31 A.L.R.2d 190 (1953).

[147] Defense Research Institute *Products Liability Trial Notebook* "Presence or Absence of Prior/Subsequent Accidents," at 51-55.

[148] *Id.*

[149] Federal Rule of Evidence 401.

[150] Federal Rule of Evidence 403.

[151] *See* §4.5.3, *supra, Subsequent Remedial Measures.*

[152] Defense Research Institute *Products Liability Trial Notebook.*

[153] *Kane v. Ford Motor Co.,* 450 F.2d 315 (3d Cir. 1971); *Manieri v. Volkswagenwerk A.G.,* 376 A.2d 1317 (N.J.Super. 1977); *Iadicicco v. Duffy,* 401 N.Y.S.2d 557 (App.Div. 1978).

subsequent or prior accidents, the courts have ruled upon the admissibility of recall letters by relying upon general rules regarding relevance[154] and probative value,[155] as well as evidence of subsequent remedial measures.[156]

In addition to general rules of relevancy, the courts have also determined the admissibility of recall letters under the hearsay rule.[157] Two of the myriad exceptions to the hearsay rule may operate in fire cases. First, the recall letter may overcome an objection that it is hearsay by denoting it as an admission.[158] Second, government agencies may open investigative files; these files may be offered under the public records exception to the hearsay rule.[159]

§4.5.4 Laws, Regulations, and Standards

Compliance or the failure to comply with statutes or regulations concerning product manufacture and design may be relevant in establishing the existence or lack of a defect.[160] This evidence may be used both to establish or rebut negligence per se in a negligence action, or evidence of product defect in a strict liability action.[161] Because strict liability actions focus on the product and not the manufacturer's conduct, the manufacturer's compliance or noncompliance with a statute or regulation is relevant, but generally is not dispositive of the issue of the product's defectiveness. It may be that the manufacturer did not comply with all statutes and regulations but

---

[154] Federal Rule of Evidence 401. *See also, Vockie v. General Motors Corp.*, 66 F.R.D. 57 (E.D.Pa.), *aff'd*, 523 F.2d 1052 (3d Cir. 1975).

[155] Federal Rule of Evidence 403. *See also Fowler v. Firestone Tire & Rubber Co.*, 92 F.R.D. 1 (N.D.Miss. 1980).

[156] Federal Rule of Evidence 407. *See also, Rozier v. Ford Motor Co.*, 573 F.2d 1332 (5th Cir. 1978).

[157] Defense Research Institute *Products Liability Trial Notebook*.

[158] Federal Rule of Evidence 801(d)(2). *See also Vockie v. General Motors Corp.*, note 154, *supra*.

[159] Federal Rule of Evidence 803(8)(c). *See also, In Re Multi-Piece Rims Products Liability Litigation*, 545 F.Supp. 149 (W.D.Mo. 1982).

[160] *See generally*, Defense Research Institute *Products Liability Trial Notebook*.

[161] *Id.*

that the product may still have been safe for its intended use. Concomitantly, full compliance will not always exculpate the product as statutes and regulations may well lag behind available technology.

In a strict liability action, however, compliance or noncompliance will likely not be sufficient to fully establish the presence or absence of a defect.[162] In addition to governmental regulations, industry standards may also be relevant, although not conclusive evidence on the issue of product defect.[163] Typically, industry standards are presented through expert testimony because the expert will be the individual most likely to have the greatest knowledge of such standards. Furthermore, experts rely upon industry standards in assessing if a product is defective, and the standard may be one of the bases for the expert's opinion.[164]

§4.6   Conflicts of Laws in Products Liability

Products liability actions pose unique conflicts of law problems. The manufacturer may be incorporated in one state, have its principal place of business in another, and its manufacturing plant in yet a third state. Wholesalers and retailers, as well as component part manufacturers, may also be found in different jurisdictions. The sale of the product may have taken place in a jurisdiction different from where the accident occurs or from where the plaintiff resides. These complexities have led to the approach adopted by the *Restatement (Second) of Conflict of Laws*. Under this formulation, the law of the state where the injury occurred will determine the rights of the parties.[165] *See generally, Annot., Modern Status of Rule That*

---

[162] *Dawson v. Chrysler Corp.*, 630 F.2d 950 (3d Cir. 1980), *cert. denied*, 450 U.S. 959 (1981); *Howard v. McCrory Corp.*, 601 F.2d 133 (4th Cir. 1979). *Simien v. S.S. Kresge Co.*, 566 F.2d 551 (5th Cir. 1978).

[163] *See generally*, Defense Research Institute *Products Liability Trial Notebook.*

[164] Federal Rules of Evidence 702-05.

[165] *Restatement (Second) of Conflict of Laws* §146 (1971), which is as follows:
    In an action for a personal injury, the local law of the state where the injury

Substantive Rights of Parties to a Tort Action Are Governed by the Law of the Place of the Wrong, 29 A.L.R.3d 603 (1970). With respect to property damage, the jurisdiction where the injury occurred also will govern the parties' rights.[166]

There may be instances, however, where another jurisdiction has a more significant relationship to the occurrence or the parties.[167] Under these circumstances, the following principles will apply in determining the applicable law:

(a)   The needs of the interstate and international systems,

(b)   The relevant policies of the forum,

(c)   The relevant policies of other interest states and the relative interests of those states in the determination of the particular issue,

(d)   The protection of justified expectations,

(e)   The basic policies underlying the particular field of law,

(f)   Certainty, predictability and uniformity of result, and

(g)   Ease in the determination and application of the law to be applied.[168]

Where the law of several jurisdictions may be applicable, consideration of which forum's law is most beneficial may be appropriate. When it is more favorable, the alternative significant relationship approach may be more useful in attempting to apply the law of another jurisdiction from that where the injury occurred. There also may be instances where the parties' contract specifically states which jurisdiction's law applies to any disputes between them.

---

occurred determines the rights and liabilities of the parties, unless, with respect to the particular issue, some other state has a more significant relationship under the principles stated in §6 to the occurrence and the parties, in which event the local law of the other state will be applied.

[166] *Restatement (Second) of Conflict of Laws* §147 (1971), which is as follows:

In an action for an injury to land or other tangible thing, the local law of the state where the injury occurred determines the rights and liabilities of the parties unless, with respect to the particular issue, some other state has a more significant relationship under the principles stated in §6 to the occurrence, the thing, and the parties, in which event the local law of the other state will be applied.

[167] *See* notes 165 & 166 *supra.*

[168] *Restatement (Second) of Conflict of Laws* §6 (1971).

These choice-of-law clauses are favored by the courts and generally upheld.

§4.7   Limitations of Action in Products Liability

In a products liability action, various dates may trigger the accrual of the applicable statute of limitations. These dates may include the date of the product's manufacture or sale, the date on which the injury was sustained, or the date the injury was discovered.[169] Typically, the statute applicable to a strict liability action will be the same statute applicable to a negligence action.[170]

In some jurisdictions, however, newly-enacted special statutes of limitations apply solely to products liability claims.[171] Many of these statutes require not only that the action be brought within a certain number of years from the date of injury, but also that the action be brought within a certain number of years from the date of manufacture or delivery.[172]

In addition to special product liability statutes, many jurisdictions have adopted what have become known as "completion statutes." These statutes apply to architects, engineers and builders, and require that actions be brought against these individuals for damage to real property within a stated number of years from the date construction is completed.[173] Repairs or modifications to property after its initial

---

[169] See generally, Annot., Products Liability: What Statute of Limitations Governs Actions Based on Strict Liability in Tort, 91 A.L.R.3d 455 (1979).

[170] Id.

[171] See, e.g., Ala. Code §6-5-502; Colo. Rev. Stat. §13-21-403; Neb. Rev. Stat. §25-224; Or. Rev. Stat. §30.905; Tenn. Code Ann. §29-28-103.

[172] Dague v. Piper Aircraft Corp., 513 F.Supp. 19 (N.D.Ind. 1980); Wilson v. Dake Corp., 497 F.Supp. 1339 (E.D.Tenn. 1980); Baird v. Electro Mart Factory Direct, Inc., 615 P.2d 335 (Or.App. 1980).

[173] Architects, Engineers and the Statute of Limitations, Defense Research Institute Monograph Vol. 1982 No. 1.

completion can be an important fact to establish, for the date of completion of the repair or modification should begin the statute running anew. These laws are delineated as statutes of repose and may destroy the right of recovery regardless of when it would accrue under a general negligence or tort statute.[174]

For breach of warranty actions the limitation period is set forth in the UCC.[175] The action must be commenced within four years after accrual.[176] Accrual is defined as the date the breach occurs, or when tender of delivery is made.[177] The four-year period may be shortened in the contract to not less than one year and may be lengthened when coupled with a warranty of future performance.[178] Warranties of future performance may include, for example, guarantees that the product will perform for a specific number of years in excess of the statutory limitation period.

Finally, many jurisdictions employ borrowing statutes which apply another jurisdiction's period of limitation, whether longer or shorter. Borrowing statutes are typically used when the claim arises in one jurisdiction and is commenced in another.[179]

---

[174] *Id.*

[175] UCC §2-725. *See generally, Annot., What Statute of Limitations Applies to Actions for Personal Injuries Based on Breach of Implied Warranty Under UCC Provisions Governing Sales (UCC §2-725(1)),* 20 A.L.R.4th 915 (1983).

[176] UCC §2-725(1).

[177] UCC §2-725(2).

[178] UCC §2-725(1)&(2).

[179] *See generally, Annot., Choice of Law As to Applicable Statute of Limitations and Contract Actions,* 78 A.L.R.3d 639 (1977).

CHAPTER 5

THE CIVIL ARSON CASE

§5.1 Introduction

Arson is the criminal act of malicious or fraudulent burning of property. In recent years, the number of reported arson fires has grown drastically. Such fires affect both life and property, and account for a large number of deaths, injuries and millions of dollars in property damage each year. Moreover, they result in increased insurance costs.

The National Fire Protection Association estimates that in 1981, property losses in the United States due to incendiary or suspicious fires amounted to 1.658 billion dollars. (Figure based on reported data submitted by fire departments for that year.) Furthermore, these fires affected some 154,500 structures and resulted in 820 deaths. As a result, numerous municipalities and government agencies have formed task forces and increased their investigative activities into the cause and origin of incendiary fires. Insurance industry task forces have been formed to increase public awareness of the arson problem and to develop measures for the detection and prevention of arson fires and insurance fraud.

Although arson is a criminal act which may result in prosecution, there are also civil aspects to arson cases. The civil case can arise in many contexts, including arson by vandals, the criminally insane, or the mentally impaired; arson for revenge; and in the context of negligence creating the occasion for arson. This last aspect of civil cases was discussed in §2.4 of this handbook. Finally, intentional tort theories, based upon acts against the arsonist or those in conspiracy with him or her, can also be applied in the above contexts.

It is "arson for profit," however, that is the most common focus of the civil case, and the focus of this chapter. These are, of course, those arson cases that arise in the context of potential insurance fraud. The factual context of such cases generally involves: 1) the submission of an insurance claim pursuant to a fire insurance policy, 2) facts indicating a deliberate burning, and 3) possible involvement of the insured or those in concert with the insured in submitting a potentially fraudulent insurance claim.

In such cases, an "arson defense" operates to relieve the insurer of liability for damage sustained by the claimant. The essential elements of a civil "arson defense" parallel the evidence utilized by a prosecutor in demonstrating violations of the criminal law. However, the elements of proof are not identical and the burdens of proof are materially different. Moreover, the courts have allowed great latitude in the introduction of circumstantial evidence which bears on the issue of "involvement" of the insured in arson and fraud.

The "arson defense" is discussed in detail in the following section.

§5.2   The "Arson Defense" in an Insurance Case

The "arson defense" to an insurance claim arises where a claim is made for insurance proceeds under circumstances where the fire investigation points to arson as the cause of the fire. Not every case involving an incendiary fire and an insurance claim gives rise to the defense. Several kinds of arson do not give rise to an inference of involvement of the insured and, therefore, will not constitute a defensible arson case. These include arson committed by third persons without connection to or participation of the insured, arson involving vandalism by juveniles or vagrants, and arson where motive or opportunity on the part of the insured are not indicated. Other circumstantial facts from which a jury can infer the involvement or participation of the insured in the incendiary fire must be present. It is only these cases in which an arson defense may be raised successfully.

An examination of the standard fire insurance policy will quickly disclose that "arson" is not explicitly mentioned anywhere in the document. "Arson" is subsumed by the larger issue of fraud. Thus, the seminal ingredient in the "arson defense" is that procuring an insurance recovery by reason of one's own wrongful act (in this circumstance — the act of arson) constitutes a fraud on the insurer and a violation of public policy. In most states, the standard fire insurance policy (also referred to as the standard "165-line" policy) provides in lines 1-6:

> This entire policy shall be void if, whether before or after a loss, the insured has willfully concealed or misrepresented any material fact or circumstance concerning this insurance or the subject thereof, or the interest of the insured therein, or in case of any fraud or false swearing by the insured relating thereto.

In most jurisdictions, reimbursement of an insured is not permitted for loss or damage incurred as a result of the insured's willful wrongdoing. Such claims generally violate public policy; accordingly, persons may not insure themselves against the economic consequences of their own intentional wrongdoing, nor utilize fraudulent means to achieve an insurance recovery. Thus arson for the purpose of achieving insurance recovery clearly falls into the category of fraudulent conduct which will render an insurance policy void from the outset.[1] Adequate proof of arson by or with the involvement of the insured is sufficient to sustain the "arson defense" and will render an insurance policy void and deny any recovery by an insured.

In addition to providing the contractual and theoretical basis for the "arson defense," this language is also utilized to defeat policy claims involving other types of fraud or false swearing by the insured. These may relate to other material facts concerning the fire, the claim itself, or the amount of the claim. Examples of these types of fraud or false swearing include the insured's willful concealment of the fire's cause and origin, claims for contents previously removed from the premises or previously claimed as lost or stolen under other insurance policies, and the insured's intentional overvaluation of the amount of the loss sustained.

The concealment and fraud language must be considered in conjunction with what is known as the "sworn statement in proof of loss." In most fire policies, standard language requires the presenta-

---

[1] *See, e.g., Miller & Dobrin Furniture Co. v. Camden Fire Insurance Co.*, 150 A.2d 276 (N.J.Super. 1959); *Ruvolo v. American Casualty Co.*, 198 A.2d 204 (N.J. 1963); *See also 7 Appleman, Insurance Law & Practice*, §4252 (1962).

tion of an insurance claim under such an oath. The "sworn statement in proof of loss" clause typically contains a representation by the insured, given under oath, which states:

> The said loss did not originate by any act, design or procurement on the part of your insured, or this affiant; nothing has been done by or with the privity or consent of your insured, or this affiant, to violate the conditions of the policy, or render it void; no articles are mentioned herein or in annexed schedules but such as were destroyed or damaged...and no attempts to deceive the said company, as to the extent of said loss, has in any manner been made....

The legal and contractual bases for the defense of arson, and the defense of fraud and of false swearing by the insured, lie in the concealment and fraud lines of the policy and in the "sworn statement in proof of loss." The insured's concealment of the act of arson; a claim for contents, not damaged; or an overvaluation of the amount of the claim on the proof of loss form constitutes fraud, under lines one to six of the standard fire policy, and false swearing as these statements by the insured are made under oath. The proof of loss form is especially important where the insureds have overvalued the claim, or claimed contents which were not damaged. The insured swears under oath that the claim is accurate and that there has been no attempt at deception.

§5.3 Other Relevant Insurance Clauses

§5.3.1 The Cooperation Clause

The standard insurance contract provides that, after a fire loss, the insurer is entitled to receive the itemized claim of the insured accompanied by a sworn statement in proof of loss. The latter item must contain representations by the insured that the claim is truthful and accurate and that the loss itself was not procured by act or design on the part of the insured or those acting under its direction or control.

To effect the submission of such a claim, the standard fire insurance policy contains what is known as a "cooperation clause."

The cooperation clause is important for several reasons. First, it gives the insurer investigative rights which are broader and more valuable than those arising from traditional discovery. The insurer is given the right, *prior* to the commencement of suit or the decision to pay or decline the claim, to examine the insured under oath, and to have the insured produce relevant documents, books and accounts as often as may reasonably be required. In doing so, the insurer may determine its liability and protect itself against potential fraudulent claims.

The language of the standard cooperation clause in effect in most jurisdictions is as follows:

> The insured, as often as may be reasonably required, shall exhibit to any person designated by this Company all that remains of any property herein described, and submit to examinations under oath by any person named by this Company, and subscribe the same; and, as often as may be reasonably required, shall produce for examination all books of account, bills, invoices and other vouchers, or certified copies thereof if originals be lost, at such reasonable time and place as may be designated by this Company or its representative, and shall permit extracts and copies thereof to be made.

The right to examine an insured under oath is an invaluable investigative tool which enables the insurer to effectively evaluate and successfully deny fraudulent claims — particularly, those which involve arson by the insured. It is also a vehicle for eliminating an initial denial of the insured's claim, prior to suit, when the facts disclose no inferences of involvement by the insured. This examination may encompass the books and accounts of the insured, as well as the right to examine the damaged property "as often as may be reasonably required."

In practice, the insurer will designate counsel or an experienced claims investigator or examiner to review the property and examine the insured. A refusal by the insured can have drastic consequences. The right to examine is contractual and is set forth in the insurance policy itself. A breach by the insured, either by refusing to submit to an examination under oath or refusing to submit its books or other records, is grounds for the insurer to deny the claim and successfully defend any subsequent lawsuit over the proceeds of the policy. Similarly, untimely or unreasonable demands for examination may result in waiver of the right by the insurer. The insurer must make its demand for an examination under oath during its investigation of the claim. A request for an examination after the insurer has decided to deny a claim will likely prove unsuccessful in a noncooperation defense in the event of a later suit by the insured. In addition, the insurer must use caution in defining the scope of its examination. Denial of a claim for refused requests for documents or other information not relevant to the claim, the insured's financial status, ownership of the property or the insured's whereabouts at the time of the fire will likely pose unsuccessful defenses. Therefore, a close evaluation of procedural regularity of these requests should be made.

The rationale behind the cooperation clause was set forth in the landmark United States Supreme Court case of *Claflin v. Commonwealth Insurance Company of Boston*, 110 U.S. 76 (1884), where the Court commented:

> The object of the provisions in the policy of insurance, requiring the assured to submit himself to an examination under oath, to be reduced to writing, was to enable the company to possess itself with all knowledge, and all information as to other sources and means of knowledge, in regard to the facts, material to its rights, to enable it to decide upon its obligations, and to protect itself against false claims. And every interrogatory that was relevant and pertinent in such an examination was material, in the sense of a true answer to it was of the substance of the obligation of the insured. A false answer as to any matter of fact, material to the inquiry, knowingly and willfully made, with intent to deceive the insurer, would be fraudulent. If it accomplished its result it would be a fraud effected; if it failed, it would be a fraud attempted. And, if the matter were material and the statement false, to the knowledge of the party making it, and willfully made, the intention to deceive the insurer would be necessarily implied for the law presumes every man to intend the natural consequences of his act.

Id. at 82.

It is generally held that, during a sworn examination under oath, the scope and latitude of an insurer's inquiry is extremely broad, and may touch upon all matters material to the loss, as well as the potential motive or opportunity of the insured to be involved in fraud or arson. Furthermore, the veracity of statements made by the insured during the examination bears heavily on the additional policy defense of fraud or false swearing outlined in lines one through six of the standard policy. As a matter of law, false statements which concern material facts knowingly and willfully made by the insured in testimony given at an "examination under oath" proceeding, in a sworn statement in proof of loss, or in schedules supporting the claim will constitute grounds to void the policy's coverage as a matter of law.[2] In the hands of the insurer, the cooperation clause can be an extremely effective tool in gathering evidence from the insured concerning possible motives for fraud, an opportunity to have committed arson or have procured someone to commit the act, and other inculpatory evidence.

§5.3.2 Fraud and False Swearing Clause

The first six lines of the standard fire insurance policy (§5.2) contain basic language concerning concealment and fraud which forms the basis of an arson defense. In addition it also forms the basis for other defenses under the policy, the first of these is the "false swearing" defense. It operates as follows. Under the cooperation clause the insurer may examine the insured under oath and obtain the insured's sworn statements concerning relevant or material facts about the loss and claim. If the insured knowingly gives false answers concerning material facts during this sworn examination, the law will

---

[2] *Gregory's Continental Coiffures & Boutique, Inc. v. St. Paul Fire & Marine Insurance Co.*, 536 F.2d 1187 (7th Cir. 1976); *Esquire Restaurant, Inc. v. Commonwealth Insurance Co. of New York*, 393 F.2d 1111 (7th Cir. 1965); *Claflin v. Commonwealth Insurance Co.*, 110 U.S. 76 (1884); *Miele v. Boston Insurance Co.*, 288 F.2d 178 (8th Cir. 1961); *Watertown Fire Insurance Co. v. Grehan*, 74 Ga. 642 (1885); *Kantor Silk Mills, Inc. v. Sentry Insurance Co.*, 228 N.Y.S. 822 (App.Div. 1928), *aff'd*, 171 N.E. 793 (N.Y. 1930); *Ellis v. Agricultural Insurance Co.*, 7 Pa.Super. 264 (1898).

imply the intent to deceive from the false statement and the insurer will be provided with a defense to the claims presented under the policy.[3]

The fraud and false swearing defense is categorized as "contractual fraud" rather than common-law fraud. The essential distinction between the two is as follows. Successful common-law fraud and misrepresentation defenses require proof of reliance on the false statement and prejudice to the victim as a result of the reliance. These two elements are not required for a successful defense of contractual fraud or false swearing under an insurance policy.

A number of cases recognize this distinction and have held that it is not necessary for the insurer to show that it was actually deceived, misled or prejudiced by false statements given by the insured in order to successfully establish the defense of false swearing under the policy. Even attempted fraud, when willfully or intentionally perpetrated, operates as a bar to an insured's claim.[4] The fraud need not be successful, it need only be attempted. For example, it is not necessary for the insurer to pay a fraudulent claim. In fact, an insurer may deny such a claim and rest on fraud as a successful defense. Under common-law fraud, payment by the insurer or a detrimental change in position would be necessary to establish the fraud defense.

The submission of exaggerated claims or deliberate overevaluation of damages is a recurring problem in false swearing cases. Pursuant to the fraud and concealment clauses of the policy, this behavior may constitute a fraud and false swearing defense under the policy.[5]

Not every overvaluation or exaggeration of an insurance claim will provide a basis for a fraud and false swearing defense; however, the extent and degree to which a claim is exaggerated or overvalued will present jury questions on the issue of intent — the greater the

---

[3] *Claflin v. Commonwealth Insurance Co.*, 110 U.S. 76 (1884); *Esquire Restaurant, Inc. v. Commonwealth Insurance Co.*, 393 F.2d 111 (7th Cir. 1965); *American Diver's Supply & Manufacturing Corp. v. Boltz*, 482 F.2d 795 (10th Cir. 1973).

[4] *Id.* See also *Ellis v. Agricultural Insurance Co.*, 7 Pa.Super. 264 (1898); *Chaachou v. American Central Insurance Co.*, 241 F.2d 889 (5th Cir. 1957).

[5] *American Diver's Supply & Manufacturing Corp. v. Boltz*, 482 F.2d 795 (10th Cir. 1973); *Esquire Restaurant, Inc. v. Commonwealth Insurance Co.*, 393 F.2d 111 (7th Cir. 1965); *Chaachou v. American Central Insurance Co.*, 241 F.2d 889 (5th Cir. 1957); *Miele v. Boston Insurance Co.*, 288 F.2d 178 (8th Cir. 1961).

overvaluation, the more persuasive the inference of intentional fraud. The overvaluation or exaggeration of the claim must be intentional, not merely a mistake or the result of negligence in computing the loss.

§5.4   Burden and Elements of Proof

§5.4.1   Burden of Proof

The "arson defense" is most often raised as an affirmative defense to a breach of contract suit brought by an insured. As is the case with most affirmative defenses, in both contract and tort litigation, the person raising the affirmative defense bears the burden of proof to establish the defense by sufficient evidence. To establish a prima facie case on the issue of liability of the insurer is relatively simple — the insured must establish the existence of a policy of insurance, and the occurrence of a covered loss (e.g., fire) which caused damage. Proofs on damages may be substantially more difficult.

Once a prima facie case of liability has been established by the insured, the burden then shifts to the insurer in establishing an arson or fraud defense sufficient to defeat the insured's claim to the policy proceeds. In this, the insurer must come forward with proofs to establish that the policy is void, suspended or restricted by virtue of the specific defense raised.[6]

§5.4.2   Elements of Proof

The insurer's affirmative defense of arson must be supported by evidence showing: 1) that the fire was of incendiary origin (deliberately set); and 2) that the insured was in some way fraudulently

---

[6] *Elgi Holding Co. v. Insurance Co. of North America*, 511 F.2d 957 (2d Cir. 1975); *Boone v. Royal Indemnity Co.*, 460 F.2d 26 (10th Cir. 1972); *Hanover Fire Insurance Co. v. Argo*, 251 F.2d 80 (5th Cir. 1958); *Greenberg v. Aetna Insurance Co.*, 235 A.2d 582 (Pa. 1967).

connected with the incendiary fire.[7] Both elements of proof may be established solely by circumstantial evidence. (See §5.5.1 *infra*.)

A succinct statement of the essential elements of proof of the arson defense appears in the case of *Mele v. All-Star Ins. Corp.*, 453 F.Supp. 1338 (E.D. Pa. 1978), where the court stated as follows:

> "In a civil matter to determine whether a reasonable inference exists that an insured is responsible for the fire which damaged the insured's property, a jury should consider the combination of evidence of: 1) an incendiary fire; 2) a motive by the insured to destroy the property; and 3) circumstantial evidence connecting the insured to the fire."
>
> *Id.* at 1341.

It is not required that the insurer prove that the fire was personally set by the insured. As long as there is sufficient circumstantial evidence to show that the fire was incendiary and that it was more probable than not that it was set by someone acting under the control of the insured, the insurer's defense can be sustained.[8] The defense will also be sustained with proofs which establish preloss knowledge by the insured of the incendiary fire or of the responsibility for causing the fire either directly or through the procurement by others.[9]

§5.4.3 Measure of Burden of Proof

The arson defense to an insurance claim is an affirmative defense to a civil suit on a contract. Accordingly, proof of this affirmative defense is no different from any other civil defense and is governed by the standard denoted as a "simple preponderance of the evidence." Most

---

[7] *Hanover Fire Insurance Co. v. Argo*, 251 F.2d 80 (5th Cir. 1958); *Mele v. All-Star Insurance Corp.*, 453 F.Supp. 1338 (E.D.Pa. 1978); *Miller & Dobrin Furniture Co. v. Camden Fire Insurance Co.*, 150 A.2d 276 (N.J.Super. 1959).

[8] *Don Burton, Inc. v. Aetna Life & Casualty Co.*, 575 F.2d 702 (9th Cir. 1978).

[9] *Boone v. Royal Indemnity Co.*, 460 F.2d 26 (10th Cir. 1972); *Fratto v. Northern Assurance Co.*, 242 F.Supp. 262 (W.D.Pa.); *aff'd*, 359 F.2d 842 (3d Cir. 1965); *Greenberg v. Aetna Insurance Co.*, 235 A.2d 582 (Pa. 1967).

jurisdictions have adopted this standard even though technically the defense imputes criminal activity to the insured.[10]

By contrast, in a criminal case of arson, the government must establish the defendant's guilt "beyond a reasonable doubt." Virtually, no American jurisdiction requires proof beyond a reasonable doubt in a civil case. However, because of the criminal overtones in a civil arson defense case, several jurisdictions have adopted the intermediate standard of proof of "clear and convincing" evidence.[11] This is clearly true in only a small minority of American courts.

To further complicate the proof picture, some courts have variously applied both the "preponderance" standard and the "clear and convincing" standard in the same factual situation. Here it has been held that proof of the arson defense and involvement of the insured in the procurement of an arson is to be judged by the simple preponderance standard; proof of fraud in connection with submission of the claim or false swearing must be proven by the clear and convincing standard.[12] Fortunately, this also is a minority view.

In those jurisdictions using the "clear and convincing" standard, the rationale is that the insurer is, in essence, attemping to set aside a written contract by reason of the fraud of one party. At common law, in order to set aside a written instrument, proof of fraud was to be established by clear and convincing evidence.

On the other hand, two Pennsylvania state cases and a Federal decision (District Court for the Eastern District of Pennsylvania) have best articulated the rationale for applying the simple preponder-

---

[10] *Goodwin v. Farmers Insurance Co.*, 646 P.2d 294 (Ariz.App. 1982); *Miller & Dobrin Furniture Co. v. Camden Fire Insurance Co.*, 150 A.2d 276 (N.J.Super. 1959); *Frizzy Hairstylists, Inc. v. Eagle Star Insurance Co.*, 392 N.Y.S.2d 554 (Misc.), *modified on other grounds*, 403 N.Y.S.2d 389 (Misc. 1977).

[11] *Rent-A-Car Co. v. Globe & Rutgers Fire Insurance Co.*, 156 A.847 (Md. 1931); *Goodman v. Poland*, 395 F.Supp. 660 (D.Md. 1975); *Hutt v. Lumbermens Mutual Casualty Co.*, 466 N.Y.S.2d 28 (App.Div. 1983).

[12] *Honeycut v. Aetna Insurance Co.*, 510 F.2d 340 (7th Cir. 1975); *Clamin v. Aetna Casualty Co.*, 501 P.2d 750 (Colo.App. 1976); *Warners Furniture, Inc. v. Commerical Union Insurance Co.*, 349 N.E.2d 616 (Ill.App. 1976); *George v. Travelers Indemnity Co.*, 265 N.W.2d 59 (Mich. 1978); *Quast v. Prudential Property & Casualty Co.*, 267 N.W.2d 493 (Minn. 1978); *Great American Insurance Co. v. Cape W. Log, Inc.*, 591 P.2d 457 (Wash.App. 1979).

ance rule. In *Greenberg v. Aetna Ins. Co., supra,* the court's rationale was stated as follows:

> Where an individual is charged with and tried for a crime, before he may be convicted, the Commonwealth must establish his guilt beyond a reasonable doubt. In civil cases, however, the extreme caution and unusual degree of precaution required in criminal cases do not obtain. Even though the result may imput a crime, the verdict should follow the preponderance of the evidence. *Somerset County Mutual Fire Ins. Co. v. Usaw,* 112 Pa. 80, 4 A.355 (1886). As the Court stated therein "in a civil issue, the life or liberty of the person whose act is sought to be proved is not involved, proof of the act is only pertinent because it is to sustain or defeat a claim for damages or respecting the right of things...The act affirmed is an incident, a fact to be proved like other pertinent facts" *Id.* at 90, 4 A. at 357. It has been advocated in many jurisdictions that wherever in a civil case a criminal act is charged as part of the case, the rule as to the standard of proof controlling in a criminal case should apply. However, this position has been repudiated in most instances. *See Wigmore on Evidence,* 3d Ed. (1940) Vol. IX, §2498. The rule generally followed in most jurisdictions in such situations is that the criminal act need only be established by a fair preponderance of the evidence. *See Wigmore on Evidence, supra.* This was the rule adopted in Pennsylvania nearly a century ago in a case where, like here, a recovery was sought on the fire insurance policy and the defense of arson was pleaded to defeat the claim...The same rule has been followed consistently in our court in similar cases ever since.

427 Pa. at 496.

This same issue was addressed by a Pennsylvania appellate court in the well-reasoned decision in *Ruttenberg v. Fire Association of Philadelphia,* 122 Pa. Super. 363, 186 A. 194 (1936):

> The defense that the fire was caused was an affirmative one, and the burden was therefore on the defendants to prove that the fire was set, or was caused to be set, by the insured. It was not necessary that the proof was beyond a reasonable doubt as is necessary when a charge is made against a defendant in a criminal case. Proof of the insured's fraudulent connection with the cause of the fire by a preponderance of the evidence was sufficient.

122 Pa. Super at 365.

Finally, then Chief Judge Lord of the United States District Court for the Eastern District of Pennsylvania discussed the quantum of proof needed to support an arson defense in *Mele v. All-Star Insurance Corp.*, 453 F. Supp. 1338, 1341 (E.D. Pa. 1978):

> The standard is whether the evidence supported a reasonable and legitimate inference that the insured fraudulently burned the building or caused it to be burned. *Ruttenberg v. Fire Association of Philadelphia*, 122 Pa. Super 363, 370, 186 A. 194, 196 (1936). A reasonable inference is one not based on speculation or conjecture but rather is a logical consequence deduced from other proven facts. *Commonwealth v. Whitman*, 199 Pa. Super 631, 634, 186 A.2d 632, 633 (1962). Although this inference must be reasoned from evidence presented, it need not be the only logical conclusion which a jury could reach. *Smith v. Bell Telephone Co.*, 397 Pa. 134, 138, 153 A.2d 477, 480 (1959).

Most jurisdictions have adopted the rationale of the Pennsylvania cases by applying the "preponderance of the evidence" test to an insurer's affirmative defense of arson. Burden of proof requirements, however, must be carefully evaluated to determine the proper quantum of proof required in a particular jurisdiction.

§5.5   Evidence in Insurance Arson Cases

§5.5.1   Use of Circumstantial Evidence in General

Experience shows that proof of an arson defense by an insurer is generally satisfied with relevant pieces of circumstantial evidence that demonstrate:
1) an incendiary fire, and
2) the insured's fraudulent connection with the fire through circumstances showing motive, opportunity or other inculpatory facts.

The authorities strongly support the insurer's right to predicate its defenses solely upon circumstantial evidence sufficient for the jury to reasonably infer either that the insured knew that a fire of incendiary origin was to take place at the premises or was fraudulently connected in some manner with setting the fire.[13] It is not necessary that the insurer prove, by direct evidence, that the insured or any specific member of his or her family or agents were the actual arsonists.[14]

Analysis of case law shows that a very broad standard is employed by the courts in determining the admissibility of circumstantial evidence in cases of this type.[15] The rule as to circumstantial evidence in an arson defense case has been stated as follows:

> Any issue may be proved by circumstantial evidence, or by a combination of direct and circumstantial evidence. A well connected train of circumstances is as cogent of the existence of a fact as any array of direct evidence, and may outweigh opposing direct testimony; and the concurrence of well authenticated circumstances has been said to be stronger evidence than positive testimony unconfirmed by circumstance.

32 C.J.S. *Evidence* at 1109-10.

The nature of arson/fraud cases often demands that circumstantial evidence is the only type of evidence available. As the court perceptively noted in *Girard v. Vermont Mutual Fire Insurance Co.,* 154 A.666, 668-69 (1931) (citations omitted):

> As we have often said, when fraud is the issue, the evidence necessarily takes a wide range...so here, though the fraud in its

---

[13] *Cora Pub v. Continental Casualty Co.,* 619 F.2d 482 (5th Cir. 1982); *Crown Colony Distributors, Inc. v. United States Fire Insurance Co.,* 510 F.2d 544 (5th Cir. 1975); *Boone v. Royal Indemnity Co.,* 460 F.2d 26 (10th Cir. 1972); *Miele v. Boston Insurance Co.,* 288 F.2d 178 (8th Cir. 1961); *Carpenter v. Union Insurance Society,* 284 F.2d 155 (4th Cir. 1960); *Stein v. Girard Insurance Co.,* 259 F.2d 764 (7th Cir. 1958).

[14] *Collins v. Fireman's Fund Insurance Co.,* 296 F.2d 562 (7th Cir. 1961); *Stein v. Girard Insurance Co.,* 259 F.2d 764 (7th Cir. 1958); *Ruttenberg v. Fire Association of Philadelphia,* 186 A.194 (Pa.Super. 1936); *Miller & Dobrin Furniture Co. v. Camden Fire Insurance Co.,* 150 A.2d 276 (N.J.Super. 1959).

[15] *McIntosh v. Eagle Fire Insurance Co.,* 325 F.2d 99 (8th Cir. 1963); *Raphtis v. St. Paul Fire & Marine Insurance Co.,* 198 N.W.2d 505 (S.D. 1972); *Girard v. Vermont Mutual Fire Insurance Co.,* 154 A. 666 (Vt. 1931).

ultimate aspects was the burning of the building, any fact or circumstance, before or after that event, which in any way indicated a purpose to accomplish that fraudulent result, was admissible. Indeed, that ultimate fact might be wholly established by circumstantial evidence...and, when such evidence is resorted to, objections to testimony on the ground of irrelevancy are not favored because the force and effect of circumstantial facts depends largely upon their relation to each other; and facts and circumstances, although wholly inconclusive when separately considered, may by their number and joint operation, be sufficient to establish the factum probandum.[16]

The rationale expressed in *Farber v. Boston Insurance Co., supra,* also clearly illustrates this point:

I cannot subscribe to the view that the court should have taken from the jury the issue as to whether the insured burned his property. To establish the rule really enunciated in Judge Arnold's opinion would, it seems to me, enable an owner to safely burn his property without fear of losing his insurance, provided he was careful enough to let no one see him burn it and arrange matters so that the fire would destroy all evidence of its origin. If the rule announced becomes the law, then in such cases, an insurance company cannot hereafter hope to successfully defend on the grounds of the owner's incendiarism, unless it is able to produce evidence that the owner or his agent was *actually* seen to apply the firebrand.

256 S.W. at 1079.[17]

---

[16] *See also,* 46 C.J.S., *Insurance* §1338 at 487-88 (1946), where it is noted:

Where the defense of incendiarism is interposed in an action on a fire insurance policy, all competent evidence tending to prove the issue is admissible. Accordingly, evidence tending to show a possible motive for the insured to destroy the property, such as evidence of the insured's financial condition, evidence of the removal of personal property from the burned building shortly before the fire occurred, and evidence of the value of the property destroyed, has been held to be admissible.

See also, 46 C.J.S., *Insurance* §1338 at 487 (1946) (footnotes omitted), wherein it is stated: "Objections on the ground of irrelevancy are not favored when circumstantial evidence is offered to show a fraudulent fire."

[17] *See Ruttenberg v. Fire Association of Philadelphia,* 186 A.194 (Pa.Super. 1936), where the court recognized that, where criminal acts are involved, they are usually committed in secret and, hence, it is difficult to obtain direct evidence. Accordingly,

The rationale behind the liberal admissability of circumstantial evidence was well-stated in *Weiner v. Aetna Insurance Co. of Hartford*, 259 N.W. 507 (Neb. 1953):

> Persons desiring to burn their property for the purpose of collecting the insurance, or for any other illegal purpose, do not discuss their intentions with others nor do they carry out such intentions in the light of day.

A typical chain of circumstantial evidence leading to a finding of arson is as follows: An increase in the amount of fire insurance coverage shortly before the fire; the insured's financial distress; secured premises at the time the police or fire fighters reach the scene; the insured is the only individual with keys; and an incendiary fire.

§5.5.2   Proof of Incendiarism

The first essential element of proof in an arson defense case is the incendiary nature of the fire — i.e., that the fire was deliberately set. Proof of incendiarism will primarily derive from the on-site investigation and examination performed by a qualified cause and origin expert.

The same general mode of fire investigation is employed in examining the scene of an incendiary fire as is used in an accidental fire. Due to the many varieties of building construction, arrangements of storage or contents, and other factors, fire patterns may vary greatly from one fire to another. There are, however, recognized differences in appearance and physical evidence which distinguish accidental fires from those which are deliberately set. For example, heavy burning near the floor as well as pour patterns on a floor indicate the use of accelerants such as gasoline.

The ingenuity employed by many arsonists and the unique nature

---

proof derived from circumstances present questions of natural presumptions which are to be found by a jury. Where the combined circumstances create inferences showing the probability of incendiarism by the insured, such matters are to be submitted for jury resolution as a matter of right.

of many arson fires prevents one from making general statements in this realm. Rather, the circumstantial facts developed by the investigator at the scene should be considered as building blocks in the construction of the arson defense. Each piece of circumstantial evidence represents a sign or an indicator of the cause and origin of the fire, and it is the cumulative sum of these facts that must be developed into the overall conclusion as to the fire's cause and origin.

The initial investigation should be primarily focused on determining the cause of the fire, whether it appears accidental or otherwise. During the course of this examination, there are certain indicia which will point to the existence of an incendiary fire. Each of these is analyzed in the italicized paragraphs that follow.

Multiple Points of Origin. Evidence indicating multiple points or areas of origin is a common indicator of incendiarism. A prime indicator of incendiarism is established when two or more points or areas of fire origin appear to be separate and unconnected, and where the on-site investigation discloses no accidental or natural cause for the separate and unconnected fires. In nearly all such cases, a complete examination of the scene is required in order to explain why the multiple fires could not have been caused by a natural phenomenon during the course of the fire. Common factors, such as radiation, conduction, convection, fire drop causing secondary fires, or flashover, must all be eliminated.

It should be remembered that, in order to destroy the property, the objective of an arsonist is to propagate the fire to as great an extent as possible. To achieve this end, it is sometimes necessary to start separate fires at strategic points within a building in order to achieve maximum destruction. Such multiple origin points are in direct contrast with the single points of origin generally associated with accidental fires. This does not mean that an incendiary fire itself cannot be one with a single point of origin. What is suggested, however, is that without adequate explanation, unconnected multiple origin points of a fire are more than likely to indicate an incendiary fire.

Accelerants. A second indicator of incendiarism is the use of accelerants to both speed up the fire and to enhance its destructive effect. Obviously, in enhancing the speed of the fire, the accelerant will cause destruction to occur before the fire fighters can extinguish the fire.

Although flammable liquids are most frequently employed by arsonists as accelerants, there are numerous other types of accelerants which can be used. Because of the natural flow of a liquid away from the point where it is poured, flammable liquid fires portray an *area* rather than a *precise point* of origin. In many instances, arsonists pour a flammable liquid over a large area, splashing or trailing it throughout a room or structure in order to achieve the desired result.

The presence of possible accelerants or flammable liquids in a structure is in itself not condusive evidence of arson. It must be determined if such accelerants are normal to the occupancy. Obviously, the presence of gasoline, kerosene or paint thinners in a living room couch or in a cabinet containing the financial records of a business would be highly suspicious. On the other hand, evidence of flammable liquids in occupancies where such substances are ordinarily stored may pose more difficult problems with regard to proof of incendiarism. However, in such circumstances, numerous other factors must be evaluated in order to reach a determination of incendiarism. These center around such indicia as burn patterns, liquid vapors, and expected damage levels.

There are various indicators which point to the use of accelerants from evidence found in fire debris and in burn patterns. For example, most accidental fires will not result in extensive charring to a floor or at the base of walls. Extensive floor damage indicating irregular shaped burn patterns may indicate the use of an accelerant. Dramatic lines of demarcation between heavily charred floor boards and unburned floor boards are further indicia that a flammable liquid has been poured in random areas in the building. By its nature, a flammable liquid will flow and settle at low points in a structure. Therefore, its presence can often be detected at locations such as corners, along the base of a wall, between floor boards, behind baseboards, or under tacked carpets.

In many instances, flammable liquid accelerants will soak into wood flooring, causing holes to burn through the floor in spots where the liquid has puddled. An examination of this type of burning can reveal that the floor boards have burned through from above by reason of the use of a flammable liquid. These physical facts, when coupled with a finding that the presence of a flammable liquid would be unusual, can be convincing proof of incendiarism.

Other indicators of the use of accelerants are closely related to the condition of fire debris and burn patterns. These include the blistering of floor tiles, and the presence of splash patterns from liquids which may appear on walls. In the latter circumstance, the liquid splashed on a wall will run downward and, when ignited, leave a distinctive pattern on the wall as compared to those remaining parts of the wall merely subjected to radiant heat. Finally, the degree of "alligatoring," which is essentially the depth of charring in wood members, may also suggest the use of an accelerant.

Second, an indication that flammable liquids may have been employed may come from the examination of low level burns on furniture or other objects extant in the area where the use of flammable liquid is suspected. Such "low burns" are often found on the undersides of furniture, the bottom edges of door frames, moldings and in other areas.

Third, it also should be noted that flammable liquids, in many instances, are not entirely consumed during the course of a fire but leave residue in materials which soak up the liquid. These include floor boards, carpeting, fabric and even concrete. Therefore, properly preserved samples of these materials can later be analyzed through gas chromatography or mass spectroscopy to identify the precise liquid used. This is even true with respect to concrete. Intense heat applied to concrete creates a condition known as "spalling." Spalling is the result of concrete reaching a very high temperature and then being rapidly cooled by water. This causes the surface to crack and loosen, resulting in a pitted appearance. Spalling may be produced by a flammable liquid; therefore, a careful examination and sampling should be done if suspected liquid burns lie solely on concrete surfaces.

A fourth indicator of the use of flammable liquid accelerants is the detection of flammable liquid vapors at the scene of the fire. This is accomplished with the use of portable vapor indicators or gas chromatographs.

A fifth indicator of incendiarism is the extent of fire damage when compared to the fire load contained in a building. Fire damage tends to become increased when a flammable accelerant is present. Evidence of extreme or extensive fire damage to an area with little or light fire load may also be suggestive of incendiarism.

Trailers and Incendiary Devices. Another common indicia of incendiarism is the use of "trailers" and incendiary devices. "Trailer" is a term commonly used by fire investigators to indicate the means used to connect separate areas of fire or to spread fire throughout a structure. The objective of the trailer is to communicate the fire from one place to another within a structure in order to enhance the damage. Many substances can be used as trailers, including paper, rope, rags, and flammable liquids.

In many instances trailers are not entirely consumed in the fire and can be retrieved and photographed. Flammable liquid trailers, in many instances, become evident when debris is cleared from floors revealing a trail of liquid from one place to another. This type of physical evidence is a strong indicator of incendiarism.

Incendiary devices are often used by arsonists to provide both the *source* of ignition and a *delay* in ignition. The main objective of an incendiary device is to facilitate ignition of other substances. Incendiary devices come in all shapes and sizes and may be simple or complex in design. Devices range from a match or a cigarette contained within a matchbook to more complicated arrangements involving the use of timers, heating coils, chemical mixtures, liquids, powders or other materials.

In many instances, simple incendiary devices, such as matchbooks with cigarettes, candles and the like are destroyed during the fire and cannot be retrieved from the fire debris. However, more complex wiring systems, batteries, heating coils, liquid-filled containers with wicks, or explosive devices can be found in the fire residue. Of course, if found they should be retained and thoroughly analyzed.

In conclusion, the attorney must remember that the kinds of circumstantial facts which point to incendiarism are as varied as the ingenuity of the arsonist. Accordingly, there is a critical need for an experienced investigator to explain each of the various indicators and its significance.

The many pieces of circumstantial evidence which may be gathered during the investigation of a fire must be evaluated for their cumulative effect and in light of the likelihood of the existence of accidental causes. Some cases will be eminently clear by reason of the combined existence of multiple unconnected origin points, the use of accelerants, the existence of incendiary devices, rapid fire progress, unexplained burn patterns and other facts. Other cases will involve more complex proofs and a more refined analysis. Since the measures of proof and the acceptance of circumstantial evidence are both wide-ranging, in most cases relevant facts can be successfully marshalled and presented to a jury for determination. The proofs of incendiarism will result in the expert opinion of the cause and origin investigator as to the incendiary origin of the fire.

§5.5.3   Motive Evidence

Evidence of motive is of critical probative value in most arson cases. As is true of much fire case evidence, motive is often ascribed circumstantially. Evidence of motive is highly significant in assessing the likelihood or reason for the insured's fraudulent connection with an incendiary fire. Was the insured motivated by profit, financial need or other facts to have set the fire to collect insurance money? The relevance of such motive proofs in a civil arson defense case is well established in the authorities.[18]

The primary motive involved in most arson insurance fraud cases is, of course, economic. Although economic motives vary considerably, generally they fall into three categories:
1)   economic gain or "arson for profit,"
2)   economic relief from financial stress or burden,

---

[18] *Harris v. Zurich Insurance Co.*, 527 F.2d 528 (8th Cir. 1975); *Elgi Holding, Inc. v. Insurance Co. of North America*, 511 F.2d 957 (2d Cir. 1975); *Boone v. Royal Indemnity Co.*, 460 F.2d 26 (10th Cir. 1972); *L & S Enterprises Co. v. Great American Insurance Co.*, 454 F.2d 457 (7th Cir. 1971); *Swindle v. Maryland Casualty Co.*, 251 So.2d 787 (La.App. 1971); *Quast v. Prudential Property & Casualty Co.*, 267 N.W.2d 493 (Minn. 1978); *Garrison v. United States Fidelity & Guaranty Co.*, 506 S.W.2d 87 (Mo.App. 1974); *Anderson v. General Accident Fire & Life Assurance Corp.*, 395 N.Y.S.2d 118 (App.Div. 1977); *Shwanga Holding Corp. v. New York Property Insurance Underwriting Association*, 394 N.Y.S.2d (App.Div. 1977).

3) a combination of the two, perhaps combined with other motives.

Each of these possibilities is discussed below.

There are numerous indicia which can establish the economic gain or "arson for profit" motive. In some circumstances, these are tied to individual situations whereas, in others, they are a part of an overall scam or scheme by professionals. In all cases, however, it is the value of the property for insurance purposes, as opposed to its true worth, that determines the extent of potential economic gain.

There are several methods used to set insurance values. In many jurisdictions, the measure of recovery under a standard fire insurance policy is "actual cash value," defined as the replacement cost of the property less depreciation.[19] In such a situation, the "actual cash value" as defined may far exceed the fair market value of the property. In a depressed real estate market, it is possible that the "actual cash value" measure of recovery under the policy would far exceed the fair market value of the property. This situation is an excellent motive for potential insurance fraud.

A second measure of recovery permits recoupment of the "replacement cost" of property, i.e., the actual cost required to replace the building with materials of like kind and quality. The availability of such "replacement cost" insurance may create a motive for arson where coverage is obtained on relatively old or obsolete buildings with a low market value and a relatively high replacement cost. In slum neighborhoods the availability of such coverage can provide the impetus for redevelopment or "rehab" schemes on the part of an unscrupulous insured. Under all measures of recovery, however, it is the relationship of the *amount* of coverage to true value that is an indicia of motive.

This syndrome takes several forms. Among these are high insurance coverage in relation to the value of buildings or contents, recently increased insurance on buildings which have been the subject of recent acquisition and resale at inflated prices, and the artificial inflation of real estate values for insurance purposes by sham

---

[19] *See generally*, 6 J. Appleman, *Insurance Law & Practice* §3823 (1972); *Annot., Depreciation as Factor in Determining Actual Cash Value for Partial Loss Under Insurance Policy*, 8 A.L.R. 4th 533 (1981).

transactions. Profit motive may also be found in over-insuring obsolete merchandise or products which have depressed values; obtaining of "valued" form insurance on low-value property; use of insurance proceeds as a substitute for capital expenditures needed for demolishment, relocation, rebuilding, remodeling, or expansion of an existing building or business. A careful and thorough investigation into the facts and details behind such overt indicia can produce persuasive evidence of profit motive when coupled with circumstantial evidence that points to an incendiary fire or other unusual circumstances.

As an economic motive for insurance fraud, relief from financial stress is far more prevalent than "arson for profit." Financial stress, of course, can be of a personal or business nature. Such phenomena as chronic unemployment, high debt, gambling debt, recorded judgments, tax liens, business failures, inability to sell obsolete or slow-moving inventory, inability to sell real estate, unfavorable lease terms, unsatisfactory business locations, and the unavailability of new capital from recognized banks or lenders are all possible motives for incendiary fires. These sources of financial stress may be documented by:

1)   an examination of employment records, corporate books and credit records,

2)   docket searches (liens, foreclosure suit), judgment searches,

3)   real estate listings,

4)   interviews with competitors and creditors, bankruptcy filings,

5)   other logical sources of information reflecting upon true financial condition or financial stress.

As is true with evidence of incendiarism, most courts have evidenced an extremely liberal attitude toward the admission of evidence which reflects the existence of financial motive when coupled with other circumstantial facts tending to show an incendiary fire and which, when considered in totality, provide reasonable inferences of the involvement of the insured.

In addition, it should be noted that there are numerous other motives connected with an incendiary fire which may or may not be relevant to a civil arson defense case. Some of these include criminal extortion; labor/management grievances; arson for the purpose of concealment of other criminal activity such as burglary or homicide, revenge, or the elimination of business competition. Each of these must be thoroughly evaluated in order to determine its relevance and application in an arson defense case. As a general principle, where there is strong evidence of economic motive, either for profit or to relieve financial stress, the case will be "strong" with a jury. Conversely, evidence of incendiarism without strong motive evidence will generally prove less convincing to a jury.

§5.5.4   Opportunity Evidence

A third type of circumstantial evidence indicative of an insured's fraudulent involvement in an incendiary fire is opportunity evidence. The question to be asked is: "What opportunity facilitated the insured's setting the fire? Here the focus is on the conditions and security of the premises at the time the fire is discovered. From the standpoint of the insurer, a fraudulent connection of the insured to the incendiary fire will be very strong if the evidence places the insured at the scene of the fire or within close proximity to the scene and within a reasonably short time of discovery of the fire. Obviously, it is rare to find eyewitness testimony to the commission of an arson fire; therefore, very few cases will have the luxury of direct evidence on this issue.

Generally, the way to establish opportunity on the part of the insured is to examine the relative security of the property at the time the fire was discovered. If the premises are found to have been secure at the time when the first witness, police or fire officials arrived, a legitimate inference can be drawn that the person or persons who set the fire had possession of the means to gain access. Careful examination and photographing of doors, windows and other means of ingress and egress are necessary.

An entirely secure building, limited distribution of keys to the premises, and no evidence of forced entry may suggest that the insured or someone to whom the insured gave keys was the most probable arsonist. On the other hand, evidence of break-in coupled with other corroborating evidence of burglary or vandalism may negate opportunity on the part of the insured.

It is well established that the opportunity for commission of an incendiary fire may be established soley by circumstantial evidence.[20] It is also well established in case law that in an arson defense it is unnecessary for an insurer to prove that the insured personally, or any specific member of the insured's business or family, was the actual arsonist.[21] For example, in *Garrison v. United States Fidelity & Guaranty Co., supra,* the court, in affirming a verdict in favor of an insurer, commented as follows:

> The person who set the fire was identified as an employee of plaintiff and her husband, and his description was essentially the same as that of Bill Hornberg, the employee described by plaintiff. Hornberg had received a key to the premises and the use of an automobile before plaintiff and her husband left for Chicago the day before the fire. When the firemen arrived to fight the fire, they found that the building was completely locked.[22]

Opportunity evidence may, of course, be rebutted at trial with the establishment of an alibi on the part of the insured. When set against such circumstantial proofs of an incendiary fire as a secured building, limited possession of keys by an insured or those under the insured's control, alibi evidence can clearly create jury questions as to the insured's direct or indirect participation in an arson incident. On the other hand, evidence of prior acts of vandalism or unsecured entryways at the time the fire fighters arrived can be used by the insured to weaken the chain of circumstantial evidence. Former disgruntled employees with keys or other means of access to the premises may also weaken any links between the insured and the fire.

---

[20] *Harris v. Zurich Insurance Co.,* 527 F.2d 528 (8th Cir. 1975); *Boone v. Royal Indemnity Co.,* 460 F.2d 26 (10th Cir. 1972); *Miele v. Boston Insurance Co.,* 288 F.2d 178 (8th Cir. 1961).

[21] *Harris v. Zurich Insurance Co.,* 527 F.2d 528 (8th Cir. 1975); *Collins v. Fireman's Fund Insurance Co.,* 296 F.2d 562 (7th Cir. 1961); *Stein v. Girard Fire Insurance Co.,* 259 F.2d 764 (7th Cir. 1958); *Garrison v. United States Fidelity & Guaranty Co.,* 506 S.W.2d 87 (Mo.App. 1974).

[22] *See also, Boone v. Royal Indemnity Co.,* 460 F.2d 26 (10th Cir. 1972); *Raphtis v. St. Paul Fire & Marine Insurance Co.,* 198 N.W.2d 505 (S.D. 1972); *People v. Lathrom,* 13 Cal.Rptr. 325 (Ct.App. 1961).

§5.5.5   Miscellaneous Connecting Evidence

In addition to proofs of the incendiary nature of the fire, motive of the insured, and opportunity, other evidence of other miscellaneous circumstances will be admitted when it supports a reasonable inference that an insured may have been fraudulently involved with the incendiary fire. The type of such miscellaneous connecting evidence will vary with the individual case, and a very broad spectrum of circumstantial facts may indicate insured involvement.

First, an examination of an insured's insurance history may prove relevant. For example, recent increases in coverage that appear to be supported by improvements to the property or normal appreciation may provide evidence that tends to exculpate the insured. The opposite conclusion is indicated where increases far exceed actual improvements or market appreciation.

Second, a common human reaction sometimes proves helpful in residential fire cases. This is the tendency for insureds to protect items of sentimental value. An examination of many fire scenes, particularly in residences, may reveal the absence of items of great monetary, personal or sentimental value owned by the insured. This may create an inference of prior knowledge by the owner that the incendiary fire would occur. Other such circumstances include the removal of pets normally left at home, an unexpected or unplanned trip, or a general inability to explain other unusual occurrences immediately preceding the occurrence of an incendiary fire.

Third, in both business and residential fires, an examination should be made of alarms, locks, detection devices, and fire protection systems. Often alarms, detection devices, sprinklers and similar equipment are found inoperable or turned off on premises where an incendiary fire has occurred. Occurrences of this nature often provide miscellaneous connecting evidence of a persuasive nature.

Fourth, a review should be made of statements made by the insured to friends, neighbors, investigating police officers and others in order to ferret out serious inconsistencies. Often, different accounts concerning the whereabouts, financial condition, and other matters are given by the insured to different people. Implausible or conflicting stories are told with respect to events immediately preceding or following a fire. Obviously, false statements concerning knowledge of the fire, the presence of accelerants, changes in insurance coverage, as well as an overly indignant reaction to legitimate requests for information can provide further supporting evidence of involvement.

Finally, the elimination of other potential motives for the fire can be a form of miscellaneous connecting evidence. For example, in examining prefire and postfire circumstances, it may be determined that there were no burglaries, no threats of labor strikes, no evidence of enemies or outside parties with motive to damage the insured's property. This may point the investigation in other, more fruitful directions. As can be seen from the foregoing analysis, a single circumstantial fact, standing alone, is likely inadequate to establish the essential requirements of the arson defense; however, when considered together, a series of facts will form a cohesive evidentiary picture implicating the insured.

§5.5.6 Evidence of Prior Fires or Insurance Claims

A question which often arises in arson defense cases is the relevance or materiality of evidence relating to an insured's prior fires or submission of prior insurance claims. Evidence of this nature is admissible depending upon the circumstances under which the evidence is offered.

For example, in *Hamman v Hartford Accident & Indemnity Co.*, 620 F.2d 588 (6th Cir. 1980), evidence was introduced showing that the insured had experienced six prior fires on various tracts of the insured's property. Four of these fires resulted in insurance recoveries. The trial judge permitted reference to the four fires which resulted in insurance recoveries but not of those fires which resulted in no recovery. Furthermore, in the first instance, the judge excluded evidence of the circumstances surrounding the four fires which involved insurance recoveries. The insured argued that even admission of the existence of the four fires was highly prejudicial and of little probative value. In affirming the trial court's admission of this evidence, the court noted:

> Here, the evidence of prior fires was properly admitted for a number of reasons: defendant attacked Hamman's credibility by establishing that he had willfully concealed several occurrences of fires from the

defendant. Second, the trial court properly instructed the jury that the fires were to be considered as bearing only on Hamman's motive. Lastly, Hartford asserted the defense of incendiarism which included evidence of Hamman's intent or knowledge of the occurrence.

Id. at 589 (citations omitted).

Accordingly, while this evidence was inadmissible to establish that the insured committed arson, it was nevertheless relevant and admissible on the issue of motive and credibility.[23]

In some cases, evidence of prior fires is offered to show the existence of a common plan or design on the part of the insured. A good analysis of this situation is found in *Hawks v. Northwestern Mutual Ins. Co.*, 461 P.2d 721 (Idaho 1969). Here the insurer sought to introduce evidence of prior fires from which the same insured had received heavy insurance benefits. The evidence was offered to prove the existence of prior fires and that they were set pursuant to a comprehensive plan, scheme or design. The court there held before such evidence could be admitted, it was necessary to show that the prior fires were incendiary in nature and that the insured owner was connected with them.

In *Harris v. Zurich Insurance Co.*, 527 F.2d 528, 532 (8th Cir. 1975), the insured argued on appeal that prejudicial error derived from remarks made by the insurer's counsel in the opening argument to the effect that evidence would be introduced showing that the insured had previous arson fires under the same or similar circumstances on which the insured collected insurance proceeds. The Eighth Circuit, after noting that timely objection was not made to the remark and, consequently, that the issue was waived as error, went on to state:

> Even assuming, arguendo, that plaintiffs had presented a question to the court below, we would find no error. Defendant proved that Harris had previously collected insurance proceeds by reason of a fire which was classified as arson by local authorities. Thus, the remark was not clearly improper.

---

[23] *See also, People v. Ferguson*, 25 Cal.Rptr. 818 (Ct.App. 1962); *People v. Miller*, 106 P.2d 239 (Cal.App. 1940); *People v. Brcgdon*, 283 P. 881 (Cal.App. 1929).

As these cases illustrate, there is no single principle governing the admissibility of "prior fires" evidence. However, depending upon the circumstances under which the evidence is offered, it may be relevant as impeaching testimony, as evidence of motive, scheme or design, or to show the insured's familiarity with insurance claims' procedures.[24] Therefore, the attorney should carefully examine the context in which the evidence is to be offered as well as the rules of evidence in the particular jurisdiction before seeking to introduce such evidence. Without proper foundation, such evidence can be highly prejudicial or denoted as reversible error on appeal, thereby seriously damaging the insurer's case.

§5.5.7   Evidence of Criminal Indictment or Conviction, Absence Thereof, or Evidence of Acquittal

Evidence of an indictment and criminal conviction for arson arising out of the same occurrence upon which a civil claim is predicated will be admissible as a bar to recovery on the insurance policy under the doctrines of collateral estoppel and res judicata. *See Imperial Kosher Catering, Inc. v. The Travelers Indemnity Co.*, 252 N.W. 2d 509 (Mich. App. 1977); *Budwit v. Herr*, 339 Mich. 265, 63 N.W. 2d 841 (1954).

On the other hand, evidence of the absence of arrest, indictment or conviction of the insured is generally held inadmissible. A reference thereto by the insured or counsel at trial may constitute ground for a mistrial.[25]

---

[24] *See generally, Aetna Insurance Co. v. Barnett Brothers*, Inc., 289 F.2d 30 (8th Cir. 1961); *Smith v. Insurance Co. of North America*, 213 F.Supp. 675 (N.D.Tenn. 1963); *Hawks v. Northwestern Mutual Insurance Co.*, 461 P.2d 721 (Idaho 1969); *Terpestra v. Niagra Fire Insurance Co.*, 256 N.E.2d 536 (N.Y. 1970); *Kramincz v. First National Bank of Green*, 302 N.Y.S.2d 22 (App.Div. 1969); *Texas Farm Bureau Insurance Co. v. Baker*, 596 S.W.2d 639 (Tex.Civ.App. 1980).

[25] *Galbraith v. Hartford Insurance Co.*, 464 F.2d 225 (3d Cir. 1972); *Kaminsky v. Blackshear*, 133 S.E.2d 441 (Ga.App. 1963); *Smith v. Federated Mutual Implement & Hardware Insurance Co.*, 185 S.E.2d 588 (Ga.App. 1971); *Miller & Dobrin Furniture Co. v. Camden Fire Insurance Co.*, 150 A.2d 276 (N.J.Super. 1959); *Greenberg v. Aetna Insurance Co.*, 235 A.2d 582 (Pa. 1967).

The rationale of this distinction rests in the differences between the criminal and civil standards of proof. It is well recognized that the preponderance of evidence standard is substantially different from the criminal standard of proof "beyond a reasonable doubt." Accordingly, the fact that the insured has not been arrested, indicted or convicted, or even criminally charged, is irrelevant and inadmissible in considering the affirmative defense of arson. It may be that a prosecutor does not feel there is sufficient evidence to convict an insured of arson, i.e., evidence beyond a reasonable doubt is not available. However, because the civil burden of proof, a preponderance of the evidence, is substantially less than the burden of beyond a reasonable doubt, the civil arson defense may be successful. In addition, because even civil arson cases appear to impute wrongdoing to the insured, the jury may be swayed by the absence of the insured's arrest, indictment or conviction by the authorities.

A good illustration of this somewhat difficult concept is found in the third circuit case of *Galbraith v. Hartford Ins. Co., supra.* In *Galbraith*, no arson charges had been filed against the insured. The insurer's defense was arson. The trial court permitted the insured's counsel to comment to the jury that the insured had not been charged by any criminal authority with the crime of arson. This remark was made in counsel's closing speech. Further, the trial court permitted the insured's counsel to elicit testimony from the insured that the insured had not been charged with the crime of arson.

On appeal, these rulings were reversed and it was established that permitting such evidence or comment in summation was improper. The court's analysis in excluding such evidence was as follows:

> The reasoning behind the exclusion of such proffered evidence is readily apparent. An acquittal in a criminal prosecution is not necessarily a judgment but merely a negative statement that the quantum of proof necessary for conviction had not been presented.
> 464 F. 2d at 227.

Accordingly, the court held that it was reversible error for the trial court to permit the jury to consider the insured's testimony and the insured's counsel's summation that no criminal charges were brought against the insured.

The New Jersey case of *Miller & Dobrin Furniture Co. Inc. v. Camden Fire Ins. Co, supra*, is in accord with this analysis. In *Miller & Dobrin*, the insured was indicted for arson, tried and acquitted. The insured's attempt to admit evidence of the acquittal was held inadmissible by the court.

> The insured's acquittal in the criminal case does not preclude the finding in this case that the fire was set by him. Even if he did not set the fire, but found it burning...and did nothing about it, this conduct, if established, could result in the defendant's insurers escaping liability under their policies because of the provisions which require the plaintiff to use all reasonable means to save and preserve the property insured.

150 A. 2d at 280.

This is not to say, however, that if the insurer presents such evidence that it would be equally inadmissable. If the evidence is adduced by the insurer's counsel, it would not be subject to the same objections.

Perhaps the best stated rationale appears in *Greenberg v. Aetna Ins. Co., supra*, where the Supreme Court of Pennsylvania squarely held that evidence of an acquittal was inadmissible as irrelevant, and highly prejudicial:

> Furthermore, the fact that the insured had been *acquitted* (by a jury) of the alleged crime or crimes in an indictment for incendiary acts is irrelevant and inadmissible in a civil case to recover damages based upon a claim under the same fire insurance policy. The reasons are obvious. First, the kind and quantum of proof required in a felony or serious criminal case—where the Commonwealth must prove the insured defendant was guilty beyond a reasonable doubt—is different from and much greater than that required in a civil suit on the insurance policy where plaintiff must prove his claim or the defendant must prove an affirmative defense (as the case may be) by a fair preponderance of the evidence.

427 Pa. at 516-17. (Emphasis in original.)[26]

---

[26] *See also, United States v. Burch*, 294 F.2d 1 (5th Cir. 1961).

These authorities recognize both the distinction between criminal and civil burdens of proof and traditional notions of identity of issues and parties used in analyzing res judicata and collateral estoppel issues.

§5.5.8   Polygraph Evidence

A question often arises in both civil and criminal arson cases concerning the admissibility of polygraph evidence on the issues of guilt or innocence, involvement, credibility or for other purposes. With very few exceptions, all recent state and federal decisions have held that polygraph evidence is inadmissible, absent an express agreement and stipulation by the parties.[27] Some courts have rejected polygraph evidence even when the parties previously agreed that it could be received in their particular case.[28]

In those cases with a prior stipulation concerning polygraph evidence, the rationale behind admission is basically one of risk rather than a recognition of the reliability or accuracy of the polygraph. In other words, it is the risk of failing the polygraph, rather than its reliability, which is recognized in these cases. Even under such circumstances, the courts have imposed stringent requirements to qualify the examiner.[29]

In general, therefore, polygraph evidence will not be admitted for purposes of evaluating credibility or establishing guilt or innocence. However, in some limited circumstances, evidence of either a refusal to take a polygraph examination or of the results of the test itself may be properly admitted. Such circumstances generally arise only in cases where bad faith on the part of the insurer is at issue.

---

[27] See *Frye v. United States,* 293 F. 1013 (D.D.C. 1923) (predicated on unreliability of polygraph testing as a scientific discipline); *Annot., Physiological or Psychological Truth and Deception Tests,* 23 A.L.R.2d 1306 (1952); *McCorskey v. United States,* 260 F.2d 377 (10th Cir. 1958), *cert. denied,* 358 U.S. 929 (1959); *United States ex rel. Szocki v. Cavell,* 156 F.Supp. 79 (W.D.Pa. 1957).

[28] *State v. Dean,* 307 N.W.2d 628 (Wis. 1981), *rev'g, State v. Stanislawski,* 216 N.W.2d 8 (Wis. 1974); *State v. Trimble,* 362 P.2d 788 (N.M. 1961).

[29] *See, e.g., Herman v. Eagle Star Insurance Co.,* 396 F.2d 427 (9th Cir. 1968), *aff'g,* 283 F.Supp. 33 (C.D.Cal. 1966); *State v. Valdez,* 371 P.2d 894 (1962).

For example, in *Moskos v. National Ben Franklin Insurance Co.*, 376 N.E.2d 388 (Ill.App. 1978), the court took into consideration the insured's conduct in a polygraph examination to determine if the insurer acted in bad faith in denying the claim. The insured had offered to take a polygraph test to prove the insured's innocence of arson. While the test was conducted, the insured apparently engaged in erratic breathing and movement, contrary to the polygrapher's express instructions. As a result of the insured's conduct, the operator was unable to secure readings to determine the truthfulness of the insured's answers. In fact, the operator stated a belief that the insured engaged in conduct to prevent the polygraph operator from determining that the insured was not telling the truth.

In discussing the polygraph evidence, the court stated:

> The results of the examination were presented to the Circuit Court, not on the issue of the plaintiff's responsibility for the fire, but on the question of the defendants' state of mind—that is, on whether the defendants, upon consideration of the results of that test, had a reasonable basis for concluding that the plaintiff was the arsonist. Whether the defendants were correct or incorrect, and whether the test results were accurate or inaccurate is irrelevant. The point is that the polygraph evidence was relevant only to establish that the defendants did not act in bad faith in contending that the plaintiff committed arson.

376 N.E.2d at 391. *Cf. deVries v. St. Paul Fire & Marine Insurance Co.*, 716 F.2d 939 (1st Cir. 1983); *Lynch v. Mid-America Fire & Marine Insurance Co.*, 418 N.E.2d 421 (Ill.App. 1981).

In a recent case tried before the Philadelphia Court of Common Pleas, the trial judge determined that polygraph evidence was admissible both as to credibility and as substantive evidence that the fire did not occur as the plaintiffs contended.[30] In *Quaker City Hide Co. v. Atlantic Richfield Co*, the plaintiffs sought damages for a fire at a warehouse they contended was caused by a defective oil furnace serviced by the defendants. Defendants, on the other hand, contended

---

[30] *Quaker City Hide Co. v. Atlantic Richfield Co.*, Phila. Ct. Common Pleas, December Term 1978 No. 4750 (Oct. 11, 1983) (appeal pending, Pa.Super. No. 2997PHL1983).

that a former employee of the plaintiff started the fire. The former employee voluntarily took a polygraph examination, at which time it was determined that the employee was practicing deception about starting the fire at the warehouse.

The court permitted the polygraph evidence to be admitted after four days of in-camera testimony of two expert polygraphers and the polygrapher who administered the examination to the plaintiff's former employee. The court based its decision to admit the polygraph evidence on the perceived reliability of the polygraph technique and in light of the evidentiary record created by the four days of in-camera testimony of the polygraphers. Despite the *Moskos* and *Quaker City* decisions, however, the prevailing authority continues to hold that polygraph testimony is inadmissible for any purpose based upon its unreliability and its lack of general acceptance in the scientific community.

§5.5.9   Arson Reporting Immunity Statutes

In arson for profit or insurance fraud cases the consequences attendant to an exchange of information between insurers and law enforcement agencies is a problem. Here the dilemma feared by insurers is that providing information of potential insurance fraud to state or federal law enforcement agencies will subject them to suits for "bad faith," malicious prosecution, libel or slander, or other intentional torts.

To counteract this fear, more than 40 states have enacted some form of reporting and immunity legislation. The primary purpose of these statutes is to facilitate the exchange of relevant information between insurers and prosecutors, and to grant immunity from civil liability to insurers, if the information is exchanged for nonmalicious purposes, i.e., to instigate a criminal investigation of the insured where one is not warranted.

There are many different forms of arson reporting immunity statutes. However, most embody the following components: 1) Circumstances under which law enforcement personnel may request information from insurers concerning nonaccidental fires; 2) those circumstances where an insurer is *required* to report nonaccidental fires to law enforcement agencies; 3) grant of civil immunity to an insurer for complying with the statutory mandate, so long as there is an absence of actual malice on the part of the insurer in providing the information; and 4) reciprocity, under certain circumstances, in providing for the exchange of prosecutorial information for an insurer's information.

The New Jersey arson reporting immunity statute is a good example of such statutes. The act[31] provides that an authorized law enforcement agency may request, in writing, that an insurer release relevant information concerning a fire loss which is under investigation by a law enforcement agency. Generally, the information sought involves all pertinent insurance policy information, including premium payment records, claims history, proofs of loss, and investigative statements. Any other relevant information which may be important to an official investigation may also be requested.

In addition, the New Jersey statutory scheme *requires* the insurer to notify the county prosecutor of a nonaccidental fire loss in which it has an interest, and requires the insurer to provide the county prosecutor with all of its investigative materials pertaining to the loss, regardless of the monetary amount of the claim. In connection with furnishing such information and unless motivated by actual malice, the insurer becomes immune from civil liability arising out of any statement made or action taken pursuant to the statute.

The information disclosed to law enforcement personnel under the New Jersey statute may be disseminated to other state or federal law enforcement officers to the extent that it is relevant to a matter which is under investigation. Insurance companies providing information to law enforcement agencies are given a reciprocal right to request relevant information, which is not otherwise privileged, from the law enforcement agencies concerning the same loss.

As noted above, the statutory scheme provides a vehicle for information exchange without fear of future civil liability. Moreover, such a scheme embodies an important spirit of cooperation in the investigation of arson for profit and other insurance fraud schemes pertaining to fire losses.

The practical operation of the information exchange is as follows.

---

[31] N.J.S.A. 17:36-14.

First, the insurer will submit a simple written report setting forth the date of the fire loss, and the name of the insured and enclosing information which is deemed to be relevant to the particular claim. In exchange, an insurer who has provided such relevant information to a law enforcement agency may make a simple request by so indicating and requesting relevant information developed by the law enforcement agency. In addition, the New Jersey statute requires that the insurer indicate that it is presently named as a party in a pending civil litigation in order to receive information from a county prosecutor's files. Information is therefore not available simply upon request prior to the commencement of a civil action. Other jurisdictions may or may not have this requirement; accordingly, an examination should be made of the specific requirements of the statute in the relevant jurisdiction.

Finally, it is important to note that information which is otherwise privileged cannot be obtained by written request under these statutory schemes. Accordingly, grand jury testimony, the statements of confidential informants or other information which, if disclosed, would jeopardize the confidentiality of an investigation, is protected from disclosure.

In order to enforce this statutory scheme, penalties can be imposed for refusal to comply with legitimate requests pursuant to the statute. The New Jersey statute provides a $250 penalty for refusals to release information, for failure to notify the prosecutor of an arson or for failure to hold the information in confidence. Many statutes also provide that law enforcement agencies may be required to testify in subsequent litigation involving the insurer where information has been exchanged.

Many jurisdictions have developed form procedures for information exchange pursuant to their arson reporting immunity statutes. A simple call to the division of state police or county prosecutor's office will assist in obtaining the necessary forms in order to facilitate the exchange of information. Arson reporting and immunity statutes can pose problems for the insurer's counsel. Many of these statutues require that the information provided by the prosecutor or law enforcement agency be held in confidence under the statute. This may pose a dilemma for insurance defense counsel when responding to discovery requests posed by the insured's counsel. Obviously, competing considerations of statutory compliance and legitimate fairness in discovery requests must be balanced by the courts in these situations. The arson reporting immunity statutes are relatively new and, accordingly, there is little case law construing the terms of these statutes. Accordingly, counsel is advised to act with discretion when faced with this problem.

The "discovery problem" is obviated to some degree by those statutes that require the insurer to inform the insured that it has provided information to law enforcement authorities (e.g., 40 Pa.C.S.A. §1610.3). This statute does not, however, require that the insured be notified of the substance of the information itself (40 Pa.C.S.A. §1610.5).

In conclusion, arson reporting and immunity statutes can be useful information-gathering tools in investigating suspected arson and fraud fires. In each jurisdiction, however, the statute must be carefully reviewed to determine the rights and obligations of law enforcement agencies, insurers and insureds.

§5.6   Conclusion

As the previous discussion indicates, arson defense cases are developed and presented through the use of circumstantial evidence. The nature of these cases and the essential elements of the defense, including the development of proofs of incendiarism and other circumstantial indicia of culpability, have led to liberal standards of admissibility on the part of the courts. Accordingly, an analysis and evaluation of *all* circumstantial facts must be conducted even though one or more of the circumstantial facts, if standing alone, are not highly probative. When the facts are considered in tandem, they can provide a strong base for the arson defense. It is the persuasive powers of the attorney, however, which determine the utlimate success of the defense.

Introduction to Section II

The focus of chapters 6-11 is on the technical and scientific aspects of fire. The attorney's point of view is used as a foundation for the presentation but it does not dominate the discussion of the technical material. No single book could fully prepare attorneys in the minutiae of scientific and engineering knowledge pertaining to fire. However, the basic principles involved in the areas which practitioners are most likely to encounter can be thoroughly presented.

A thorough examination of Section II will not make the reader a fire expert, but Section II will impart knowledge and insights into fire and fire hazards and aid in establishing a vocabulary through which the litigator can come to more fully know the subject matter. This knowledge can then be used as an aid in forming basic theories of liability and defense, analysis of fact patterns, basic analysis of evidence, and a foundation for questioning—whether that questioning be of a fact or an expert witness. Few, if any, significant fire cases can be properly litigated without an expert. It is likely that each side will be armed with at least one expert to bolster its case. Well prepared attorneys should have in their arsenal at least a basic personal knowledge of the subject matter. Without this, they run the risk of being at the mercy of the experts. For those moments of analysis, investigation, and interrogation when attorneys must act alone, they need some command of the subject.

Chapter 6 presents fire loss figures derived from the ongoing fire data collection and statistical analysis conducted by the National Fire Protection Association. The NFPA has long been the leading organization in compiling this information which quantifies and leads to understanding of the national fire experience. Data are presented here in summary fashion so that they will be readily available when the right statistic is needed to assist the busy litigator in the final stages of trial preparation. Other data sources are also mentioned to provide further references in the event they are needed, and NFPA issues new statistical reports annually.

Chapter 7 discusses the characteristics and behavior of fire. This is a basic scientific presentation of available knowledge about fire. The discussion is not exhaustive, nor is it overly technical, but it provides a basis from which the non-engineer lawyer can work.

Chapter 8 discusses the nature and properties of flammable gases and liquids, which are among the most common fuels for fire. A few of the more common flammable gases and liquids are given particular treatment, while descriptions of about forty others are taken directly from a referenced NFPA standard.

Since most fires that are involved in litigation occur in buildings, Chapters 9, 10, and 11 use NFPA standards as well as other technical sources to discuss various aspects of fire hazards in buildings. Chapter 9 looks at the structure and covering of a building from a firesafety point of view. Chapter 10 is based on the NFPA Life Safety Code®, one of the most widely adopted safety codes in the country. The chapter deals with the particular aspects of a structure that involve the safety of occupants. Chapter 11 is concerned with the walls and hidden areas of a building, including examination of the operating systems that turn the cold, dark shell of a structure into a well-lighted, heated, ventilated and air conditioned building where human activities take place.

Finally, a glossary is included to facilitate the understanding of some of the important commonly used terms which occur in the text and in the field of fire investigation and analysis.

Even a complete knowledge of all the material in Section II will not replace a good expert. However, unless there is some basic understanding of the technical forces involved, an attorney will be controlled by the expert instead of leading the expert.

Section II Table of Contents (detailed)

CHAPTER 6

FIRE-LOSS DATA

§6.1 Introduction

This chapter presents statistics which provide numbers to quantify the United States fire problem and experience. The statistics are drawn from a study conducted by the National Fire Protection Association (NFPA) in cooperation with the Federal Emergency Management Agency (FEMA). Like all statistical studies, the numbers presented herein are taken from large samplings, followed by mathematical calculations to project these samples to the entire country. These figures are probably the most thoroughly researched available into the United States fire experience. All figures are estimated to the extent that each incident has not been individually verified.

Figures are drawn from studies of 1982 and from five-year studies of the years 1978 - 1982.

§6.2 Total Fires

For 1982, NFPA figures show 2,538,000 total reported fires.

Over the 1978 - 1982 period, fire departments in the United States responded to an average of 2,816,500 fires annually according to NFPA estimates. The high point of this period was 1980 with 2,988,000 reported incidents. Total fires decreased in 1981 and also in 1982, when they reached the low of that five-year period at 2,538,000.

Table 6-1 shows the estimates of the total reported fires in the United States in 1978 - 1982. The table is broken down according to residential and nonresidential fires, with residential fires being further broken down by the various types of structures. The predominance of fires in one- and two-family dwellings among all types of structural fires is clearly observed.

TABLE 6-1
Estimates of Number of Fires Attended
by Fire Departments by Major Property Use
in the United States by Year (1978-82)

	1978	1979	1980	1981	1982
Residential (Total)	730,500	721,500	757,500	733,000	676,500
One- and Two-Family Dwellings*	569,000	550,500	590,500	574,000	538,000
Apartments	137,500	146,000	143,500	137,000	116,500
Hotels and Motels	12,000	11,500	11,500	11,500	10,500
Other Residential	12,000	13,500	12,000	10,500	11,500
Public Assembly	29,500	30,500	29,000	28,500	28,000
Schools and Colleges	20,500	22,500	19,500	19,500	16,500
Institutions	39,000	29,500	28,000	27,000	29,500
Stores and Offices	63,500	64,000	62,000	65,500	56,000
Industry, Utility, Defense**	55,000	58,000	55,000	49,000	42,000
Storage in Structures	59,500	64,500	65,500	62,000	54,000
Special Structures	64,500	46,000	48,500	43,000	44,000
TOTAL STRUCTURE FIRES	1,062,000	1,036,500	1,065,000	1,027,500	946,500
Outside Fires (with value)	110,000	80,500	86,500	81,000	54,000
Vehicles***	503,000	495,500	471,500	466,500	443,000
Brush, Grass, Wildland (no value)	582,500	614,500	718,500	711,000	522,500
Rubbish	296,000	342,000	397,000	341,000	309,500
All Other	264,000	276,500	249,500	266,500	262,500
TOTAL FIRES	2,817,500	2,845,500	2,988,000	2,893,500	2,538,000

Estimates are based on data reported to the NFPA by public fire departments that responded to the NFPA's Annual Survey of Fire Departments for U.S. Fire Experience (1978-1982). No adjustments were made for fires not attended by the public fire service.

*Includes mobile homes.

**Since some incidents for these property uses were handled only by private fire brigades or fixed suppression systems and were not reported to the NFPA, the results represent only a portion of U.S. fire experience.

*** This category includes highway vehicles, trains, boats, ships, aircraft, farm vehicles, and construction vehicles.

§6.3 Civilian Fire Deaths and Injuries

§6.3.1 Civilian Fire Deaths

For 1982, NFPA statistics show 6,020 civilian (other than fire service) deaths due to fire. This was a significant decrease of 680, or 10.1 percent from the 1981 level of 6,700.

Over the 1978 - 1982 period, an average of 6,900 civilians died annually from fire; 84.9 percent of these, or 5,790, died in structural fires. Examining only the structural fire deaths, note that an average of 5,575 persons died in one- and two-family dwellings and apartments, 96.2 percent of all structural fire deaths. The number of fire deaths in the home peaked in 1978 with a total of 6,015 deaths, then declined to 4,820 deaths in 1982.

Table 6-2 shows the estimates of United States civilian fire deaths from 1978 - 1982 broken down by selected properties.

TABLE 6-2
Estimates of US Civilian Fire Deaths by Selected Property Breakdowns (1978-82)

	1978	1979	1980	1981	1982
Structure (total)	6,350	5,970	5,675	5,760	5,200
Residental[1]	6,185	5,765	5,446	5,540	4,940
Nonresidential[2]	165	205	229	220	260
Vehicle[3]	1,255	1,535	740	840	695
Other[4]	105	70	90	100	125
Total:	7,710	7,575	6,505	6,700	6,020

Estimates are based primarily on data reported to the NFPA by fire departments that responded to the NFPA's Annual Survey of Fire Departments for US Fire Experience (1978-82); some of the estimates are based on data reported to the NFPA in the FIDO (Fire Incident Data Organization) System.

[1] Includes one- and two-family dwellings, apartments, hotels and motels, and other residential properties.

[2] Includes public assembly, educational, institutional, stores and offices, industry, utility, storage, and special structure properties.

[3] Includes highway and other vehicles (e.g., trains, ships). Estimates for 1978 and 1979 were based on data reported by fire departments but unverified, while estimates for 1980-82 were based on reported data by fire departments that was verified. Because of this factor, vehicle estimates for 1978 and 1979 may be on the high side, and should not be compared to estimates for 1980-82.

[4] Includes outside properties with value, brush, rubbish, and other.

§6.3.2   Vehicular Fire Deaths

The figures in Table 6-2 for vehicular fire deaths show significant differences among succeeding years. This may, in part, be explained by the difficulty in determining when a vehicular death is due to a fire and when it is due to the impact of a collision or other accident. Efforts have been made since 1980 to verify the cause of death in each case as being due to fire. Such efforts were not made in prior years.

§6.3.3   Civilian Fire Injuries

For the year 1982, the number of reported fire injuries to civilians was almost 31,000. Figures compiled by NFPA in Table 6-3 show 30,525 injuries, a slight increase of 0.3 percent over 1981. A five-year average for the years 1978 - 1982 shows approximately 30,460 fire-related injuries per year. It is widely believed that this figure is on

TABLE 6-3
Estimates of U.S. Civilian Fire Injuries by Selected
Property Breakdowns and Year (1978-82)

	1978	1979	1980	1981	1982
Structure (total)	24,985	24,850	24,725	25,700	25,575
Residential[1]	21,260	20,450	21,100	20,375	21,100
Nonresidential[2]	3,725	4,400	3,625	5,325	4,475
Vehicle[3]	3,720	5,175	4,075	3,400	3,425
Other[4]	1,120	1,300	1,400	1,350	1,525
TOTAL	29,825	31,325	30,200	30,450	30,525

Estimates are based on data reported to the NFPA by fire departments that responded to the NFPA's Annual Survey of Fire Departments for U.S. Fire Experience (1978-82); adjustments were made based on data reported to the NFPA in the FIDO (Fire Incident Data Organization) System.

[1]Includes one- and two-family dwellings, apartments, hotels and motels, and other residential properties.
[2]Includes public assembly, educational, institutional, stores and offices, industry, utility, storage, and special structure properties.
[3]Includes highway and other vehicles (e.g., trains, ships).
[4]Includes outside properties with value, brush, rubbish, and other.

the low side due to the large number of fire-related injuries that are unreported. The trends observed in fire deaths regarding the predominance of deaths in structures, and of these the predominance of fire deaths in the residential setting, are reflected in the injury statistics.

§6.4 Fire Fighter Deaths

The incidents of U.S. fire fighter deaths have been well documented and carefully followed with regard to total numbers, type of injuries leading to death, and the nature of the specific duty which was involved at the time of death. Of course, the numbers are smaller than among the civilian population. Summary statistics are presented below.

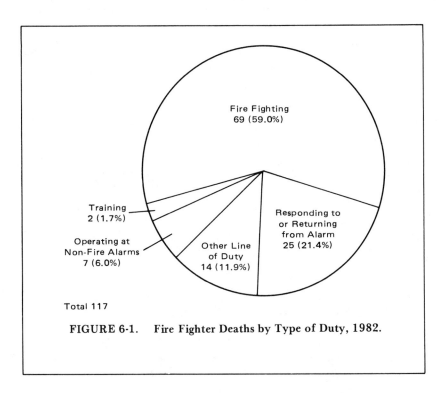

FIGURE 6-1.   Fire Fighter Deaths by Type of Duty, 1982.

If further information is needed, sources listed at the end of this chapter should be consulted.

In 1982, 117 fire fighters were killed in the line of duty, a decrease of 4.9 percent from 1981. Of the 117 deaths, 49 were career fire fighters and 68 were volunteers. Heart attacks were the predominant cause of fire fighter deaths, accounting for 54, or 46.1 percent, of all deaths recorded in the line of duty. Of course, fire fighting was the leading type of duty which brought about death, but there was a significant percentage, 21.4 percent, of deaths caused by responding to or returning from an alarm. There were 23 deaths as a result of motor vehicle accidents.

Figures 6-1, 6-2, and 6-3 show the 1982 causes of fire fighter deaths, both the number and percentage, broken down by type of duty, cause of injury, and the nature of fatal injuries.

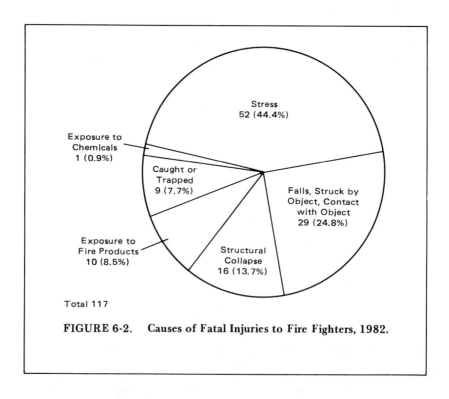

Total 117

FIGURE 6-2.   Causes of Fatal Injuries to Fire Fighters, 1982.

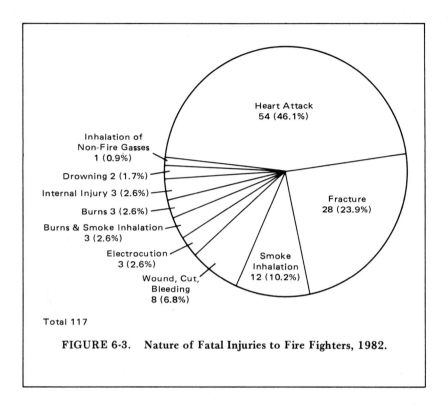

Inhalation of
Non-Fire Gasses
1 (0.9%)

Drowning 2 (1.7%)

Internal Injury 3 (2.6%)

Burns 3 (2.6%)

Burns & Smoke Inhalation
3 (2.6%)

Electrocution
3 (2.6%)

Wound, Cut,
Bleeding
8 (6.8%)

Heart Attack
54 (46.1%)

Fracture
28 (23.9%)

Smoke
Inhalation
12 (10.2%)

Total 117

FIGURE 6-3.   Nature of Fatal Injuries to Fire Fighters, 1982.

§6.5   *Property Loss*

Financial figures cannot be put on the value of human lives lost due to fire, but property damage can be quantified. For the 1978-1981 period, total fire property damage increased 49 percent from an estimated $4,022,000,000 in 1978 to $5,976,000,000 in 1981. In 1982, structure property loss decreased to $5,731,000,000, the result of the significant decrease in structure fires. Most of the dollar increase, both in total property damage and in structures lost, can be attributed to the effects of inflation. When an inflation adjustment factor is used, there is very little change in the value of property damage to structures over the period. Fires in structures accounted for an average of 88 percent of fires during the period, 45 percent of these structural fires being in one- and two-family dwellings, making them by far the leading type of structural fire loss.

Table 6-4 and Figure 6-4 show the dollar amounts of property damage for 1982, broken down by major categories, compared with 1981 figures.

For the year 1982, there was a downturn of 7.9 percent in the total number of structural fires. Stores and offices showed a 14.5 percent loss in the number of reported fires. Another category, institutional fires, showed a 9.3 percent gain. The effects of inflation, however, caused property-loss dollar estimates to climb in most categories. These figures appear in Table 6-4 and Figure 6-4 that give estimates for 1982 and show the percentage gain or loss from the same figures in 1981.

An estimated 676,500 residential fires, accounting for 71.4 percent of all structural fires, occurred in 1982, a highly significant decrease of 7.7 percent from the previous year.

TABLE 6-4
Estimates of 1982 United States Structure Fires
and Property Loss by Property Use

Property Use	Structure Fires		Property Loss[1]	
	Estimate	Percent Change from 1981	Estimate	Percent Change from 1981
Public Assembly	28,000	−1.8	$ 381,000,000	+7.0
Educational	16,500	−15.4†	161,000,000	−12.5
Institutional	29,500	+9.3	17,000,000	−55.3‡
Residential (Total)	676,500	−7.7‡	3,253,000,000	−0.2
One- and Two- Family Dwellings[2]	538,000	−6.2	2,794,000,000	+1.9
Apartments	116,500	−15.0	353,000,000	−14.9
Hotels and Motels	10,500	−8.7	84,000,000	−15.2
Other	11,500	+9.5	22,000,000	−31.2†
Stores and Offices	56,000	−14.5‡	510,000,000	−20.6
Industry, Utility, Defense[3]	42,000	−14.3†	663,000,000	−14.5
Storage in Structures[3]	54,000	−12.9‡	584,000,000	−5.2
Special Structures	44,000	+2.3	162,000,000	+52.8
	946,500	−7.9‡	$5,731,000,000	−4.1

The estimates are based primarily on data reported to the NFPA by fire departments that responded to the 1982 National Fire Experience Survey.

[1] Includes overall direct property loss (loss to contents, structure, a vehicle, machinery, vegetation or anything else involved in a fire), and does not include indirect losses, e.g., business interruption or temporary shelter costs; no adjustment was made for inflation in the year-to-year comparison.

[2] Includes mobile homes.

[3] Since some incidents for these property uses were handled only by private fire brigades or fixed suppression systems and were not reported to the NFPA, the results represent only a portion of US fire experience.

† Change was statistically significant at the 0.05 level.

‡ Change was statistically significant at the 0.01 level.

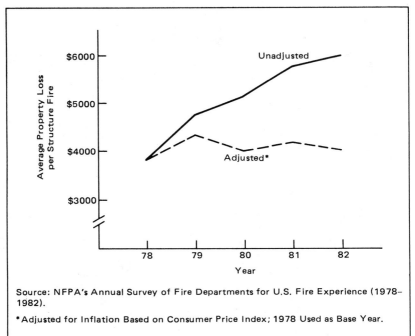

Source: NFPA's Annual Survey of Fire Departments for U.S. Fire Experience (1978–1982).

*Adjusted for Inflation Based on Consumer Price Index; 1978 Used as Base Year.

FIGURE 6-4.   Average Property Loss per Structure Fire in the United States (1978–82).

§6.6 Causes of Fire, Fire Deaths, Fire Injuries, and Property Loss

The cause of every single fire cannot be fully determined and reported. Therefore, whatever figures are used, there will be a significant number of reported fires of unknown origin. However, since the cause of most fires can be ascertained, some very meaningful statistics can be derived. The emphasis in this chapter is on residential occupancies because of the much larger number of variables which come into play. In the nonresidential setting, i.e., commercial factories and the outdoors, the causes of fire are generally fewer in number. Also, the vast majority of fire injuries and deaths occur in residences.

The figures used in Table 6-5 are from a 1981 FEMA study. Leading the list of causes of residential fires in 1981 were the factors of heating, cooking, incendiary/suspicious, and electrical distribution. The table shows that heating, the most predominate cause of fires in one- and two-family dwellings, is not the most predominate cause in apartments. Cooking and smoking, which caused 8.3 percent of total residential fires, was listed as the cause of 25.5 percent of the fire deaths. Interpretation of these statistics leads to numerous, interesting theories. It is hoped that the theories will lead to ways of preventing fires in the future.

Table 6-5 shows the estimates of the causes of fires, fire deaths, and fire injuries in residential occupancies for 1981, and the estimates of the causes of property loss in various residential occupancies.

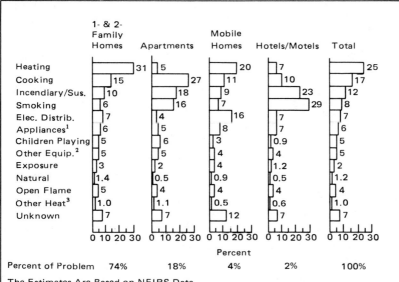

The Estimates Are Based on NFIRS Data.
[1] Includes All Items Listed as Appliances in NFPA 901, Along with Air Conditioners.
[2] Includes Special Equipment, Processing Equipment, and Service or Maintenance Equipment.
[3] Includes Rekindled Fire, Fireworks, Explosives, Heat or Spark from Friction, Molten or Hot Material, and Other Heat from a Spark or Hot Object.

TABLE 6-5.   Causes of Fires in Residential Occupancies.

Hopefully, the continued decrease in fire deaths is the result of recent fire safety efforts in the home, including numerous local ordinances and state laws that require the installation of smoke detectors in residential properties.

The recent decrease in civilian fire deaths should encourage continued firesafety education in the home on topics such as common causes of fatal fires in the home, the use and maintenance of smoke detectors, and escape planning.

The text and tables which have been presented in this chapter are not exhaustive of all statistics kept. Each year FEMA and NFPA jointly do a large-scale study of the fire problem. Statistics are printed in various issues of the National Fire Protection Association's *Fire Journal®*. Another NFPA periodical, *Fire Command®*, occasionally carries articles directly concerning the death and injury experience of the Fire Service. The NFPA *Fire Almanac* contains up-to-date information in all areas of the fire field.

In addition to the mentioned sources, there are other organizations that maintain and process statistics dealing with the national fire experience. Many of these organizations are more specialized than the NFPA or FEMA, and the information gathered may be more distinctive and, perhaps, more useful for litigation purposes.

The Public Safety Officer's Benefits Program located at 633 Indiana Ave., N.W., Washington, DC 20531, (202) 724-7620, was established by the Public Safety Officer's Benefits Act of 1976. The act provides for a $50,000 federal benefit to the survivors of state and local fire fighters and law enforcement officers who die as the direct or proximate result of personal injury sustained in the line of duty. The program is interested in the cause of fire fighter deaths and requires a significant quantity of investigative and medical information to support any claim. It is administered by the Department of Justice. As a resource, this program is more likely to be useful in specific incidents, rather than overall statistics.

Another source of statistics on fire fighter deaths and injuries is the International Association of Fire Fighters, 1750 New York Ave., N.W., Washington, DC 20006, commonly known as the IAFF. They commonly keep statistics on fire fighter injuries and fatalities for members of their association.

The FBI maintains and publishes its annual Uniform Crime Report, which includes national statistics, particularly arson statistics. The FBI's figures are helpful, principally because of the inherent credibility given to the FBI by most jurors. If for no other reason, the Uniform Crime Report should be checked if quantifying or distinguishing the arson problem becomes important. Another possible source for arson information is the Bureau of Alcohol, Tobacco and Firearms, located in Washington, D.C.

Trade associations, such as the American Petroleum Institute and the Insurance Services Office, provide statistics geared to more particular situations. There are a large number of trade associations with an interest in fire protection matters. Many of these associations are listed in the appendix of the 15th edition of the NFPA *Fire Protection Handbook*. Lists of trade associations can be found in most libraries. Credibility, along with the limited precise nature of trade association statistics, would make their use advisable for some situations.

CHAPTER 7

CHARACTERISTICS AND
BEHAVIOR OF FIRE

§7.1 Introduction

In order to litigate a fire case in the best manner, the attorney should have some basic knowledge of the complete interactions that constitute what has come to be classified under the single name of fire. The terms *fire* and *combustion* are often used to mean generally the same thing. In fact, a fire is a form of combustion which includes, by definition, varying intensities of heat and light.

Four basic definitions are presented here, the understanding of which will facilitate the related discussion of fire principles.

Combustion - a rapid exothermic oxidation process accompanied by continuous evolution of heat and usually light.

Fire - rapid self-sustaining oxidation, accompanied by the evolution of varying intensities of heat and light.

Oxidation - a chemical reaction in which electrons are transferred. Oxidation and reduction always occur simultaneously, and the substance which gains the electrons is called the oxidizing agent.

Reduction - a chemical process in which oxygen is removed from a compound or environment.

§7.2 Small Fires

Scientific research on the properties of fire has been confined to laboratory experiments, small fires, and theoretical work. There has been little work done to determine the complex reactions that go on in a large-scale fire burning out of control and consuming large amounts of property as its fuel. A scientifically accurate prediction or completely precise after-the-fact reproduction is, therefore, impossible. However, the basic science of fire can be described from what has been learned.

§7.3 Basics of Combustion and the Traditional Fire Triangle

The attorney can see from the definitions presented that fire and combustion are practically the same, with the exception that fire will always emit heat and light while combustion will emit heat and usually light. Fire is actually a more intense form of combustion. Both fire and combustion are exothermic reactions in that they produce substances with less energy than was in the reacting materials so that energy is released from the reaction. The energy that is released from fire is the heat and light which are seen and felt in the area of a fire. Fire is a self-sustaining reaction in that, once ignited, it will continue to burn as long as the factors which led to ignition remain available.

It is common to depict fire as a triangle of three equal legs consisting of oxygen, fuel, and heat (temperature).

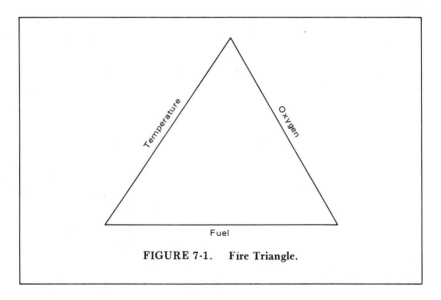

FIGURE 7-1.   Fire Triangle.

The diagram shows that a fire has to have three components in order to ignite and survive: oxygen, fuel, and heat. It is also true that the three components must be in a balanced state; too much of any one of them may cause a different reaction among the remaining two which may negate the possibility of ignition or sustaining of the fire.

Oxygen is the first leg of the triangle. While a necessary factor, it actually does not burn; rather, it supports combustion. If the oxygen supply is removed or cut off, such as by placing a cover over a pan containing a grease fire, the combustion will stop once the available oxygen is consumed. In order to sustain combustion there must be at least 16 percent oxygen in the surrounding atmosphere. An atmosphere of less than 16 percent may support a momentary ignition, but a sustained reaction will fail. Since air at sea level contains 21 percent oxygen, there is generally sufficient oxygen for combustion in all but a chemically contaminated or an intentionally oxygen-starved atmosphere. Under normal circumstances there is sufficient oxygen for fire in almost every exposed atmosphere. In certain instances where oxygen is not available from the atmosphere, there may be other chemicals which will break down as a result of heat and, in breaking down, release oxygen into enough of the surrounding atmosphere to allow ignition and a sustained reaction. These chemicals are called oxidizing agents, and their reaction is necessary for ignition in otherwise apparently sealed off areas or areas contaminated with other chemical gases and vapors which lower oxygen below the 16 percent threshold.

Fuel is the second leg of the traditional triangle. Fuel is the material that burns and is actually consumed by the fire. In a more technical sense, this can be considered the reducing agent in that it reduces an oxidizing agent. Almost any material will serve as a fuel as long as sufficient heat and oxygen are present. However, it is generally required that a material be in a vapor or gaseous state before it will serve as fuel. Given sufficient heat, most materials will give off the necessary vapors and gases to serve as fuel for combustion. Of course, different materials require different amounts of heat before they will vaporize, and this explains, in part, differences in fuel characteristics. Materials already in a gaseous state will, for the most part, serve as the best fuels. Propensity to vaporize is but one factor in determining the value of a particular material or fuel for combustion. Other characteristics, such as flash point, ignition temperature, and flammable limits, will become involved in this determination. As an example of the vaporizing of a solid in order to act as a fuel, and as an example of many of the chemical or physical processes involved in fires, the appendix of this chapter describes the ignition, burning, and eventual extinction of a wood slab in a typical situation such as a fireplace.

An important factor about fuel, of interest to the litigator, is that the amount of available fuel, as well as its nature, location, and existence of any protection, can be the key to the development and spread of a fire. Since a fire can generate its own heat, and oxygen can come directly from the atmosphere, a factor often controllable before the fire begins is the amount, position, and protection of any material which will serve as fuel.

Heat is the third leg of the traditional fire triangle. While heat is a necessary by-product of fire, it also is necessary prior to ignition to raise the fuel to the point where vapors are given off which, when mixed with oxygen, will allow ignition to occur. Heat is also the form of energy that will cause ignition to occur. Thus, heat plays a multifaceted and continuing role in the development of the self-sustaining nature of fire. The heat causes the fuel to vaporize into the oxygen atmosphere and then to ignite. Once the fire, as such, has started, heat is given off as a product of the exothermic reaction. The heat will then continue to vaporize fuel into the oxygen-containing environment, which will then ignite the vaporized fuel to keep the fire going.

It is because of this interaction that fire has its self-sustaining nature. When one of the three factors is removed, the fire will go out. The most common and traditional method of fire suppression, the use of water, is a method of reducing the heat which will prevent vaporizing of the fuel, thus breaking the reaction that sustains the fire. Other examples would be the mentioned covering of a pan fire to remove the oxygen supply, or shutting off the gas in a gas-fueled fire.

Modern fire theory has added a fourth interesting element, thus often depicting the traditional triangle as a tetrahedron. The fourth factor necessary for fire to sustain itself and increase in size is a chemical chain reaction between the fuel and the oxidizing agent. It could be argued that this is not a wholly separate factor. Rather, it is the existing interaction among the other factors. The chemical chain reaction, being viewed as a factor of its own, results from an increased understanding of the nature of the action of molecules when heated and ignited in the atmosphere. A full description of the chain reaction would require an understanding of atomic structure and molecular activity which is beyond the scope of this book. However, a basic understanding of the necessary chain reaction can be presented. As a fire burns, fuel molecules are reduced within the flame to simpler molecules. As the process of combustion continues, the increase in temperature causes additional oxygen to be drawn into the flame area. More fuel molecules will then break down, enter into the chain reaction, draw additional oxygen, and continue the chain reaction. This process will continue until all the substances involved reach a cooler area of the flame.[1] As long as there is sufficient heat, fuel, and oxygen, this chain reaction will continue to propagate the combustion process.

§7.4   Heat Source and Transferal

Fuel for combustion and burning can be almost any substance which contains gases that will be given off when heated and will themselves burn. Since oxygen is usually present or is supplied by the breakdown of some other substance (perhaps the fuel) which acts as an oxidizing agent, the remaining variable needed to prove a cause and origin of fire is the transfer or development of heat. The question of cause and origin is often of interest in a fire case. However, in numerous situations the precise cause may be of less than critical importance. Nevertheless, if the full development of the fire is to be understood, a reasonable and documented theory of causation should be available.

What, then, is heat and how is it generated and transferred? Heat can be defined as energy that is associated with and proportional to molecular motion that can be transferred from one body to another by

[1] Gold, David T., *Fire Brigade Training Manual*, National Fire Protection Association, Quincy, MA, 1982.

radiation, conduction, and convection. Thus, the capacity to be transferred and the methods by which that transferal can take place are parts of the definition of heat. There are numerous chemical and physical processes which can cause the necessary increase in molecular activity that results in heat. Below are the more common of these processes, along with a description of each.

§7.5   *Chemical Heat Energy*

§7.5.1   Heat of Combustion

The heat that is part of the exothermic reaction of fire is called heat of combustion. It is more closely defined as the amount of heat released during the complete oxidation of a substance. Expressions such as joules per gram, Btus per pound, or calories per gram are often used to describe the heat of combustion of a quantity of a particular fuel.

§7.5.2   Spontaneous Heating

Spontaneous heating is the process which will lead to what has become known as spontaneous combustion. Spontaneous heating is a process by which the temperature of a material is increased without drawing heat from its surroundings. Spontaneous heating is common among agricultural products as a result of bacteriological action and among oily rags due to chemical reactions. In order for spontaneous heating to ignite into a fire, there must be sufficient air to supply oxygen, but not enough air so as to carry away the heat that is produced. Spontaneous heating is generally the result of an oxidation reaction which, in itself, is very common among all materials, but which in some cases produces heat faster than it can be carried away. Thus, storage and surroundings are more important than the reacting materials themselves for the ignition of combustion due to spontaneous heating.

§7.5.3 Heat of Decomposition

Although not a common phenomenon, heat of decomposition involves the heat released in the breaking down of a compound which required the addition of heat for its original formulation. Since most compounds are produced by exothermic reactions, heat produced by decomposition is not common. Many of the compounds necessary for heat of decomposition are particularly unstable and often form the basis of explosions. As a reaction, decomposition is an important cause of explosions and will be discussed later in this chapter under explosions.

§7.5.4 Heat of Solution

The heat released when a substance is dissolved in a liquid is called heat of solution. While most materials release heat when dissolved, the amount is usually of little significance. Some chemicals, however, such as sulfuric acid, may release enough heat through this process to be of importance.

§7.6 Electrical Heat Energy

§7.6.1 Resistance Heating

The most common form of heating due to the flow of electrical current through a conductor and the resistance of the conductor to that flow is called resistance heating. The rate of heat generated is proportional to the resistance and the square of the current. The light of an incandescent bulb is based on this resistance, but ignition is prevented by evacuating oxygen from the bulb and using a filament with a very high melting point.

§7.6.2 Leakage Current Heating

Because all insulators of electric wire are to some degree imperfect, there is some current flow when these insulators are subjected to substantial voltages. This is commonly referred to as leakage current and in most cases is insignificant. However, leakage current may become important if the insulator or its installation is improper, because the insulator may break down.

§7.6.3 Heat from Arcing

Arcing occurs when an electrical circuit that is carrying current is interrupted intentionally or accidentally. The resulting arc may be one of very high temperature so that it can ignite combustible or flammable material in the vicinity. Under certain circumstances, the arc may be of sufficient heat and power to melt the conductor, with the result that molten metal is scattered and can cause ignition of available fuel. It should be noted that arcs, when they occur, are often in wires located within walls or other areas containing significant available fuel.

§7.6.4 Static Electricity

When two surfaces come together, one will take on a positive charge and the other a negative charge. If enough potential has built up when they are separated, a spark will discharge between them. Ordinarily, this spark is of short duration and not of sufficient energy or heat to cause combustion. However, flammable gases and vapors have been known to ignite as a result of heat developed by a static charge.

§7.6.5   Lightning

An electrical charge of lightning may cause ignition as it passes between clouds or between a cloud and the ground.

§7.7   *Mechanical Heat Energy*

§7.7.1   Frictional Heat

All bodies possess resistance to motion. The mechanical energy used to overcome this resistance when two solids are rubbed together is called friction. Any form of friction will generate heat. The danger to the ignition of surrounding fuel depends upon the available mechanical energy, the rate at which the heat is generated, and its rate of dissipation.

§7.7.2   Friction Sparks

These sparks are largely a product of friction resulting from the impact of two hard surfaces, at least one of which is usually metal. Heat generated by the initial friction will heat the separated particle which will bring about oxidation, and the heat from this will increase the temperature until it is incandescent. Usually, the particle is so small that its heat is quickly lost; thus, it does not have enough total heat energy to cause combustion. But under certain conditions and with the presence of a specific fuel, ignition is possible.

§7.7.3   Heat of Compression

The heat released when a gas is compressed is called heat of compression. It is often called the diesel affect because it has its most common application in the diesel engine. In that engine, diesel fuel is sprayed into a cylinder in which air has been compressed. The heat released by the compression of the air is sufficient to ignite the diesel fuel.

There are other particular methods of heating not mentioned here, but the general chemical, electrical, and mechanical categories are the areas of concern to the litigator. Nuclear heat energy is an area which, while very much a part of the mechanical world, does not figure in a significant number of fires to be of interest to the litigator. Future technology may make this a more significant source of accidental fire ignition.

Further treatment of the scientific nature of and sources of heat and heat energy would involve an extended discussion into atomic structure and molecular activity which is beyond the scope of this book as well as the interest of litigators. What is worthy of further examination for the practitioner is the way in which heat is transferred from its source of generation to the point of ignition.

§7.8   Heat Transferal

Another important aspect of heat is its capacity to be transferred from one body to another. According to the original definition, this can be done in three ways — by conduction, radiation, and convection. The discussion of the fire triangle explained that heat is often the operative outside force which, in effect, triggers the ignition. Given this importance, it is necessary to understand not only how heat can be originally generated, but also how heat can travel from the source of generation to the point of ignition. If fuel and oxygen are present in the required amounts at the source of the heat, ignition may take place in the immediate vicinity and there will be little need for transferal. However, since ignition may take place outside of the immediate proximity of the heat source, the method of heat transferal may require explanation. For a complete understanding of fire development, the litigator may then have to consider the three following means for the transferal of heat.

§7.8.1 Conduction

Conduction is a method of transfer that takes place when the objects transferring heat are in direct contact with each other. For many materials their conduction rates have been timed and their other heat-conducting capabilities noted. Conduction cannot be stopped by any heat-insulating material; the best these materials can do is lower the normal conductivity rate. The best heat insulation should have some way of venting air which will carry the heat away, thus dissipating its energy. Allowing a method for heat energy to be carried away is an insulation technique far superior to simple reliance on the thickness of the insulating material. Given sufficient time and intensity, heat will work its way even through concrete to ignite wood or other flammable material.

§7.8.2 Convection

Convection is the process by which heat transferal takes place through a circulating medium. The circulating medium is usually air, although it may be a liquid. Heat generated in a stove is heated initially by conduction as it comes in contact with heated coils or surfaces. The heated air then warms the room by convection as it flows throughout. Since hot air rises naturally, heating by convection often begins with an upward air flow, although this can be altered by means of a blower, a fan, or any other means by which air is forced in another direction.

§7.8.3 Radiation

Radiation is a form of energy by which heat passes through space like an electromagnetic wave in the same way that light, radio waves, or X rays travel through space as waves. In a vacuum, these waves travel at the speed of light. Upon coming into contact with an object, they are either absorbed, reflected, or transmitted. Heat carried by radiation is, in most cases, more dangerous than heat carried by convection. The basis of this increased danger is that a stationary surface near a fire will absorb nearly all the radiated heat to which it is exposed, while the heat traveling by convection will largely be carried off in the continuing flow of air.

§7.9   Ignition Temperature

In order for fire to ignite, the heat, regardless of its source and manner of transmission, must come in contact with the fuel in the presence of oxygen. As was noted earlier in this chapter, oxygen does not burn but only supports combustion. The temperature at which a substance will ignite is known as its ignition temperature. Once ignition has occurred, the combustion reaction will continue as long as fuel and oxygen are present. The terms piloted ignition and autoignition are often used to refer to the source of the flame, spark, or glowing object which brings about ignition. If the source is from outside the object, it is called piloted ignition. If ignition occurs without the assistance of an external pilot source, it is called autoignition. The precise temperature at which a solid or a liquid will ignite is influenced by a number of factors, including the rate of air flow, the rate of heating, and the size and shape of the mass of the solid or the body of liquid. For gaseous fuels, these factors are likewise important, along with the nature and purity of the gas and the pressure at which it is being held, as well as the shape of the containing vessel.

§7.10   Foreign Substances

A material's propensity toward ignition, either from an outside source or otherwise, may also be affected by the presence of other foreign substances within its composition. These foreign substances fall into three categories known as catalysts, inhibitors, and contaminants. Catalysts are substances that hasten the rate of chemical reactions, though the catalyst itself is not affected. An inhibitor is a chemical which, when added to a potential fuel, will prevent a vigorous reaction such as fire. As previously mentioned, there are no perfect inhibitors that can be added to fully prevent a fire from occurring. Contaminants are any foreign substances not normally found in a material. Of course, the presence, quantity, and nature of any contaminant may cause a material to react differently once ignited. The presence of any of these three types of foreign substances in any material under examination as a possible fuel in a fire sequence should always be explored.

§7.11 Flash Point

Because of the importance of liquids used as fuels for fire, particularly those fires involved in litigation, there is an often-used measurement with which the litigator should be familiar that indicates the propensity of the liquid toward ignition. That measurement is called the flash point. The flash point is the minimum temperature at which a liquid gives off vapor of sufficient concentration to form an ignitable mixture with air near the surface of the liquid in a laboratory vessel. This is a laboratory measurement calculated precisely for most hazardous liquids. This definition does not deal directly with the ignition of fire. That will take place at the fire point which is not precisely calculated but is usually a few degrees higher than the flash point. Flash point is used to indicate a danger point at which the state of a given liquid is changed to a vapor where it becomes flammable. When dealing with flammable liquids, the flash point is often used as a measure of the relative danger involved. Simply by reaching its flash point, a liquid that has turned to vapor will not automatically ignite, but in the vapor state, it will ignite quickly if sufficient heat, a spark or other stimulus is present. Chapter 8 deals in greater depth with flash point and other factors involved with hazardous liquids.

§7.12   Explosions

Perhaps the most spectacular manifestation of combustion comes in the combustion explosion. Not all explosions are as a result of combustion. However, the vast majority of building explosions are combustion explosions.

The terms "explosion" is not one that is easily defined. The scientific community cannot agree with any degree of precision on a definition. For the purpose of this book, it is best to think of explosions in terms of the effect that they cause — the destruction of a vessel or building, or the powerful expansion of gases which causes some visible damage to impacted objects. What causes this explosion, i.e., the effect, is the sudden, violent expansion of gases. What causes these gases to suddenly and violently expand is the real focus of this section.

In order for a combustion explosion to occur, a quantity of flammable gas/vapor-air mixture must accumulate in an enclosure. Then, when this mixture is ignited by a spark or other source, the pressure developed must exceed the strength of the enclosure. If the strength of the enclosure is such that it will contain the pressure, there will be no explosion because the criteria of the mechanical effect being caused will not be met. Therefore, realistically, it is the strength of the container or enclosure that determines whether an explosion takes place.

The stages of development of such an explosion are as follows:

1.   A flammable gas, the liquid phase of a liquefied flammable gas, or a flammable liquid is released from its container, either intentionally or unintentionally. If the escaping material is a liquefied gas, it rapidly vaporizes and produces the *potentially* large quantities of vapor associated with the transfer of this liquid to vapor.

2.   The gas (or liquid turned to vapor) mixes with the air.

3.   With certain proportions of gas or vapor and air, known as the flammable or combustible range (see Chapter 8), the gas/vapor-air mixture is ignitable and will burn.

4.   When ignited, the flammable mixture burns rapidly and produces heat rapidly.

5.   The heat is absorbed by everything in the vicinity of the flame and the very hot gaseous combination products.

6.   As a result of this absorbed heat, all of the materials will expand. This expansion is characteristic of nearly all materials and is especially pronounced in air which will expand to double its original volume for every 459° F that it is heated.

7.   If the heated air is not free to expand because it is confined in a room, vessel, or other container, there is a resulting rise of pressure in the room, vessel, or other container.

8.   If the containing structure is not strong enough to withstand the pressure, some part of the structure, either the walls of a room or portions of some vessel, will suddenly and abruptly move from their original position and a bang, woosh, boom, or other noise will be heard. This activity, in part, describes an explosion. Note that the explosion is considered to be the effect of suddenly moving some part of a containing structure. Because the source of the pressure which causes this movement is combustion, it is called a combustion explosion.

The idea that the strength of the enclosure will determine whether or not a particular event occurs may seem unusual, but it reinforces the concept that a combustion explosion is simply a very rapid form of combustion. The rapidity is based on the intimate premixing of a sufficient quantity of flammable vapor and air or oxygen, prior to the introduction the ignition source.

An enclosure full of a flammble gas/vapor-air mixture at atmospheric pressure would have to withstand pressure of 60 to 110 psi in order to remain intact. This pressure could be even higher if a reactive flammable gas or vapor is involved or if the mixture is oxygen-rich. Conventional building structures are built to withstand pressures on the order of ½ of 1 psi. This wide disparity between what a structure is intended to withstand and the pressures that may be generated by a combustion explosion demonstrate why an explosion may occur if an enclosure is far less that full of the flammable gas-air mixture. It has been estimated that most combustion explosions occur when less than 25 percent of the enclosure is occupied by the flammable mixture.

Two terms which are often associated with explosions are

deflagration and detonation. Each of these describes processes which may be involved in an explosion, though every explosion is not necessarily a detonation or a deflagration. A full description of each of these terms could become quite technical. Thus the two reactions are briefly described and distinguished below.

A deflagration is an exothermic reaction which propagates from the burning gases to the unreacted material by conduction, convection, or radiation. The combustion explosion previously described is a type of deflagration. An exothermic reaction, as mentioned previously, is a type of reaction that produces substances with less energy than was in the reacting materials so that energy is released from the reaction. The point that is important is the propagation by the normal methods of heat transfer.

A detonation is also an exothermic reaction, the propagation of which is characterized by a shock wave in the material which establishes and maintains the reaction. The force causing the temperature increase in the affected objects is the intensity of the shock wave, rather than some method of thermal conduction, as is the case with a deflagration. In principle, any material that is capable of releasing energy may support an initiating shock; in practice, the rate of energy release must be sufficiently great to overcome energy losses at free boundaries.

§7.13   Products of Combustion

The complex reaction or series of reactions known as combustion results in four products: heat, flame, fire gases, and smoke. For the litigator, an understanding of these is important because many more persons are killed or injured by one of these products than by the combustion itself. Victims of fires are often directly affected in physiological ways by one or more of these products, causing injury or death. Also, particularly in building fires, fire gases and smoke may be found far from the scene of the fire and are often the first signs individuals have of the existence of a fire. These signs can contribute to panic, obscure vision, and slow reflexes which will hamper or prevent movement to exits or other life-saving actions.

A common scenario in fatal occupancy fires is a fire that is itself quickly extinguished. However, injuries and death are caused by smoke and gases carried throughout the building and especially to upper floors by natural forces or a working air-conditioning or air-handling system.

Because of its practical significance, smoke will be treated last and the theory of its movement in buildings will also be discussed.

§7.13.1 Heat

The importance of heat to the development of ignition and the spread of fire, as well as the manner in which heat is transferred, has already been discussed. Heat also must be understood to be a product of combustion.

Perhaps the first human impression of fire is that it burns; in other words, the nearness of a surface made hot by fire will cause pain to that part of the body that makes contact with it. The common system of classifying burns to the body of a victim involves three degrees: first, second, and third. First-degree burns involve the outer skin layer and will result in abnormal redness, pain, and sometimes an accumulation of fluid below the skin. This is often the case in a sunburn. Second-degree burns involve a deeper penetration below the surface area, and will result in a burned area that is moist and pink with blisters and considerable fluid accumulation below the surface of the skin. Third-degree burns involve a more serious and deep burn which leaves the affected skin area charred pearly white.

The degree of severity of any burn will depend on numerous external factors such as time, temperature, and moisture of the air, as well as internal factors such as age, infirmity, and effectiveness of natural cooling mechanisms. It has been shown that exposure to heat at 111°F for six hours will produce second-degree burns but that a slight increase in temperature will result in burns in a greatly reduced exposure time. Temperatures of 131°F can cause second-degree burns in 20 seconds and temperatures of 158°F can cause such burns in one second.[2]

[2] "Effects of Extreme Heat on Man," *Journal of the American Medical Association*, Vol. 144, No. 9, October 28, 1950.

Air temperature does not immediately equate to skin temperature because the body has natural cooling mechanisms, principally perspiration and blood circulation. Increased heat naturally leads to perspiration which, when evaporating, can counteract air temperatures up to 60° F. This is most effective in dry air. Moisture in the air will not only transmit heat to the body more efficiently through conduction and convection, but it will also deter the cooling effect of evaporation. Blood circulation can carry away by convection some measure of the heat which gets below the skin, but this has the adverse effect of carrying heat throughout the body and causing a rise in the entire body temperature.

Burning is not the only adverse effect of heat on the body. Inhalation of hot air can cause damage to respiratory organs, including an increase of fluid in the lungs, and falling blood pressure, which may result in the collapse of a capillary blood vessel. The total heat environment can overcharge the body's heat balance leading to hyperthermia, a condition where the body absorbs heat faster than it can be discharged, causing a breakdown of the central nervous system.

In order to portray the damage done to human skin by heat energy, the scientific community usually measures the absorbed calories per square centimeter per second. A calorie is defined as the amount of heat required to raise the temperature of 1 gram of water 1° C. This will show what has actually happened to the skin which, in the case of most victims, is the key point of evidence. To try to establish accurate tables showing what happens to skin at different temperatures would be to ignore the many variables involved, including, among others, how much skin is exposed, the intensity of the heat, and the condition of the ambient air. Therefore, you have to view the skin damage from the inside and look at the quantity of heat absorbed, not just the temperature of the air.

Naturally, there is a correlation among the amount of time that the skin is exposed, the severity of the burn or limit of tolerance, and the quantity of energy absorbed. This relation is shown in Figure 7-2.[3]

Figure 7-2 shows for the indicated times, between one and forty seconds, the quantity of heat skin can tolerate for pain or what will

[3] Dr. James Veghte of Biotherm, Inc., Beavercreek, Ohio.

188

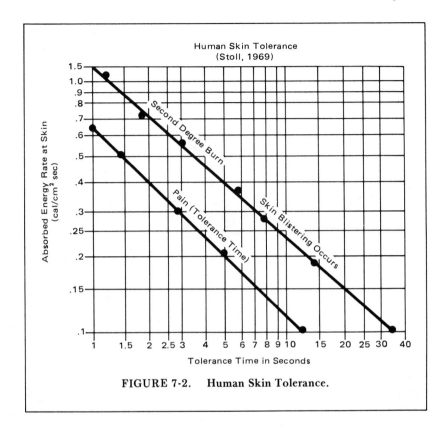

FIGURE 7-2.   Human Skin Tolerance.

occur to skin under the related heat and time conditions. Note that significant damage is done in very brief spans of time, much of it in the first five seconds. At the same time, since the exposed skin area is a key variable, the larger the exposed area, the less will be the damage unless heat intensity is, likewise, raised.

Like all charts depicting physiological response to outside conditions, Figure 7-2 should be understood not to be precise for all individuals, but it is accurate enough to be scientifically valid.

§7.13.2   Flame

The most widely recognized symbol of fire and probably the most vividly remembered factor of any fire is that of flame. Not all combustion reactions produce flames, but flame is characteristic of those reactions known as fire. To understand what a flame is, keep in mind that the burning of most materials is an exothermic, chemical oxidation process. As an exothermic process, energy is evolved and released as part of the reaction. This energy is evolved as heat which possesses convective and radioactive components. The radioactive component is represented by energy released in the visible and infrared portions of the spectrum and is seen as flame or luminosity. Therefore, flame is a visible form of heat energy, released as a result of the combustion reaction.

As a visible form of heat energy, flame is unlikely to play any singular role in injury to persons or property. Whether or not the energy is released in the visible range will not by itself greatly affect any damage that is done. Therefore, the most important injurious impact of flame is likely to be in the fact that when it is observed, it confirms the presence of fire which may lead to the creation of panic conditions. An increase in heat can come from numerous sources, but the presence of a flame, by definition, means the presence of fire.

The color of the flame can be used as an indication of the intensity and heat of the fire. This is not a fully scientific measurement and should be used only within broad limits to judge the heat caused by the fire. Time, availablity of oxygen, fuel, and fuel load are all important contributing factors toward the intensity and heat of a fire manifested by the flame color. General guidelines for correlating intensity and flame colors are as follows:

Reddish Glow   — about 500°C
Light Red   — about 1,000°C
Yellow   — about 1,100°C
Blinding White — reaches 1,500°C[4]

An important point for the litigator is that in the absence of some accelerant, it is unlikely that most common materials would reach the blinding white temperature. Whether the accelerant was used in the ignition of an arson fire, or whether it exists as a natural part of the structure, such as an oil tank, is another question.

---

[4]"Arson and Fraud Fires," Max S. Swertzman, *Forum* 12: 827-59, Spring, 1977.

§7.13.3   Fire Gases

Every type of combustion gives off some gases which are the invisible matter resulting from decomposition of the fuel and the interaction of the oxygen and other gases found in the ambient atmosphere. Fire gases are those which remain when the products of combustion are cooled to normal temperature. What gases are formed in a fire will depend on many variables, the principal ones being the chemical composition of the burning material, the amount of oxygen available for combustion, and the temperature.[5] Probably the most common gas formed is carbon monoxide. It is a combination of carbon, which is found in most fuels, and oxygen formed in a situation of imperfect combustion that occurs when there is not sufficient oxygen for complete combustion. If the combustion is complete, carbon dioxide will be formed. Other important gases include sulfur dioxide, ammonia, hydrogen cyanide, nitrogen oxides, halogen acids, isocyanates, acrolein, and volatile hydrocarbons.

For the litigator these gases can be all-important. Gases from a fire are the most common cause of death and injury to individuals. The ability of these gases to move through heating ducts and any other open passageways, and to exist long after the fire itself is put out, makes them particularly important in high-rise and other large-occupancy buildings. In any case involving fire death or injury, the litigator should, as fully as possible, investigate the possible effect of fire gases on the victim. The litigator also should be able to trace the movement of the gas from its source to the victim and explain the toxicological events which took place to bring about the injury. A full explanation of how the various gases react on the body to bring about damage is beyond the scope of this book.

Definitive statements are impossible about which burning material will emit what gases to bring about precisely what effects. Indeed, presently there is considerable controversy within the scientific community about the factor of what gases are emitted by various burning objects. The controversy is centered mostly among plastics,

---

[5] McKinnon, Gordon, et al, *Fire Protection Handbook*, 15th ed., National Fire Protection Association, Quincy, MA, 1981, pp. 3-19.

particularly polyvinyl chloride (PVC). To date, there is no definitive test that can precisely measure the quantity of gas released or the toxicological effects of burning plastic within a building. Expert opinions can be obtained for different sides of the controversy. For the litigator, it is enough to note at this point that the toxicological effect of gases on a fire victim is an item which must be carefully investigated.

As mentioned, the most common example of a toxic fire gas is carbon monoxide. Since, in most cases, there is not sufficient oxygen for complete combustion, carbon monoxide will form. It affects victims due to its propensity to combine with hemoglobin, the oxygen-carrying constituent of blood. Carbon monoxide lowers the oxygen-carrying capacity of the blood, thus making the normal gaseous exchange functions of the blood impossible. This result is poisoning by a form of asphyxiation. Relatively low concentrations of carbon monoxide can bring about severe consequences from 100 ppm, an allowable exposure for several hours, to 4,000 ppm, which can be fatal when inhaled for less than one hour, and 10,000 ppm, which can be fatal when inhaled for one minute.

Carbon dioxide (CO_2), which forms from the complete combustion of a carbon-containing fuel in the presence of oxygen, is not normally toxic. However, concentrations of carbon dioxide can create an inverse breathing rate that may cause a victim to inhale toxic gases and other irritants at a much greater rate than normal.

Another common fire gas which has a similar effect is hydrogen cyanide. This gas, like carbon monoxide, will combine readily with the blood and, due to enzyme inhibition, prevent normal cellular metabolism from occurring. Again, the body is not allowed normal metabolic processes, with death being a possibility if exposure is of sufficient length.

Another effect caused by the presence of fire gases is the using up of oxygen that is needed to sustain life. Particularly in a closed area, oxygen may be consumed by other gases to the point where it goes below the percentage needed for an individual to function effectively, and may drop to the point where life cannot be sustained. Keep in mind, too, that the presence of carbon dioxide or exertion experienced in fighting a fire or escaping from a fire may increase the need for oxygen just at the time when it is being depleted.

TABLE 7-1. Toxicological Effects of Fire Gases[6]

Toxicant	Sources	Toxicological Effects	Estimate of short-term (10-minute) lethal concentration (ppm)
Hydrogen cyanide (HCN)	From combustion of wool, silk, polyacrylonitrile, nylon, polyurethane, and paper.	A rapidly fatal asphyxiant poison.	350
Nitrogen dioxide (NO₂) and other oxides of nitrogen	Produced in small quantities from fabrics and in larger quantities from cellulose nitrate and celluloid.	Strong pulmonary irritant capable of causing immediate death as well as delayed injury.	>200
Ammonia (NH₃)	Produced in combustion of wool, silk, nylon, and melamine, concentrations generally low in ordinary building fires.	Pungent, unbearable odor; irritant to eyes and nose.	>1000
Hydrogen chloride (HCl)	From combustion of polyvinyl chloride (PVC), and some fire-retardant treated materials.	Respiratory irritant; potential toxicity of HCl coated on particulate may be greater than that for an equivalent amount of gaseous HCl.	>500, if particulate is absent
Other halogen acid gases (HF and HBr)	From combustion of fluorinated resins or films and some fire-retardant materials containing bromine.	Respiratory irritants.	HF ~400 HBr >500
Sulfur dioxide (SO₂)	From materials containing sulfur.	A strong irritant, intolerable well below lethal concentrations.	>500
Isocyanates	From urethane polymers; pyrolysis products, such as toluene-2,4-diisocyanate (TDI), have been reported in small-scale laboratory studies; their significance in actual fires is undefined.	Potent respiratory irritants; believed the major irritants in smoke of isocyanate-based urethanes.	~100 (TDI)
Acrolein	From pyrolysis of polyolefins and cellulosics at lower temperatures (~400 °C).	Potent respiratory irritant.	50 to 100

[6] Petajan, J.H., Voorhees, K.J., Packham, S.C., Baldwin, R.C., Einhorn, I.N., Grunnet, M.L., Dinger, B.G., and Birkey, M.M., "Extreme Toxicity from Combustion Products of a Fire-Retarded Polyurethane Foam," *Science*, No. 187, 1975, p 742.

§7.13.4 Smoke

Smoke and suspended liquid droplets, known as aerosols, are the visible particulate matter which are produced by fire. As a result of incomplete combustion, most carbon-based materials will give off visible particulates. Thus, while not all fires will be evidenced by smoke, some kind of visible particulates are usually given off, and the existence of smoke, like that of flame, can serve as evidence of a fire. Petroleum products in particular, when subject to incomplete combustion, are prone to giving off great quantities of smoke.

Unlike fire gases which were described earlier, smoke, being visible, often surfaces as the first warning of a fire. Smoke can set off panic or other reactions in persons confined within a fire area. Fire gases are generally carried with the smoke, however, and add to its injurious potential. Smoke is made up of fine particles of matter as opposed to the radiated energy of flames. Thus, smoke will travel far from the fire source and may remain after the fire is extinguished or controlled. Smoke being carried by an air-handling system many floors above the fire floor is a recognized factor in a hotel, office building or other high-occupancy fire location.

Another highly injurious effect of smoke is that it obscures the vision of persons looking for an exit, thus adding to the potential of individuals being trapped. Tests and victim reports have shown the rapid rate at which smoke can be generated and by which vision is obscured. A series of school fire tests performed in Los Angeles established that in nearly every case, the smoke in corridors reached untenable proportions before the heat did.[7]

The impression should not be given that smoke is unharmful in and of itself. Particulates are often irritating to the eyes and may cause more serious injuries to the respiratory system when ingested. The particulates and the toxic chemicals they may carry can be inhaled deeply into the lungs, causing further damage. Because most deaths from fire are caused by the presence of fire gases, and because these invisible gases are also airborne along with smoke, there has been incomplete research on the negative effects of smoke.

---

[7] Cloudy, W.D., *Respiratory Hazards of the Fire Services*, 1st ed., National Fire Protection Association, Boston, 1951.

One aspect of smoke, other than its potential for direct injury, makes it an important area of consideration for litigators explaining or reconstructing a fire scene. That aspect is the identifiable nature of smoke present after a fire which can indicate the location and transmission pattern of not only the smoke itself, but the fire gases as well. Also, because smoke is particulate matter, its actions have been studied and follow some generally understood patterns. Therefore, after a fire is extinguished, the visible effects of smoke can provide evidence of the distribution of the fire gases, perhaps explaining death or injury far from the source of the fire. Also, because smoke follows generally understood patterns, this flow can be predicted, explained, and tested in the event of a gap in the visible evidence.

Turning this point of view around, an understanding of smoke patterns may allow the litigator to question an opposing theory of fire development and spread of injury. The litigator should start with the generation of smoke and the fact that it generally moves like air. In small buildings there is not the conceptual problem of smoke control and effect as there is in large or tall buildings.

Given the numerous factors involved in smoke generation — nature of fuel, size of fire, size of room or area, ventilation, and heat, to name a few — it is not possible to accurately predict the total quantity of smoke that will be generated by a fire, nor is it possible to quantify what had been produced after a fire.

It has been theorized that since every rising plume of smoke is made up mostly of entrained air rather than actual products of combustion, it is possible to determine the amount of smoke generated by the amount of air drawn into the fire. An equation has been developed by which this can be estimated. However, the equation is dependent on variables which will change in the course of a large fire; thus it would probably have little value to a litigator. Another theory advanced is that the larger the fire, the greater the volume of smoke generated.[8] This theory may be too simplistic and again of little value to a litigator.

Of more importance to the litigator is the manner in which smoke moved through a building and what this can tell in an after-the-fact examination.

[8] McKinnon, Gordon, *Op. cit.*, pp. 3-30.

The quantity of particulate matter that makes up smoke is much less than the air which is entrapped with smoke as it moves from a fire. Because the greater quantity of smoke is actually entrained air, the smoke will move similar to normal air patterns in the same space. There is thus, generally speaking, no special movement of smoke apart from the normal movement of air. Expanding gases is one factor that does effect smoke movement apart from normal air.

A room fully involved in fire may have a temperature of 1,200° F to 2,000° F or higher. Given the physical nature of gases, the rise in temperature will cause the gases to expand in volume. The air pressure will also rise. Temperature increases of 1,200° F to 2,000° F can cause a three-fold expansion of air volume and pressure. The increase in volume and pressure will bring about some degree of air movement away from the fire. However, the temperature will usually recede quickly as the gas moves away from the heat source, thus bringing down the expanded volume and pressure. It is likely that in many large buildings the air heated by the fire will be only a small part of the total mass. Therefore, while pressure may have some influence on smoke movement, it is unlikely to predominate.

Ventilation, outside air movement, and the equal expansion of gases are the predominant factors controlling smoke movement in small low-rise buildings or houses where smoke movement is an issue. However, in many cases the size of the structure may lead to total or nearly total involvement of the structure in the fire; thus neither gas movement as a cause of injury nor fire development as a explanation of structural damage will be an issue. For those structures in which this is not the case, particularly for tall, modern buildings, four predominating factors can be examined. These are: 1) the expansion of gases due to temperature; 2) the stack effect; 3) the influence of external winds; and 4) the forced air movement within the building.

A full technical analysis of all of these factors would be necessary for any given building to explain the rate or cause of smoke movement in any particular direction. This work describes those factors in a general way to give the litigator a feel for their existence.

The stack effect is of major importance in tall buildings because it involves the natural movement of air from the ground to the roof. This movement is based on the difference in weight between the column of air inside the building and a corresponding column of air outside the

building. This vertical movement will take place when the temperature outside the building is less than the temperature inside. The
reverse will be true when the temperature inside the building is less
than the temperature outside.

While the walls and partitions of a modern office building will
undoubtedly have an effect on the actual flow of smoke in a given
situation, it is easiest to conceptualize the stack effect in terms of an
empty box with an opening near the bottom and one near the top, as
shown in Figure 7-3.

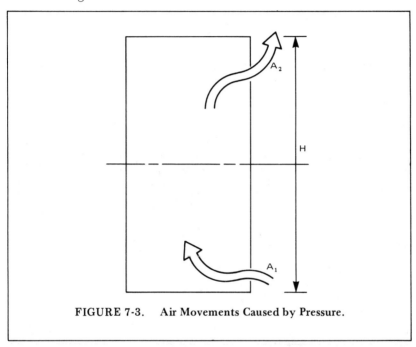

FIGURE 7-3.   Air Movements Caused by Pressure.

If $T_o \leq T_1$, the exterior pressure will be greater than the interior
pressure. Given this greater pressure, air will flow into the building at
the lower opening and out of the building at the higher opening. It is
assumed that the pressure distribution between these two openings is
linear. Because of this linear relationship, it is possible to divide the
building into two zones: one of positive pressure forcing air into the
building, and one of negative pressure forcing air out of the building.
The center line will be one of neutral pressure where air is forced
either in or out. Any openings in the walls which occur in the

respective zones will be characterized by air and smoke movement in or out in relation to the respective zone. Naturally, any openings that occur will alter the neutral pressure zone because the openings will no longer be in the same location. See Figures 7-4a and 7-4b.

Because few buildings are as clear of obstructions and as solid-walled as the Figures 7-4a and 7-4b, the general principle of stack effect should be understood, although its total application to any building is questionable. The concept of a neutral pressure plane will have importance in judging the natural movement of smoke in a building even though there are numerous ventilation holes allowing for air movement. The litigator may seek to use this concept to show either how smoke should have traveled without the intervention of some other effect, or to demonstrate how smoke actually did travel to cause injury to affected persons.

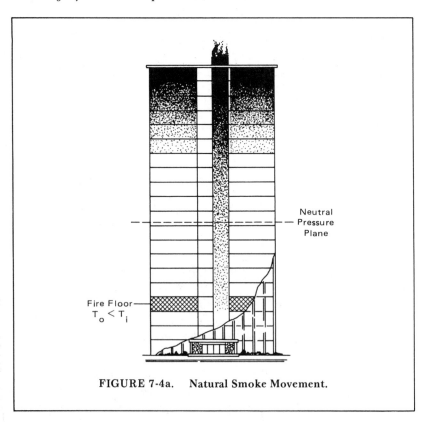

FIGURE 7-4a.   Natural Smoke Movement.

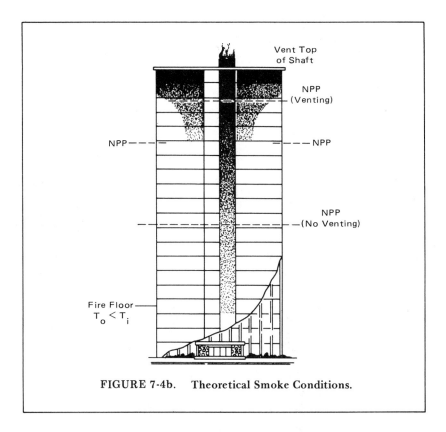

FIGURE 7-4b.   Theoretical Smoke Conditions.

Modern air-handling equipment should be designed to shut down in case of fire in order to slow the spread of deadly smoke and fumes. When this does not fully happen either through malfunction or otherwise, this failure is usually of sufficient strength to mitigate theories about stack effect on air movement. However, to the extent that air-handling systems fail to function or do not cover a particular area, stack effect will have a great influence on smoke movement.

Outside air may well play a part in smoke movement, though not to the extent of some of the other factors mentioned in regard to interior smoke. If the litigator is dealing with exterior smoke and gases, or with a building where the walls have ceased to provide protection, the wind may become the predominant factor.

Wind is best thought of as a mass of air moving around and over the structure. A tall narrow building will react differently with the wind being forced more around the building than onto the roof. This will cause comparatively stronger forces on the sides and comparatively less on the roof.

§7.14   Conclusion

Obviously, every fire case starts with a fire. Fire itself is well within the experience of every judge or juror. That fire is a complex reaction with a very definite structure, with necessary elements for ignition and growth, and often measurable outputs, is a factor not fully appreciated by most jurors who are used to striking a match or lighting a pilot light. Every juror knows that a fire can be contained by a brick wall and put out by water. Why this happens probably has not been the subject of much thought. Most jurors will assume that when a victim dies in a fire, it is inevitably because the victim has been bound up in the inferno of flame.

The point is that a jury will approach a fire without a clean slate and will not judge a theory of recovery or defense based solely on the evidence presented. Rather, each juror has his or her own experience with fire from the very outset of trial. Only if the litigator plausibly explains each point of a theory based on scientific data or visible facts will the juror resist the temptation to introduce his or her personal theory of how a phenomenon took place. Any point missed by the litigator in presenting evidence will be filled in by each juror's personal understanding of a complex reaction, an understanding which in fact may be erroneous, and at the very least, unknown to the litigator.

Therefore, the principles covered in this chapter — from the prerequisites of oxygen, fuel, and heat, to ignition and self-sustaining reaction, to the products of combustion and their effects — must be explicitly presented and fully supported in any litigation. Some points will require expert testimony, others will be made obvious from the facts of the incident. It is important for the litigator to cover all the bases. Any area left untouched will give ample ground for either the litigator's opponent to cast doubt on a theory or for jurors to fill in the facts with their own perceptions, whatever they may be.

§7.15   Appendix

As an illustration of the physical and chemical processes involved in fires, this appendix describes the ignition, burning, and eventual extinction of a wood slab in a typical situation, such as a fireplace.

1.   Suppose a wood slab is initially heated by thermal radiation. As its surface temperature approaches the boiling point of water, gases (principally steam) slowly evolve from the wood. These initial gases have little, if any, combustible content. As the slab temperature increases above the boiling point of water, the drying process penetrates deeper into the wood interior.

2.   With continued heating, the wood surface begins to discolor when the surface temperature approaches 575° F. This discoloration is visible evidence of pyrolysis, the chemical decomposition of matter through the action of heat. When wood pyrolyzes, it releases combustible gases while leaving behind a black carbonaceous residue called char. The pyrolysis process penetrates deeper into the wood slab interior as the heating continues.

3.   Soon after active pyrolysis begins, combustible gases typically evolve rapidly enough to support gas-phase combustion. However, combustion will occur only if there is a pilot flame or some other source of chemically activated molecules sufficient to support a piloted ignition. If no such pilot flame is present, the wood surface often must be heated to a much higher temperature before autoignition occurs.

4.   Once ignition occurs, a diffusion flame rapidly covers the pyrolyzing surface. The diffusion flame shields the pyrolyzing surface from direct contact with oxygen. Meanwhile, the flame heats the fuel surface and causes an increase in the rate of pyrolysis. If the original radiant heat source is withdrawn at ignition, the burning will continue provided the wood slab is thin enough (less than ¾ of an inch). Otherwise the flames will go out because the slab surface loses too much heat by thermal radiation and thermal conduction into its interior. If there is an adjacent parallel wood surface (or insulating material) facing the ignited slab, it can recapture and return much of the surface radiation loss, so that the ignited slab can continue burning even after the withdrawal of the initial heat source. This explains why a single large log cannot be burned in a fireplace and why several logs must be used to recapture the radiant heat losses.

5. As the burning continues, a char layer builds up. This char layer, which is a good thermal insulator, restricts the flow of heat to the wood interior, and consequently tends to reduce the rate of pyrolysis. The pyrolysis rate will also decrease when the supply of unpyrolyzed wood runs out. When the pyrolysis rate decreases to the point of not being able to sustain gas-phase combustion, oxygen from the air will come into direct contact with the char, permitting it to undergo direct glowing combustion, provided the radiant heat losses are not too large.

6. The above discussion presumes an ample (but not excessive) supply of air (oxidant) for combustion. If there is insufficient oxidant to burn the available fuel vapor, the excess vapors will travel with the flow and possibly burn where they eventually find sufficient oxidant. For example, this happens when fuel vapors emerge and burn outside a window of a fully involved but underventilated room fire.

If, on the other hand, a forced draft is imposed onto the pyrolyzing surface, the oxidant supply may exceed that required for complete combustion of the fuel vapors. In this case, the excess oxidant can cool the flames sufficiently to suppress their chemical reaction and extinguish them, as happens, for example, when blowing out a match or a candle. In the case of larger fires with an ample supply of fuel vapors, blowing on them simply increases their rate of burning by increasing the flame to fuel-surface heat transfer which in turn enhances the fuel supply rate.

7. Following the ignition of a certain portion of the wood slab, the flames are likely to spread over the entire fuel area. Flame spread can be thought of as a continuous succession of piloted ignitions where the flames themselves provide the heat source. Upward flame spread is much more rapid than downward or horizontal flame spread. This is because hot flames generally travel upward and contribute their heat over a greater area in an upward direction. Thus each successive upward ignition adds a much greater burning area to the fire than a corresponding downward or horizontal ignition.

Generally, materials which ignite easily (rapidly) also propagate flames rapidly. The ignitability of a material is controlled by its resistance to heating (thermal inertia) and the temperature rise required for it to start pyrolyzing. Materials with low thermal inertia, such as foamed plastics or balsa wood, heat up rapidly when subjected to the intensity of heat over a given area, called heat flux. These materials are often easy to ignite and can cause very rapid flame spread. On the other hand, dense materials, such as ebony wood, tend to have relatively high thermal inertias and are difficult to ignite.

8. The burning rates of larger, more hazardous fires are principally governed by the radiant heat transfer from the flames to the pyrolyzing fuel surface[9]. This flame radiation comes primarily from the luminous soot particles in the flames. Combustibles which tend to produce copious amounts of soot or smoke (such as polystyrene) also tend to support more intense fires, despite the fact that their fuel vapors burn less completely, as evidenced by their higher smoke output.

Generally, the sootiness of fire flames increases strongly with the oxygen content of the surrounding atmosphere. Fires in vitiated (i.e., oxygen-depleted) atmospheres produce considerably less soot, whereas fires in oxygen-enriched atmospheres can be very sooty. This is true even for fuels that produce negligible soot when burned in normal air. This sensitivity to oxygen content can strongly influence both flame radiation and burning rates.

[9] Petajan, J.H., et al., *Op. cit.*

CHAPTER 8

FIRE HAZARDS OF LIQUIDS
AND GASES

§8.1 Introduction

The previous chapter dealt with the basic science of fire and its effects. This chapter examines those inherent properties that render certain substances hazardous, especially some of the more common substances. Of principal concern are the fire hazard characteristics of a substance, as opposed to other hazardous properties like health or reactivity. This chapter is limited to liquids and gases, and does not cover hazardous solids.

Among liquids and gases, liquids are more prevalent due to their wide distribution and variety of uses. For example, gasoline, although not the most flammable liquid, is perhaps the most widely transported and used hazardous liquid due to its widespread dissemination, easy and comparatively cheap availability, and its common use as a motor fuel. Gasoline is regularly a factor in probably more fires than any other single hazardous material.

The material in this chapter is arranged to provide information on hazardous liquids and gases in a usable manner for the litigator. The initial discussion explains some of the terminology used to describe fire hazard properties of liquids and gases. Sections 8.3.1 through 8.3.7 offer more detailed information concerning seven commonly used substances, followed by a listing excerpted from NFPA 49, Hazardous Chemicals Data, providing pertinent information on many different hazardous liquids and gases. In addition, the NFPA hazardous materials identification system is described in some detail, thereby offering the litigator a variety of vantage points on the different aspects of the subject.

For more information on any of the materials discussed in this chapter, reference can be made to a number of chemistry texts listed in the references to several NFPA standards, such as NFPA 49, Hazardous Chemicals Data; 325M, Fire Hazard Properties of Flammable Liquids, Gases, and Volatile Solids; or 491M, Manual of Hazardous Chemical Reactions. These standards, along with the *Flash Point Index of Trade Name Liquids* and NFPA 704, Recommended System for the Identification of the Fire Hazards of Materials, are published by the NFPA in one volume entitled *Fire Protection Guide on Hazardous Materials*.

§8.2   Properties of Hazardous Materials

§8.2.1   Flash Point

One of the principal properties used to determine the relative hazard of a liquid is its flash point. The flash point is the lowest temperature at which a liquid gives off vapor in sufficient concentration to form an ignitable mixture with air near the surface of the liquid. Ignitable mixture means a mixture within the flammable range (between the upper and lower limits) that is capable of the propagation of flame away from the source of ignition when ignited.[1] One does not ascribe a flash point to flammable gases, since it is the nature of a gas to exist in the vapor state. Flash point, then, is a characteristic normally associated with liquids, and the temperature value assigned is derived by a laboratory test. At the flash point temperature, combustion will not be sustained. The point at which combustion will be sustained is called the fire point, and it is generally several degrees higher than the flash point. Once a liquid begins to burn, the flash point is of no significance in determining its burning characteristics.

---

[1] NFPA 325M, *Fire Hazard Properties of Flammable Liquids, Gases, and Volatile Solids*, National Fire Protection Association, Quincy, MA.

§8.2.2 Flammable Limits and Flammable Range

The flammable limits of any hazardous liquid define the upper and lower bounds of the flammable range. This range is often referred to as the explosive range. What this means is that for any ignitable gas or vapor there is a minimum and maximum concentration of the vapor with oxygen above or below which ignition will not occur. If there is too great a concentration of the vapor for the quantity of oxygen, the mixture will be above the flammable limit for that material, and propagation of flame will not occur as the mixture is considered to be too rich. On the other hand, if there is too great a concentration of oxygen for the quantity of vapor, propagation of flame will not occur as the mixture is considered to be too lean.

The flammable limits and, therefore, the flammable range of each material is different. Also, the flammable limits used in most tables are based upon normal atmospheric temperature and pressure. Altered conditions will generally have an effect on the actual limits. A general rule is that an increase in temperature or pressure will lower the lower limit of flammability and raise the upper limit, thus increasing the flammable range. A decrease in temperature or pressure will decrease the flammable range.

§8.2.3 Ignition Temperature

The ignition temperature of any material is the minimum temperature required to initiate or cause self-sustained combustion independent of the heating or heated element.[2] The previous chapter explained that one of the unique characteristics of combustion is its self-sustaining quality. Once initiated, the reaction will continue until one of three essential elements is no longer available in sufficient quantity to keep the reaction going.

Because a change of condition can have a dramatic effect, ignition temperatures should be looked upon only as approximations. In

[2] NFPA 325M, *Fire Hazard Properties of Flammable Liquids, Gases, and Volatile Solids*, National Fire Protection Association, Quincy, MA.

reality, changes in the percentage composition of gas-air mixtures, the size of the containing vessel, the duration of the heating source, and the purity of the gas or oxygen are all factors which can affect the ignition temperature. Solids are also capable of burning and sustaining combustion, and they also have ignition temperatures. However, there are no solids which are so susceptible to ignition that they are considered hazardous enough to be included in this chapter.

Autoignition temperature is a phrase often used in describing the properties of a particular gas. It refers to the temperature at which self-sustaining combustion will be initiated without the assistance of an external pilot source.

§8.2.4   Flammable Liquids and Combustible Liquids

Hazardous liquids are often referred to as being either flammable or combustible. There is a difference between the two designations, although it is one of definition rather than concept. A combustible liquid is one which has a flash point of 100° F or higher. A flammable liquid is one which has a flash point of less than 100° F.

Because of their lower flash points, flammable liquids are the more dangerous. However, all hazardous liquids should be handled with extreme caution because a temperature of 100° F is easily exceeded unless precautions are taken during transportation and storage. A temperature of 100° F will quickly be exceeded once ignition has occurred.

A full technical analysis or discussion of hazardous materials, particularly liquids, would require an examination of other factors such as specific gravity, vapor density, and boiling point. These are not mentioned in this chapter because it is less likely that a litigator will need this information about hazardous materials. However, some of these terms are covered in the previous chapter and are contained within the glossary. Reference should be made to either of these sources if factors other than those mentioned become important in litigating a case.

§8.3 Characteristics of Some Common Hazardous Materials

Section 8.4 lists a large number of hazardous gases and liquids along with some of their important characteristics. The seven that are singled out here for specific treatment — acetylene, ammonia, butane, fuel oil, gasoline, methanol, and propane — represent the more common of the hazardous gases and liquids which the litigator is likely to encounter. In any given case, further research will likely be needed to effectively present or refute a theory of recovery based on the involvement of one of these materials.

§8.3.1 Acetylene

Acetylene is a colorless gas that is odorless in its pure form, but that often has a garliclike odor in practice. Acetylene is derived in a number of different ways, but particularly by the cracking of petroleum hydrocarbons with steam or separating it from natural gas by partial oxidation. Storage can be a major problem since acetylene cannot be stored alone without the possibility of disassociating into carbon and hydrogen, with a resulting release of energy.

Acetylene's most widespread use is in welding and cutting torches because of the very high heat achieved when it is combined with oxygen. The temperature may reach 5,400°F when pure acetylene and oxygen are combined. Because of its wide use, many fires have been started by the flames from an acetylene torch, although these fires were really not started by the gas itself.

Acetylene is also an unstable gas and has a much broader flammable range than most gases. This characteristic contributes to its hazard. The lower flammable limit is 25 percent while the upper flammable limit is 80 percent concentration in air. Thus acetylene can ignite and burn over a range that is much wider than most gases. The flash point is 0°F while the autoignition temperature is 531°F. An ever-dangerous fire risk, acetylene also will form explosive compounds with silver, mercury, and copper. Each of these three metals should be excluded from any transmission system for acetylene.

A moderate fire hazard, but one which may affect fire victims, is that when mixed with oxygen in preparation of 40 percent or more, acetylene will act as a mild asphyxiant.

§8.3.2   Ammonia

Ammonia is a colorless gas that will liquify under pressure. Ammonia's principal characteristic is its highly irritating qualities, well known through everyday experience. While not a significant fire hazard in itself, the irritating qualities can hamper effective fire fighting or escape by fire victims. Because ammonia is used in numerous compounds, it is a potential irritant at any fire scene. The presence of ammonia should always be investigated in a case involving a victim's inability to reach safety or a fire fighter's inability to carry out effective fire fighting tactics.

As a hazardous or flammable gas, ammonia has a narrow explosive range with a lower explosive limit of 16 percent and an upper explosive limit of 25 percent concentration in air. Ammonia will burn when exposed to air, although it is not likely to be involved in spontaneous ignition because it has an autoignition temperature of 1,204° F.

§8.3.3   Butane

Butane is a colorless gas with a natural gas odor, which is often artificially produced by the addition of an odorous sulphur compound. Butane is very stable; has no corrosive action on metals; does not react with moisture; and is soluble in water, alcohol, and chloroform. Butane is often used in various mixtures with isobutane, propane, and pentrane.

As a fire hazard, butane is extremely flammable and can provide a constant risk of fire and explosion if not properly handled. The lower explosive limit is 1.9 percent and the upper explosive limit is 8.5

percent concentration in air. Butane has a very low flash point of -76° F and an autoignition temperature of 761° F.

Butane is a heavier hydrocarbon of low vapor pressure and is among those materials commonly classified as liquefied petroleum gas which are all derived from petroleum. Because butane will readily liquefy, it can be stored and shipped as a liquid, although it is most commonly used in its gaseous phase as a fuel. The vaporizing action of the liquid can cause freezing or chilling to skin tissue even at normal temperatures. The vaporizing action may also cause the formation of frost on any container where there is a leak. The frost often provides a telltale sign of such leaks.

§8.3.4 Fuel Oil

Fuel oil is a general classification term for any liquid petroleum product that is burned in a furnace for heat or used in an engine for generating power, excepting oils having a flash point below 100° F or oils used in burners with cotton or wool wicks. There are several grades of fuel oil; the grades generally refer to the degree of refinement of the particular oil from the crude oil. Fuel oil of a particular grade may also be made from residues from the distillation or refining process or a combination of two or more of these. The particular requirements of each of the six grades of fuel oil recognized by the American Society for Testing and Materials (ASTM) are shown in Table 8-1.

All fuel oils are combustible, with the degree of combustibility dependent on the grade of refinement. Table 8-1 should be consulted for the particulars concerning each grade of fuel oil. Because all fuel oils are combustible to some degree, they are considered a moderate fire hazard especially when they come into contact with or are exposed to heat, flame, or oxidizers.

Improper storage of fuel oil is often involved in fires; therefore, in any litigation where fuel oil is considered a factor, the storage should be carefully checked. Also keep in mind that the quantities of fuel oil stored in the normal course of business may exceed that of any other hazardous material. The quantity of oil stored can be a serious problem unless adequate precautions are taken. Reference should be

TABLE 8-1   Detailed Requirements for Fuel Oils*

Grade of Fuel Oil	Flash Point, deg F (deg C) Min	Pour Point, deg F (deg C) Max	Water and Sediment, percent by volume Max	Carbon Residue on 10 percent Bottoms, percent Max	Ash, percent by weight Max	Distillation Temperatures, deg F (deg C) 10 percent Point Max	90 percent Point Min	90 percent Point Max	Saybolt Viscosity, sec Universal at 100°F (38°C) Min	Universal Max	Furol at 122°F (50°C) Min	Furol Max	Kinematic Viscosity, centistokes At 100°F (38°C) Min	At 100°F Max	At 122°F (50°C) Min	At 122°F Max	Gravity, deg API Min	Copper Strip Corrosion Max	Sulfur, percent Max
No. 1: A distillate oil intended for vaporizing pot-type burners and other burners requiring this grade of fuel	100 or legal (38)	0§	trace	0.15	—	420 (215)	—	550 (288)	—	—	—	—	1.4	2.2	—	—	35	No. 3	0.5 or legal
No. 2: A distillate oil for general purpose domestic heating for use in burners not requiring No. 1 fuel oil	100 or legal (38)	20§ (—7)	0.05	0.35	—	—	540§ (282)	640 (338)	(32.6)#	(37.93)	—	—	2.0§	3.6	—	—	30	—	0.5§ or legal
No. 4: Preheating not usually required for handling or burning	130 or legal (55)	20 (—7)	0.50	—	0.10	—	—	—	45	125	—	—	(5.8)	(26.4)	—	—	—	—	‖
No. 5 (Light): Preheating may be required depending on climate and equipment	130 or legal (55)	—	1.00	—	0.10	—	—	—	150	300	—	—	(32)	(65)	—	—	—	—	‖
No. 5 (Heavy): Preheating may be required for burning and, in cold climates, may be required for handling	130 or legal (55)	—	1.00	—	0.10	—	—	—	350	750	(23)	(40)	(75)	(162)	(42)	(81)	—	—	‖
No. 6: Preheating required for burning and handling	150 (65)	—	2.00**	—	—	—	—	—	(900)	(9000)	45	300	—	—	(92)	(638)	—	—	‖

* It is the intent of these classifications that failure to meet any requirement of a given grade does not automatically place an oil in the next lower grade unless in fact it meets all requirements of the lower grade.

This table reprinted from ASTM D-396-69. See complete specification ASTM D-396-69.

§ Lower or higher pour points may be specified whenever required by conditions of storage or use. When pour point less than 0°F is specified, the minimum viscosity shall be 1.8 cSt (32.0 sec. Saybolt Universal) and the minimum 90 percent point shall be waived.

‖ The 10 percent distillation temperature point may be specified at 440°F (226°C) maximum for use in other than atomizing burners.

Viscosity values in parentheses are for information only and not necessarily limiting.

** The amount of water by distillation plus the sediment by extraction shall not exceed 2.00 percent. The amount of sediment by extraction shall not exceed 0.50 percent. A deduction in quantity shall be made for all water and sediment in excess of 1.0 percent.

made to NFPA 30, Flammable and Combustible Liquids Code, which sets standards for fuel storage and the containers used in that storage, and to NFPA 31, Installation of Oil Burning Equipment.

§8.3.5   Gasoline

While not the most flammable of all liquids, gasoline is the most commonly used of all hazardous materials. Gasoline is stored in a variety of containers and vessels, and improper storage is a major cause of fire. Gasoline is involved annually in more fires than any other flammable liquid.

Like fuel oil, gasoline is a hydrocarbon and is refined from petroleum. Gasoline is the chief product of petroleum distribution systems. Gasoline has a flash point of -45° F, and since it will vaporize readily at room temperature, it constitutes a constant hazard unless properly stored and handled. For purposes of contrast, it is worth noting the difference in higher flash points between gasoline and kerosene (100° F and higher). The autoignition temperature is between 635° F and 853° F. The lower explosive limit is 1.4 percent and the upper explosive limit is 7.6 percent concentration in air. This upper explosive limit illustrates that a confined space can easily become saturated with gasoline so that it will be too rich to burn.

§8.3.6   Methanol

Methanol, also called methyl alcohol, is a colorless liquid at ambient temperatures. It is the principal ingredient in wood alcohol with which it is often confused. Methanol is derived from the high-pressure catalytic synthesis of carbon monoxide and hydrogen or the partial oxidation of natural gas hydrocarbons.

The flash point of methanol is 52° F, its autoignition temperature is 725° F, with a lower explosive limit of 6.0 percent and an upper explosive limit of 36 percent concentration in air.

One property of methanol is its high toxicity, particularly involving the nervous system. Once absorbed into the body, methanol is only very slowly eliminated. Thus a significant exposure to pure methanol can cause serious effects, including blindness and coma. Death can result from extended exposure. Blurring vision from the effects of methanol on the optic nerve may be the first symptom with more severe reactions following.

§8.3.7 Propane

Propane is a colorless gas derived from petroleum and natural gas. Along with butane, methane and ethane, propane is a component of liquefied natural gas, often called LNG. Propane on its own is highly flammable and constitutes a dangerous fire risk. Its explosive range is between 2.4 percent and 9.5 percent of air. The flash point is -156° F and the autoignition temperature is 842° F. Therefore, while flammable vapor will be given off at an extremely low temperature, it will take a moderately high temperature for the gas to self-ignite. Thus spontaneous ignition is not a great risk with propane. Among its other properties, propane is noncorrosive and nontoxic. Also it is soluble in ether or alcohol and slightly soluble in water.

Propane is becoming more widely used as a fuel, and is sold commercially in bottled form for use in barbecues and other common places where such a fuel is needed. Propane is now used as an automobile fuel, and this use is growing. Propane-fueled automobiles have moved out of the experimental stage and into some measure of mass use. A labeling requirement has recently been added to NFPA 54, National Fuel Gas Code, so that in the case of accident or other potential danger, fire service personnel and civilians can be warned of the presence of this dangerous gas and act accordingly.

§8.4 Other Hazardous Materials

Among the materials already discussed were some of the more prevalent hazardous materials involved in the greatest number of fires which the litigator is likely to encounter. However, there are thousands of other hazardous liquids and gases which, though less common, may still have importance in a particular case.

A description follows of other hazardous liquids and gases referenced in NFPA 49, Hazardous Chemicals Data. Only 40 chemicals from that standard are included within this list. Preference has been given to those chemicals which have a flammability hazard of 3 or greater, although other chemicals, important either because they are commonly used or possess a high health or reactivity hazard, are included. Reference should be made to the complete standard or to other referenced works if a particular gas or liquid is not listed here. (See §8.5, NFPA Fire Hazard Labeling System, for explanation of diamond-shaped figures in the following material.) The "Explanatory" referenced under some liquids and gases appears on page 255.

ACETALDEHYDE CH_3CHO

DESCRIPTION: Colorless liquid at temperatures below 69° F. but rapidly volatilizes at this temperature; suffocating, fruity odor.

FIRE AND EXPLOSION HAZARDS: Reactive and flammable liquid which rapidly volatilizes at 69° F. Vapor forms explosive mixtures with air over a wide range. Flammable limits, 4% and 60%. Flash point, minus 36° F. Ignition temperature, 365° F. Liquid is lighter than water (specific gravity, 0.8). Vapors are heavier than air (vapor density, 1.5), and may travel a considerable distance to a source of ignition and flash back. Very reactive and can be oxidized or reduced readily. Combines with halogens and amines, and forms a great number of condensation products with alcohols, ketones, acid anhydrides, phenols and similar compounds. Hydrogen cyanide, hydrogen sulfide and anhydrous ammonia react with acetaldehyde readily. Acetaldehyde oxidizes readily in air to unstable peroxides that may explode spontaneously. Easily undergoes polymerization which is accompanied by evolution of heat. All of these reactions can be violent. Vapor oxidizes readily with air and may form highly explosive and unstable peroxides. Acetaldehyde is soluble in water.

LIFE HAZARD: Eye, skin and respiratory irritant. Capable of producing serious eye burns. Prolonged inhalation may have a narcotic effect, resulting in drowsiness.

PERSONAL PROTECTION: Wear self-contained breathing apparatus; wear goggles if eye protection not provided.

FIRE FIGHTING PHASES: In advanced or massive fires, fire fighting should be done from a safe distance or from a protected location. Use dry chemical, "alcohol" foam, or carbon dioxide.

Water may be ineffective (see Explanatory), but water should be used to keep fire-exposed containers cool. If a leak or spill has not ignited, use water spray to disperse the vapors. If it is necessary to stop a leak, use water spray to protect men attempting to do so. Water spray may be used to flush spills away from exposures and to dilute spills to nonflammable mixtures.

USUAL SHIPPING CONTAINERS: One-quart glass pressure bottles, 5- to 55-gallon metal drums, insulated tank cars and insulated tank trucks, tank barges.

STORAGE: Protect against physical damage. Store bulk quantities outside in detached tanks provided with refrigeration and inert gas blanket, such as nitrogen, in void space above liquid level. Smaller container storage should be in a detached noncombustible building, provided with cooling facilities, adequate ventilation and free of sources of ignition; no alkaline materials (such as caustics, ammonia, amines), halogens, alcohols, ketones, acid anhydrides, phenols, nor oxidizing materials, permitted in storage room. Inside storage should be in a standard flammable liquids storage room or cabinet. Isolate from other storage.

REMARKS: Electrical installations in Class I hazardous locations, as defined in Article 500 of the National Electrical Code, should be in accordance with Article 501 of the Code. If explosionproof electrical equipment is necessary, it shall be suitable for use in Group C. See Flammable and Combustible Liquids Code (NFPA No. 30), National Electrical Code (NFPA No. 70), Static Electricity (NFPA No. 77), Lightning Protection Code (NFPA No. 78), Fire-Hazard Properties of Flammable Liquids, Gases and Volatile Solids (NFPA No. 325M), and Chemical Safety Data Sheet SD-43 (Manufacturing Chemists' Association, Inc.).

ACETYLENE CH⋮CH

Dissolved in
Acetone in
Closed Cylinder

DESCRIPTION: Colorless gas with slight garliclike odor.

FIRE AND EXPLOSION HAZARDS: Flammable gas. Forms explosive mixtures with air over a very wide range. Flammable limits, 2.5% and 100%. Ignition temperature is comparatively low and varies according to mixture composition, pressure, water vapor content and initial temperature; minimum ignition temperature is about 571° F. Lighter than air (vapor density, 0.9). Acetylene not dissolved in acetone is unstable at high pressures and may decompose into hydrogen and carbon with explosive violence. Generation, distribution through hose or piping, or utilization of acetylene should be maintained at a pressure less than 15 psi gage. Under certain conditions, acetylene forms explosive compounds with copper, silver and mercury. Also forms spontaneously explosive acetylene chloride with chlorine.

LIFE HAZARD: Nontoxic but can cause asphyxiation by exclusion of oxygen.

FIRE FIGHTING PHASES: Stop flow of gas. Use water to keep fire-exposed containers cool and to protect men effecting the shut-off.

USUAL SHIPPING CONTAINERS: Steel cylinders, 10 to 300 standard cubic feet capacity, containing a porous material and acetone.

STORAGE: Protect against physical damage. Outside or detached storage is preferred. Isolate from oxidizing gases, especially chlorine. Store in cool, well-ventilated, noncombustible place, away from all possible sources of ignition and combustible materials. Protect against lightning and static electricity. Store cylinders upright.

REMARKS: Electrical installations in Class I hazardous locations, as defined in Article 500 of the National Electrical Code, should be in accordance with Article 501 of the Code. If explosionproof electrical equipment is necessary, it shall be suitable for use in Group A. See Standard for the Installation and Operation of Gas Systems for Welding and Cutting (NFPA No. 51), Explosion Venting Guide (NFPA No. 68), National Electrical Code (NFPA No. 70), Static Electricity (NFPA No. 77), Lightning Protection Code (NFPA No. 78), Fire-Hazard Properties of Flammable Liquids, Gases and Volatile Solids (NFPA No. 325M) and Chemical Safety Data Sheet SD-7 (Manufacturing Chemists' Association, Inc.).

AMMONIUM PERCHLORATE NH_4ClO_4

Nonfire   Fire

DESCRIPTION: White crystalline substance.

FIRE AND EXPLOSION HAZARDS: Powerful oxidizing material. Becomes an explosive when mixed with finely divided organic materials. Exhibits the same explosive sensitivity to shock as picric acid; sensitivity to shock and friction may be great when contaminated with small amounts of some impurities such as sulfur, powdered metals and carbonaceous materials. May explode when involved in fire.

LIFE HAZARD: Yields toxic products of combustion.

PERSONAL PROTECTION: In fire conditions wear self-contained breathing apparatus.

FIRE FIGHTING PHASES: Fight fires with water from an explosion-resistant location. In advanced or massive fires, the area should be evacuated. If fire occurs in the vicinity of this material water should be used to keep containers cool.

USUAL SHIPPING CONTAINERS: Bags, bottles, wooden drums or kegs, or fiber drums.

STORAGE: Protect containers against physical damage. Isolate from other materials, especially combustibles, sulfur, carbonaceous materials, finely divided metals, organic or other readily oxidizable materials and mineral acids. Never store in an acute fire hazard area, in the vicinity of constant sources of ignition or where subject to elevated temperatures. Immediately remove and dispose of any spilled perchlorate.

REMARKS: See Code for the Storage of Liquid and Solid Oxidizing Materials (NFPA No. 43A).

BENZENE C_6H_6

DESCRIPTION: Colorless liquid with aromatic odor.

FIRE AND EXPLOSION HAZARDS: Flammable liquid. Vapor forms explosive mixtures with air. Flammable limits, 1.3% and 7.1%. Flash point, 12° F. Ignition temperature, 1044° F. Liquid is lighter than water (specific gravity, 0.9). Vapor is heavier than air (vapor-air density at 100° F., 1.4) and may travel considerable distance to a source of ignition and flash back. Not soluble in water.

LIFE HAZARD: Breathing of high concentrations of benzene may cause acute poisoning and death. Repeated inhalation of low concentrations often results in severe or fatal anemia. Also a skin and eye irritant.

PERSONAL PROTECTION: Wear self-contained breathing apparatus.

FIRE FIGHTING PHASES: Use dry chemical, foam, or carbon dioxide. Water may be ineffective (see Explanatory), but water should be used to keep fire-exposed containers cool. If a leak or spill has not ignited, use water spray to disperse the vapors and to protect men attempting to stop a leak. Water spray may be used to flush spills away from exposures.

USUAL SHIPPING CONTAINERS: Small glass bottles, one-gallon cans, 5- to 55-gallon metal drums, tank cars and tank trucks. Tank barges.

STORAGE: Protect against physical damage. Outside or detached storage is preferable. Inside storage should be in a standard flammable liquids storage room or cabinet. Separate from oxidizing materials.

REMARKS: Electrical installations in Class I hazardous locations, as defined in Article 500 of the National Electrical Code, should be in accordance with Article 501 of the Code. If explosionproof electrical equipment is necessary, it shall be suitable for use in Group D. See Flammable and Combustible Liquids Code (NFPA No. 30), National Electrical Code (NFPA No. 70), Static Electricity (NFPA No. 77), Lightning Protection Code (NFPA No. 78), Fire-Hazard Properties of Flammable Liquids, Gases and Volatile Solids (NFPA No. 325M), and Chemical Safety Data Sheet SD-2 (Manufacturing Chemists' Association, Inc.).

3-BROMOPROPYNE CH:CCH_2Br

DESCRIPTION: Colorless to light-amber liquid; strong lachrymator.

FIRE AND EXPLOSION HAZARDS: Flammable liquid. Vapor forms flammable mixtures with air. Lower flammable limit, 3.0%. Flash point, 50° F (cc). Ignition temperature, 615° F. Liquid is heavier than water (specific gravity, 1.57). Melting point, minus 76° F. Boiling point, 185° F. 3-bromopropyne is an acetylenic compound which may be decomposed by mild shock. When heated under confinement, such as in an exposure fire, the material decomposes with explosive violence and may detonate. When suitably diluted, however, as with 20–30% by weight of toluene, its explosive properties are practically eliminated.

LIFE HAZARD: Vapors are very toxic. Strong lachrymator and skin irritant.

PERSONAL PROTECTION: Wear full protective clothing.

FIRE FIGHTING PHASES: Fight fires from an explosion-resistant location. In advanced or massive fires, the area should be evacuated. If fire occurs in the vicinity of this material, water should be used to keep containers cool. Tanks or drums containing 3-bromopropyne should not be approached directly after they have been involved in a fire or heated by exposure fires. Clean-up or salvage operations should not be attempted until the 3-bromopropyne is cooled.

USUAL SHIPPING CONTAINERS: Glass carboys and metal drums.

STORAGE: Unstabilized material should be stored like an explosive. Material properly stabilized by dilution should be stored like a flammable liquid. Outside or detached storage of the latter is preferred. Inside storage should be in a standard flammable liquids storage room or cabinet. Separate from oxidizing materials.

REMARKS: Electrical installations in Class I hazardous locations, as defined in Article 500 of the National Electrical Code, should be in accordance with Article 501 of the Code; and electrical equipment should be suitable for use in atmospheres containing 3-bromopropyne vapors. See Flammable and Combustible Liquids Code (NFPA No. 30), National Electrical Code (NFPA No. 70), and Static Electricity (NFPA No.77).

BUTADIENE-1,3 $CH_2:CHCH:CH_2$

DESCRIPTION: A colorless, mildly aromatic gas.

FIRE AND EXPLOSION HAZARDS: Flammable gas. Forms explosive mixtures with air. Flammable limits, 2% and 12%. Ignition temperature, 804° F. Heavier than air (vapor density, 1.9) and may travel a considerable distance to a source of ignition and flash back. Usually contains inhibitors to prevent self-polymerization (which is accompanied by evolution of heat) and to prevent formation of peroxides. At elevated temperatures, such as in fire conditions, polymerization may take place. If the polymerization takes place in a container, there is possibility of violent rupture of the container.

LIFE HAZARD: Slightly toxic but may cause asphyxiation by exclusion of oxygen. Slight respiratory irritant. Direct expansion on skin may cause freeze burns.

PERSONAL PROTECTION: Wear self-contained breathing apparatus.

FIRE FIGHTING PHASES: Stop flow of gas. Use water to keep fire-exposed containers cool and to protect men effecting the shut-off. Butadiene vapors are uninhibited and may form polymers in vents or flame arresters of storage tanks, resulting in stoppage of vents.

USUAL SHIPPING CONTAINERS: Liquefied in steel pressure cylinders, tank cars, tank barges.

STORAGE: Protect against physical damage. Outside or detached storage is preferred. Inside storage should be in a cool, well-ventilated, noncombustible location, away from all possible sources of ignition. Store cylinders vertically and do not stack. Do not store with oxidizing material.

REMARKS: Electrical installations in Class I hazardous locations, as defined in Article 500 of the National Electrical Code, should be in accordance with Article 501 of the Code. If explosionproof electrical equipment is necessary, it shall be suitable for use in Group B. Group D equipment may be used if such equipment is isolated in accordance with Section 501–5(a) by sealing all conduit $\frac{1}{2}$-inch size or larger. See Explosion Venting Guide (NFPA No. 68), National Electrical Code (NFPA No. 70), Static Electricity (NFPA No. 77), Lightning Protection Code (NFPA No. 78), Fire-Hazard Properties of Flammable Liquids, Gases and Volatile Solids (NFPA No. 325M), and Chemical Safety Data Sheet SD-55 (Manufacturing Chemists' Association, Inc.).

n-BUTYLAMINE $C_4H_9NH_2$

DESCRIPTION: Colorless, volatile liquid with ammonia-like odor.

FIRE AND EXPLOSION HAZARDS: Flammable liquid. Vapor forms explosive mixtures with air. Flammable limits, 1.7% and 9.8%. Flash point, 10° F. Ignition temperature, 594° F. Boiling point, 172° F. Liquid is lighter than water (specific gravity, 0.74). Vapor is heavier than air (vapor-air density at 100° F., 1.5) and may travel a considerable distance to a source of ignition and flash back. Liquid is soluble in water, alcohol and ether.

LIFE HAZARD: Moderately toxic. Eye, skin and respiratory irritant.

PERSONAL PROTECTION: Wear self-contained breathing apparatus; wear goggles if eye protection not provided.

FIRE FIGHTING PHASES: Use dry chemical, "alcohol" foam, or carbon dioxide; water may be an ineffective agent (see Explanatory), but water should be used to keep fire-exposed containers cool. If a leak or spill has not ignited, use water spray to disperse the vapors and to protect men attempting to stop a leak. Water spray may be used to flush spills away from exposures and to dilute spills to nonflammable mixtures.

USUAL SHIPPING CONTAINERS: 1-gallon cans; 5- and 55-gallon drums; tank cars.

STORAGE: Protect against physical damage. Separate from oxidizing materials and sources of heat. Outside or detached storage is preferable. Inside storage should be in a standard flammable liquids storage room or cabinet.

REMARKS: Electrical installations in Class I hazardous locations, as defined in Article 500 of the National Electrical Code, should be in accordance with Article 501 of the Code; and electrical equipment should be suitable for use in atmospheres containing butylamine vapors. See Flammable and Combustible Liquids Code (NFPA No. 30), Fire Hazard Properties of Flammable Liquids, Gases and Volatile Solids (NFPA No. 325M), and National Electrical Code (NFPA No. 70).

CALCIUM CARBIDE CaC_2

DESCRIPTION: Grayish-black irregular lumps.

FIRE AND EXPLOSION HAZARDS: Not flammable in dry state but produces acetylene gas on contact with water or moisture. Will generate sufficient heat on contact with small amount of water to ignite the acetylene gas formed.

LIFE HAZARD: Dust is an eye and respiratory irritant, and can cause skin burns.

FIRE FIGHTING PHASES: Do not use water, vaporizing liquids or foam. Carbon dioxide is ineffective. Smother with suitable dry powder.

USUAL SHIPPING CONTAINERS: Steel drums and cans.

STORAGE: Protect against physical damage. Store in dry, noncombustible, well-ventilated place without sprinkler protection and exclude possible sources of ignition of acetylene gas. Isolate from other materials.

REMARKS: See Standard for the Installation and Operation of Gas Systems for Welding and Cutting (NFPA No. 51), and Chemical Safety Data Sheet SD-23 (Manufacturing Chemists' Association, Inc.).

CARBON DISULFIDE (Carbon Bisulfide) CS₂

DESCRIPTION: Clear, colorless to faint yellow liquid with strong, disagreeable odor.

FIRE AND EXPLOSION HAZARDS: A flammable liquid which gives off flammable vapors even at low temperatures which will form explosive mixtures in air over a wide range. Flash point, minus 22° F. Flammable limits are 1.3% and 50%. Ignition temperature is dangerously low, 194° F (90° C). Vapor is heavier than air (vapor-air density at 100° F., 2.2) and may travel a considerable distance to a source of ignition and flash back. Boiling point is low, 115° F. Liquid is heavier than water (specific gravity, 1.3), and is not soluble in water. Vapors may be ignited by contact with an ordinary light bulb. Not soluble in water.

LIFE HAZARD: Toxic by oral intake, inhalation or prolonged contact with skin.

PERSONAL PROTECTION: Wear self-contained breathing apparatus.

FIRE FIGHTING PHASES: Foam is ineffective. Use dry chemical, carbon dioxide or other inert gas. Cooling and blanketing with water spray is effective in case of fires in metal containers or tanks to help prevent reignition by hot surfaces.

USUAL SHIPPING CONTAINERS: Small glass or metal containers packed inside fiber or wood boxes and barrels, 5-gallon cans, metal drums, tank trucks and tank cars, tank barges.

STORAGE: Protect against physical damage. Store well detached and isolated from other buildings, other materials and possible sources of ignition, preferably in a building of noncombustible, or better, construction with floor level ventilation. Avoid direct sunlight. During hot weather, spray drums with water to keep vapor pressure down. Tanks should be submerged in water or located over concrete basins containing water of sufficient capacity to hold all of the tank contents in addition to the water. Water or inert gas should be provided over the carbon disulfide in all tanks. No electrical installations or heating facilities should be permitted in or near storage area. Protect against lightning and static electricity.

REMARKS: Recent tests have shown that carbon disulfide, because of its low ignition temperature and because of the extremely small joint clearance required to arrest its flame, cannot be included in any of the atmospheric groups in Section 500-2 of the National Electrical Code. Carbon disulfide should never be transferred by means of air; use pump, water, or inert gas. Do not use spark-producing tools or devices where stored, handled or used. Use wood measuring stick for measuring contents of containers and tanks. Do not dispose of carbon disulfide by pouring it on the ground; provide a safe place for burning it.

See the Standard for Carbon Dioxide Extinguishing Systems (NFPA No. 12), Flammable and Combustible Liquids Code (NFPA No. 30), National Electrical Code (NFPA No. 70), Static Electricity (NFPA No. 77), Lightning Protection Code (NFPA No. 78), Fire-Hazard Properties of Flammable Liquids, Gases and Volatile Solids (NFPA No. 325M), Chemical Safety Data Sheet SD-12 (Manufacturing Chemists' Association, Inc.), and Handbook of Industrial Loss Prevention, Chapter 52 (Factory Mutual Engineering Division).

CROTONALDEHYDE $CH_3CH:CHCHO$

DESCRIPTION: Water-white to straw-colored liquid, characterized by extreme degree of fluidity, and possessing a penetrating, pungent and suffocating odor. Commercial grade stabilized with water; solid phase separates out at 23° F.

FIRE AND EXPLOSION HAZARDS: Flammable liquid. Vapor forms explosive mixtures with air. Flammable limits, 2.1% and 15.5%. Flash point, 55° F (cc) and 127.4° F (oc) for commercial grade (93%); 55° F (oc) for anhydrous. Liquid is lighter than water (specific gravity, 0.9). Vapor is heavier than air (vapor-air density at 100° F, 1.10), and may travel considerable distance to a source of ignition and flash back. Slightly soluble in water. At elevated temperatures, such as in fire conditions, polymerization may take place. If the polymerization takes place in a container, there is possibility of violent rupture of the container. Readily converted by oxygen to hazardous peroxides and acids. Extremely violent polymerization reaction results when in contact with alkaline materials such as caustics, ammonia, or amines.

LIFE HAZARD: Extreme eye, respiratory and skin irritant. Very dangerous to eyes, can cause corneal damage. Regarded as an acute irritant on very short exposure to small quantities. Flush eyes and skin with copious quantities of water and obtain medical attention.

PERSONAL PROTECTION: Wear full protective clothing.

FIRE FIGHTING PHASES: In advanced or massive fires, fire fighting should be done from an explosion-resistant location. Use dry chemical, foam, or carbon dioxide. Water may be ineffective (see Explanatory), but water should be used to keep fire-exposed containers cool. If a leak or spill has not ignited, use water spray to disperse the vapors. If it is necessary to stop a leak, use water spray to protect men attempting to do so. Water spray may be used to flush spills away from exposures.

USUAL SHIPPING CONTAINERS: 10-gallon boxed carboys.

STORAGE: Protect against physical damage. Outside or detached storage is preferred. Inside storage should be in a standard flammable liquids storage room or cabinet; no alkaline materials, such as caustics, ammonia or amines, or oxidizing materials permitted in storage room or cabinet.

REMARKS: Electrical installations in Class I hazardous locations, as defined in Article 500 of the National Electrical Code, should be in accordance with Article 501 of the Code; and electrical equipment should be suitable for use in atmospheres containing crotonaldehyde vapors. See Flammable and Combustible Liquids Code (NFPA No. 30).

CYANOGEN NCCN

DESCRIPTION: Colorless gas with pungent penetrating odor.

FIRE AND EXPLOSION HAZARDS: Flammable gas. Flammable limits, 6.6% and 32%. Heavier than air (vapor density, 1.8), and may travel considerable distance to a source of ignition and flash back.

LIFE HAZARD: Highly toxic. In addition, when heated to decomposition or on contact with acid, acid fumes, water, or steam, it will react to produce highly toxic fumes.

PERSONAL PROTECTION: Wear special protective clothing.

FIRE FIGHTING PHASES: Stop flow of gas. Use water to keep fire-exposed containers cool and to protect men effecting the shut-off.

USUAL SHIPPING CONTAINERS: High-pressure metal cylinders not over 125 pounds water capacity.

STORAGE: Protect containers against physical damage. Prevent shock. Keep cylinders away from any source of heat. Isolate from acids, acid fumes, or water. Store in cool, well-ventilated area of noncombustible construction away from any source of ignition. Outside or detached storage is preferred.

REMARKS: Electrical installations in Class I hazardous locations, as defined in Article 500 of the National Electrical Code, should be in accordance with Article 501 of the Code; and electrical equipment should be suitable for use in atmospheres containing cyanogen. See National Electrical Code (NFPA No. 70) and Fire-Hazard Properties of Flammable Liquids, Gases and Volatile Solids (NFPA No. 325M).

DIBORANE B₂H₆

DESCRIPTION: Colorless gas; repulsive, sweet odor.

FIRE AND EXPLOSION HAZARDS: Highly reactive and flammable gas. Forms flammable mixtures with air over a wide range. Flammable limits, 0.8% and 88%. Ignition temperature is low, 100° to 125° F. Vapor is lighter than air (vapor density, 0.96). Diborane will ignite spontaneously in moist air at room temperature. It reacts spontaneously with chlorine and forms hydrides with aluminum and lithium which may ignite spontaneously in air. Reacts with many oxidized surfaces as a strong reducing agent. Reacts violently with vaporizing liquid-type extinguishing agents.

LIFE HAZARD: Extremely toxic due to respiratory irritation and lung congestion.

PERSONAL PROTECTION: Wear self-contained breathing apparatus.

FIRE FIGHTING PHASES: Fire fighting should be done from an explosion-resistant location. Use water from unmanned monitors or hoseholders to keep fire-exposed containers cool. If it is necessary to stop the flow of gas, use water spray to protect men effecting shut-off. Personnel should be evacuated immediately.

USUAL SHIPPING CONTAINERS: Steel pressure cylinders.

STORAGE: Protect against physical damage. Storage should be in detached, refrigerated (less than 68° F.) and well-ventilated place. Containers must be clean, dry, and free of oxygen. Separate from halogens and other oxidizing agents, and check periodically for decomposition. Protect against electrical sparks, open flames, or any other heat source. Dry nitrogen purge should be used in any transfer.

REMARKS: Electrical installations in Class I hazardous locations, as defined in Article 500 of the National Electrical Code, should be in accordance with Article 501 of the Code; and electrical equipment should be suitable for use in atmospheres containing diborane. See National Electrical Code (NFPA No. 70), Static Electricity (NFPA No. 77), and Fire-Hazard Properties of Flammable Liquids, Gases and Volatile Solids (NFPA No. 325M).

DIBUTYL ETHER (Normal) $C_4H_9OC_4H_9$

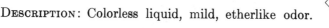

DESCRIPTION: Colorless liquid, mild, etherlike odor.

FIRE AND EXPLOSION HAZARDS: Flammable liquid. Vapor forms explosive mixtures with air. Flammable limits, 1.5% and 7.6%. Flash point, 77° F. Ignition temperature, 382° F. Liquid is lighter than water (specific gravity, 0.8). Vapor is heavier than air (vapor-air density at 100° F., 1.1) and may travel a considerable distance to a source of ignition and flash back. Tends to form explosive peroxides especially when anhydrous. Not soluble in water.

LIFE HAZARD: Irritating to eyes, nose, throat, respiratory or gastrointestinal system. Wear goggles and self-contained breathing apparatus.

FIRE FIGHTING PHASES: Use dry chemical, "alcohol" foam, or carbon dioxide. Water may be ineffective agent (see Explanatory), but water should be used to keep fire-exposed containers cool. If a leak or spill has not ignited, use water spray to disperse the vapors and to protect men attempting to stop a leak. Water spray may be used to flush spills away from exposures. Wear gogles and self-contained breathing apparatus.

USUAL SHIPPING CONTAINERS: 1-gallon cans; 5- and 55-gallon drums; tank cars.

STORAGE: Protect against physical damage. Outside or detached storage is preferable. Inside storage should be in a standard flammable liquids storage room or cabinet. Separate from oxidizing materials.

REMARKS: Electrical installations in Class I hazardous locations, as defined in Article 500 of the National Electrical Code, should be in accordance with Article 501 of the Code; and electrical equipment should be suitable for atmospheres containing butyl ether vapors. See Flammable and Combustible Liquids Code (NFPA No. 30), National Electrical Code

(NFPA No. 70), Static Electricity (NFPA No. 77), Lightning Protection Code (NFPA No. 78), Fire-Hazard Properties of Flammable Liquids, Gases and Volatile Solids (NFPA No. 325M).

1,2-DICHLOROETHENE ClCH:CHCl

DESCRIPTION: Colorless liquid with an ethereal, slightly acrid odor.

FIRE AND EXPLOSION HAZARDS: Flammable liquids. Vapors form explosive mixtures with air. Flammable limits, 9.7% and 12.8%. Flash points: cis, 39° F.; trans, 36° F. Liquid is heavier than water (specific gravity, 1.3). Vapor is heavier than air (vapor-air density at 100° F., 2.1) and may travel a considerable distance to a source of ignition and flash back. Not soluble in water.

LIFE HAZARD: Toxic by inhalation, skin contact, or if swallowed. Prolonged, excessive, or repeated exposures in any form are hazardous. Also an eye irritant and can cause serious damage.

PERSONAL PROTECTION: Wear self-contained breathing apparatus; wear goggles if eye protection not provided.

FIRE FIGHTING PHASES: In advanced or massive fires, fire fighting should be done from a safe distance or from a protected location. Use dry chemical, foam, or carbon dioxide. Water may be ineffective (see Explanatory), but water should be used to keep fire-exposed containers cool. If a leak or spill has not ignited, use water spray to disperse the vapors. If it is necessary to stop a leak, use water spray to protect men attempting to do so. Water spray may be used to flush spills away from exposures.

USUAL SHIPPING CONTAINERS: Glass bottles, metal cans, drums, tank trucks, and tank cars.

STORAGE: Protect against physical damage. Outside or detached storage is preferable. Inside storage should be in a standard flammable liquids storage room or cabinet. Separate from oxidizing materials.

REMARKS: Electrical installations in Class I hazardous locations, as defined in Article 500 of the National Electrical Code, should be in accordance with Article 501 of the Code; and electrical equipment should be suitable for use in atmospheres containing 1,2-dichloroethene. See Flammable and Combustible

Liquids Code (NFPA No. 30), National Electrical Code (NFPA No. 70), Static Electricity (NFPA No. 77), and Fire-Hazard Properties of Flammable Liquids, Gases and Volatile Solids (NFPA No. 325M).

DIETHYL ETHER (Ether; Ethyl Ether)
$C_2H_5OC_2H_5$

DESCRIPTION: Colorless liquid with characteristic anesthetic odor.

FIRE AND EXPLOSION HAZARDS: Flammable liquid. Even at low temperatures, vapor forms flammable mixtures with air over a wide range. Flammable limits, 1.9% and 36%. Flash point, minus 49° F. Ignition temperature is relatively low, 356° F. Boiling point, 95° F. Vapor is heavier than air (vapor density, 2.6) and may travel considerable distance to a source of ignition and flash back. In presence of oxygen, or on long standing, or exposed to sunlight in bottles, unstable peroxides sometimes form which may explode spontaneously or when heated. Ether may be readily ignited by static electricity. Not soluble in water.

LIFE HAZARD: Medical anesthetic low concentration of vapor in air rapidly causes unconsciousness.

PERSONAL PROTECTION: Wear self-contained breathing apparatus.

FIRE FIGHTING PHASES: Use dry chemical, foam, or carbon dioxide. Water may be ineffective (see Explanatory), but water should be used to keep fire-exposed containers cool. If a leak or spill has not ignited, use water spray to disperse the vapors and to protect men attempting to stop a leak. Water spray may be used to flush spills away from exposures.

USUAL SHIPPING CONTAINERS: Glass bottles or cans inside boxes, steel drums, tank barges.

STORAGE: Protect against physical damage. Detached outside storage is preferred. Inside storage should be in a standard flammable liquids storage room or cabinet. Isolate from other combustible materials. Avoid direct sunlight. Protect against static electricity and lightning. For large quantity storage rooms, protect with automatic sprinkler systems and total flooding carbon dioxide systems. The reactivity hazard may be increased to 3 on long standing due to peroxide formation. Separate from oxidizing materials.

REMARKS: Electrical installations in Class I hazardous locations, as defined in Article 500 of the National Electrical Code, should be in accordance with Article 501 of the Code. If explosion-proof electrical equipment is necessary, it shall be suitable for use in Group C. See Flammable and Combustible Liquids Code (NFPA No. 30), Standard for the Use of Inhalation Anesthetics (NFPA No. 56A), National Electrical Code (NFPA No. 70), Static Electricity (NFPA No. 77), Lightning Protection Code (NFPA No. 78), Fire-Hazard Properties of Flammable Liquids, Gases and Volatile Solids (NFPA No. 325M), and Chemical Safety Data Sheet SD-29 (Manufacturing Chemists' Association, Inc.).

DIMETHYL ETHER CH_3OCH_3

DESCRIPTION: Colorless gas with an ethereal odor. Liquid below minus 11° F (minus 23.9° C).

FIRE AND EXPLOSION HAZARDS: Flammable gas. Forms explosive mixtures with air. Flammable limits, 3.4% and 27%. Ignition temperature, 662° F (350° C). Vapor is heavier than air (vapor density, 1.6). Presence of oxygen, long standing, or exposure in bottles to sunlight may result in formation of unstable peroxides which may explode spontaneously or when heated. Soluble in water.

LIFE HAZARD: Gas possesses irritative and narcotic properties. Absorption of excessive quantities by inhalation and skin may lead progressively to a state of intoxication, loss of consciousness and death due to respiratory failure.

PERSONAL PROTECTION: Wear self-contained breathing apparatus.

FIRE FIGHTING PHASES: Stop flow of gas. Use water to keep fire-exposed containers cool and to protect men effecting the shutoff.

USUAL SHIPPING CONTAINERS: 25-, 50-, 100- and 150-pound cylinders.

STORAGE: Protect against physical damage. Outside or detached storage is preferred. Inside storage should be in cool, well-ventilated, noncombustible location away from all possible sources of ignition.

REMARKS: Electrical installations in Class I hazardous locations as defined in Article 500 of the National Electrical Code should be in accordance with Article 501 of the Code; and electrical equipment should be suitable for use in atmospheres containing dimethyl ether vapors.   See National Electrical Code (NFPA No. 70), Fire-Hazard Properties of Flammable Liquids, Gases and Volatile Solids (NFPA No. 325M), and Manual of Hazardous Chemical Reactions (NFPA No. 491M).

ETHENE $CH_2:CH_2$

DESCRIPTION: Colorless gas; sweet odor and taste.

FIRE AND EXPLOSION HAZARDS: Flammable gas.   Forms flammable mixtures with air over a wide range.   Flammable limits, 2.7% and 36%.   Ignition temperature, 914° F.   Vapor density is approximately the same as air.   Spontaneously explosive in sunlight with chlorine.   Can react vigorously with oxidizing materials.

LIFE HAZARD: Medical anesthetic, moderate concentration in air causes unconsciousness.

PERSONAL PROTECTION: Wear self-contained breathing apparatus.

FIRE FIGHTING PHASES: Stop flow of gas.   Use water to keep fire-exposed containers cool and to protect men effecting the shut-off.   If a burning cylinder is mounted on an anesthetic machine or truck, move the cylinder to a safe place.

USUAL SHIPPING CONTAINERS: Steel pressure cylinders, tank barges.

STORAGE: Protect against physical damage.   Isolate from oxygen chlorine, combustible, organic and oxidizing materials.   Store in cool, well-ventilated area, of noncombustible construction, away from possible sources of ignition.   Protect against static electricity and lightning.   Outside or detached storage is preferred.

REMARKS: Electrical installations in Class I hazardous locations, as defined in Article 500 of the National Electrical Code, should be in accordance with Article 501 of the Code.   If explosionproof electrical equipment is necessary, it shall be suitable for use in Group C.   See Standard for the Use of Inhalation Anesthetics (NFPA No. 56A), Explosion Venting Guide (NFPA No. 68), National Electrical Code (NFPA No. 70), Static Electricity (NFPA No. 77), Lightning Protection Code (NFPA No. 78), and Fire-Hazard Properties of Flammable Liquids, Gases and Volatile Solids (NFPA No. 325M).

ETHYLAMINE $C_2H_5NH_2$

DESCRIPTION: Colorless liquid (boiling point 62° F.) having strong ammoniacal odor. Furnished also as aqueous solutions containing 33 or 70 per cent ethylamine.

FIRE AND EXPLOSION HAZARDS: Flammable liquid. Vapor forms flammable mixtures with air. Flammable limits, 3.5% and 14%. Flash point, less than 0° F. Ignition temperature, 724° F. Liquid is lighter than water (specific gravity 0.7), and is soluble in water in all proportions. Vapor is heavier than air (vapor density 1.6) and may travel a considerable distance to a source of ignition and flash back.

LIFE HAZARD: Eye, skin and respiratory irritant. Direct contact can cause burns.

PERSONAL PROTECTION: Wear full protective clothing.

FIRE FIGHTING PHASES: For anhydrous ethylamine in steel cylinders, stop flow of vapor. Use water to keep fire-exposed containers cool and to protect men effecting the shutoff. Use carbon dioxide, or dry chemical on fires involving anhydrous ethylamine. Water may be ineffective (see Explanatory). Use carbon dioxide, or dry chemical on fires involving aqueous solutions; water spray may be ineffective as an extinguishing agent (see Explanatory). If a leak or spill has not ignited, use water spray to disperse the vapors and to protect men attempting to stop a leak. Water spray may be used to flush spills away from exposures and to dilute spills to nonflammable mixtures.

USUAL SHIPPING CONTAINERS: For anhydrous ethylamine; steel cylinders. For water solutions of ethylamine: steel drums.

STORAGE: Protect against physical damage. Outside or detached storage is preferred. Separate from oxidizing materials and sources of heat. Inside storage of water solutions should be in a standard flammable liquids storage room or cabinet. Store cylinders containing anhydrous liquid in cool, well-ventilated, noncombustible location away from all possible sources of ignition.

REMARKS: Electrical installations in Class I hazardous locations, as defined in Article 500 of the National Electrical Code, should be in accordance with Article 501 of the Code; and electrical equipment should be suitable for use in atmospheres containing ethylamine vapors. See Flammable and Combustible Liquids Code (NFPA No. 30), National Electrical Code (NFPA No. 70), and Fire Hazard Properties of Flammable Liquids, Gases and Volatile Solids (NFPA No. 325M).

ETHYLENE OXIDE CH₂OCH₂

DESCRIPTION: Colorless gas at ordinary temperatures; liquid below 51° F.; has an ether-like odor.

FIRE AND EXPLOSION HAZARDS: Flammable liquid with boiling point of 51° F. Vapor forms explosive mixtures with air over wide range. Flammable limits, 3% and 100%. Flash point, less than 0° F. Ignition temperature in air, 804° F.; ignition temperature of 100% ethylene oxide, 1,058° F. Liquid is lighter than water (specific gravity, 0.9). Vapor is heavier than air (vapor density, 1.5.) and may travel considerable distance to a source of ignition and flash back. Dangerously reactive; may rearrange chemically and/or polymerize violently with evolution of heat, when in contact with highly active catalytic surfaces such as anhydrous chlorides of iron, tin and aluminum, pure oxides of iron and aluminum, and alkali metal hydroxides. Although soluble in water, solutions will continue to burn until diluted to approximately 22 volumes of water to one volume of ethylene oxide.

LIFE HAZARD: Moderately toxic by inhalation; eye, skin and respiratory irritant; prolonged contact with skin may result in delayed burns.

PERSONAL PROTECTION: Wear self-contained breathing apparatus.

FIRE FIGHTING PHASES: Fire fighting should be done from an explosion-resistant location. Use water from unmanned monitors or hoseholders to keep fire-exposed containers cool. If it is necessary to stop the flow of gas, use water spray to protect men effecting shut-off.

USUAL SHIPPING CONTAINERS: Steel cylinders, drums, insulated tank cars, tank barges.

STORAGE: Protect against physical damage. Should be kept cool, below 86° F. Should be stored outside, away from buildings and other materials, in insulated tanks or containers, shielded from sun-heat, provided with cooling facilities and protected by a properly designed water-spray system. Adequate diking and drainage should be provided in tank area to confine and dispose of liquid in case of tank rupture. Avoid pits and depressions. Inside storage should be held to a minimum and confined to a standard fire-resistive flammable liquids storage room, provided with continuous ventilation and free of sources of ignition. Do not permit chlorides, oxides, acids, organic bases, alkali metal hydroxides, metallic potassium or other combustible materials in storage room.

REMARKS: Electrical installations in Class I hazardous locations, as defined in Article 500 of the National Electrical Code, should be in accordance with Article 501 of the Code. If explosionproof electrical equipment is necessary, it shall be suitable for use in Group B, except that Group C equipment may be used if such equipment is isolated in accordance with Section 501–5(a) by sealing all conduit ½-inch size or larger. See Flammable and Combustible Liquids Code (NFPA No. 30), National Electrical Code (NFPA No. 70), Static Electricity (NFPA No. 77), Lightning Protection Code (NFPA No. 78) and Fire-Hazard Properties of Flammable Liquids, Gases and Volatile Solids (NFPA No. 325M).

ETHYL METHYL ETHER $CH_3OC_2H_5$

DESCRIPTION: Colorless liquid with characteristic anesthetic odor and sweet taste.

FIRE AND EXPLOSION HAZARDS: Flammable liquid. Vapor forms explosive mixtures with air. Flammable limits, 2% and 10.1%. Flash point, minus 35° F. Ignition temperature is relatively low, 374° F. Boiling point, 51.4° F. Vapor is heavier than air (vapor density, 2.1) and may travel considerable distance to a source of ignition and flash back. In presence of oxygen, or on long standing, or on exposure to sunlight in bottles, unstable peroxides sometimes form which may explode spontaneously or when heated. Ethyl methyl ether may be readily ignited by static electricity. Soluble in water.

LIFE HAZARD: Strongly anesthetic.

PERSONAL PROTECTION: Wear self-contained breathing apparatus.

FIRE FIGHTING PHASES: Use dry chemical, "alcohol" foam, or carbon dioxide; water may be ineffective (see Explanatory), but water should be used to keep fire-exposed containers cool. If a leak or spill has not ignited, use water spray to disperse the vapors and to protect men attempting to stop a leak. Water spray may be used to flush spills away from exposures and to dilute spills to nonflammable mixtures.

USUAL SHIPPING CONTAINERS: Glass bottles or cans inside boxes, and steel drums.

STORAGE: Protect against physical damage. Detached outside storage is preferred. Inside storage should be in a standard flammable liquids storage room or cabinet. Isolate from either strongly oxidizing or strongly reducing materials. Avoid direct sunlight. Protect against static electricity and lightning. For large quantity storage rooms, protect with automatic sprinkler systems or total flooding carbon dioxide systems.

REMARKS: Electrical installations in Class I hazardous locations, as defined in Article 500 of the National Electrical Code, should be in accordance with Article 501 of the Code; and electrical equipment should be suitable for use in atmospheres containing ethyl methyl ether. See Flammable and Combustible Liquids Code (NFPA No. 30), Standard for the Use of Inhalation Anesthetics (NFPA No. 56A), National Electrical Code (NFPA No. 70), Static Electricity (NFPA No. 77), Lightning Protection Code (NFPA No. 78), and Fire Hazard Properties of Flammable Liquids, Gases and Volatile Solids (NFPA No. 325M).

ETHYL NITRITE C₂H₅ONO

DESCRIPTION: Colorless liquid.

FIRE AND EXPLOSION HAZARDS: Flammable liquid. Flash point minus 31° F. Flammable in air over a wide range from 4.1% to greater than 50%. Boiling point, 63° F. Ethyl nitrite tends to decompose on exposure to light or heat. Spontaneously decomposes at 194° F. The decomposition can be violently explosive. Vapors are heavier than air and may travel a considerable distance to a source of ignition and flash back. Not soluble in water.

LIFE HAZARD: Decomposition forms irritating and toxic oxides of nitrogen. Heavy exposure to ethyl nitrite itself can result in increased pulse rate, decreased blood pressure, and unconsciousness.

PERSONAL PROTECTION: Wear self-contained breathing apparatus.

FIRE FIGHTING PHASES: Fight fires from an explosion-resistant location. In advanced or massive fires, the area should be evacuated. If a fire occurs in the vicinity of this material water should be used to keep containers cool.

USUAL SHIPPING CONTAINERS: Glass carboys and bottles.

STORAGE: Protect against physical damage. Outside or detached storage is preferred. Separate from oxidizing materials. Protect from light. Inside storage should be in a standard flammable liquids storage room or cabinet. For large quantity storage rooms, protect with automatic sprinkler systems or total flooding carbon dioxide systems.

REMARKS: Electrical installations in Class I hazardous locations as defined in Article 500 of the National Electrical Code, should be in compliance with Article 501 of the Code. Electrical equipment should be suitable for use in atmospheres containing ethyl nitrite vapors. See Flammable and Combustible Liquids Code (NFPA No. 30), National Electrical Code (NFPA No. 70), Static Electricity (NFPA No. 77), and Fire-Hazard Properties of Flammable Liquids, Gases and Volatile Solids (NFPA No. 325M).

FORMALDEHYDE HCHO

DESCRIPTION: Colorless gas with a highly irritating odor. The commercial product is a colorless water solution containing 37% to 55% formaldehyde, stabilized against polymerization by 0.05% to 15% methyl alcohol.

Water Solution

Gas

FIRE AND EXPLOSION HAZARDS: The gas vaporizes readily from solution and is flammable in air. Flammable limits, 7% and 73%. Ignition temperature, 806° F. Boiling points, minus 3° F. for pure liquid; 214° F. for 37% solution. Flash points for 37% solution range from 185° F. with 0.05% methyl alcohol to 122° F. with 15% methyl alcohol. Soluble in water.

LIFE HAZARD: Eyes, skin, and respiratory irritant.

PERSONAL PROTECTION: Wear self-contained breathing apparatus; wear goggles if eye protection not provided.

FIRE FIGHTING PHASES: Use water spray, dry chemical, "alcohol" foam, or carbon dioxide. Use water to keep fire-exposed containers cool. If a leak or spill has not ignited, use water spray to disperse the vapors and to protect men attempting to stop a leak. Water spray may be used to flush spills away from exposures and to dilute spills to nonflammable mixtures.

USUAL SHIPPING CONTAINERS: Insulated tank cars and tank trucks; 5- to 55-gallon metal drums, carboys, and bottles. Tank barges.

STORAGE: Protect against physical damage. Separate from oxidizing and alkaline materials. Indoor storage should be in areas having floors pitched toward a trapped drain or in curbed retention areas. Minimum storage temperatures to prevent polymerization range from 83° F for 37% formaldehyde containing 0.05% methyl alcohol to 29° F. for formaldehyde containing 15% methyl alcohol.

REMARKS: Electrical installations in Class I hazardous locations, as defined in Article 500 of the National Electrical Code, should be in accordance with Article 501 of the Code; and electrical equipment should be suitable for use in atmospheres containing formaldehyde vapors. See Flammable and Combustible Liquids Code (NFPA No. 30), National Electrical Code (NFPA No. 70), Static Electricity (NFPA No. 77), Lightning Protection Code (NFPA No. 78), Fire-Hazard Properties of Flammable Liquids, Gases and Volatile Solids (NFPA No. 325M), and Chemical Safety Data Sheet SD-1 (Manufacturing Chemists' Association, Inc.).

HYDROGEN (Liquefied) H₂

DESCRIPTION: Colorless, odorless liquid.

FIRE AND EXPLOSION HAZARDS: Flammable liquefied gas that burns with a practically invisible flame. Liquefied hydrogen is usually shipped and stored at a temperature slightly above its boiling point at atmospheric pressure. Boiling point, minus 423.3° F. Vapors form flammable mixtures with air. Flammable limits, 4 and 74% at atmospheric pressure. Ignition temperature, 1022° F. Vapor density at 32° F, 0.078. Unlike hydrogen at normal temperatures, the cold gas as it comes from the liquid is slightly heavier than air (vapor density at minus 423.3° F, 1.04) and may remain near ground level until it warms up. Fog formed when the cold gas comes in contact with the moisture of the air will indicate where the gas is spreading. However, flammable mixtures can exist outside the visible fog. A fireball is formed if the gas cloud is ignited immediately after the first flash evaporation. Liquefied hydrogen is lighter than water (specific gravity at boiling point, 0.07).

LIFE HAZARD: Nontoxic, but the gas can cause asphyxiation by displacement of air in a closed space. The liquid can cause severe frostbite or "burns" to the skin or other body tissues.

PERSONAL PROTECTION: Wear special protective clothing designed to prevent liquefied hydrogen or the cold vapors from coming in contact with the body.

FIRE FIGHTING PHASES: If a spill has not ignited, use water spray to direct flammable gas-air mixtures away from sources of ignition. If it is desirable to evaporate a spill quickly, water spray may be used to increase the rate of evaporation, if the increased vapor evolution can be controlled. Do not discharge solid streams into liquid.

Because of danger of reignition, hydrogen fires normally should not be extinguished until the supply of hydrogen has been shut off. If liquid hydrogen has ignited, use water to keep fire-exposed containers cool and to protect men stopping the source of a spill. If it is necessary to extinguish small hydrogen fires, use dry chemical, carbon dioxide, or halogenated extinguishing agent.

USUAL SHIPPING CONTAINERS: Insulated containers; insulated tank trucks, tank cars.

STORAGE: Protect against physical damage. Outdoor storage is preferred.   See NFPA No. 50B, Standard for Liquefied Hydrogen Systems at Consumer Sites.

REMARKS: See Standard for Liquefied Hydrogen Systems at Consumer Sites (NFPA No. 50B), Fire-Hazard Properties of Flammable Liquids, Gases and Volatile Solids (NFPA No. 325M).

HYDROGEN CHLORIDE (Hydrochloric Acid)
HCl

DESCRIPTION: Hydrogen chloride is a colorless gas. Hydrochloric acid a water solution of hydrogen chloride, is a clear, colorless or slightly yellow, fuming liquid with irritating pungent odor.

FIRE AND EXPLOSION HAZARDS: Not combustible but contact with common metals produces hydrogen which may form explosive mixtures with air. Soluble in water.

LIFE HAZARD: Toxic. Eye, skin and respiratory irritant. Inhalation of concentrations of about 1500 parts per million in air are fatal in a few minutes.

PERSONAL PROTECTION: Wear full protective clothing.

FIRE FIGHTING PHASES: Use water, neutralize with chemically basic substances such as soda ash or slaked lime.

USUAL SHIPPING CONTAINERS: Aqueous solutions in glass bottles, carboys, rubber-lined tank cars. Anhydrous hydrogen chloride in steel cylinders and tank barges.

STORAGE: Protect against physical damage. Store in cool, well-ventilated place, separated from all oxidizing materials.

REMARKS: See Chemical Safety Data Sheet SD-39 (Manufacturing Chemists' Association, Inc.).

HYDROGEN CYANIDE (Hydrocyanic Acid)
HCN

DESCRIPTION: Clear, colorless liquid (boiling point 79° F.) or gas, with faint odor of bitter almonds.

FIRE AND EXPLOSION HAZARDS: Flammable liquid. Vapor forms explosive mixtures with air over a wide range. A 96% mixture of hydrogen cyanide and water has the following properties: flammable limits, 6% and 41%; flash point 0° F.; ignition temperature, 1000° F. Lighter than air (vapor density 0.9). Hydrogen cyanide liquid is lighter than water (specific gravity 0.68). May become unstable and subject to explosion if stored for extended time. Soluble in water.

LIFE HAZARD: Extremely toxic. Few breaths may cause unconsiousness and death. Breathing apparatus alone not considered complete protection in atmospheres containing over 100 parts per million as toxic amounts can be absorbed through the skin.

PERSONAL PROTECTION: Wear special protective clothing.

FIRE FIGHTING PHASES: Extremely toxic. In advanced or massive fires, fire fighting should be done from a safe distance or from a protected location. Use dry chemical "alcohol" foam, or carbon dioxide. Water spray may be ineffective as an extinguishing agent (see Explanatory), but water should be used to keep fire-exposed containers cool. If a leak or spill has not ignited, use water spray to disperse the vapors. If it is necessary to stop a leak, use water spray to protect men attempting to do so. Water spray may be used to flush spills away from exposures and to dilute spills to nonflammable mixtures.

USUAL SHIPPING CONTAINERS: Steel cylinders or completely absorbed in inert material in cans inside wooden boxes, tank cars.

STORAGE: Protect against physical damage. Outside or detached storage is preferred. Inside storage should be in a standard flammable liquids storage room or cabinet. Isolate from other storage and all possible sources of ignition. Individual containers should not remain in storage for more than 90 days or not longer than recommended by supplier.

239

REMARKS: Electrical installations in Class I hazardous locations, as defined in Article 500 of the National Electrical Code, should be in accordance with Article 501 of the Code; and electrical equipment should be suitable for use in amospheres containing hydrogen cyanide vapors. See Flammable and Combustible Liquids Code (NFPA No. 30), Fire-Hazard Properties of Flammable Liquids, Gases and Volatile Solids (NFPA No. 325M), and Chemical Safety Data Sheet SD-67 (Manufacturing Chemists' Association, Inc.).

HYDROGEN SULFIDE H$_2$S

DESCRIPTION: Colorless gas; offensive strong odor similar to rotten eggs.

FIRE AND EXPLOSION HAZARDS: Flammable gas. Forms explosive mixtures with air over a wide range. Flammable limits, 4.3% and 45%. Ignition temperature, 500° F. Heavier than air (vapor density, 1.2) and may travel considerable distance to a source of ignition and flash back. Dangerously reactive with fuming or strong nitric acid and powerful oxidizing materials.

LIFE HAZARD: Highly toxic. Irritating to eyes and respiratory tract. High concentrations cause almost immediate death.

PERSONAL PROTECTION: Wear self-contained breathing apparatus; wear goggles if eye protection not provided.

FIRE FIGHTING PHASES: Stop flow of gas. Use water to keep fire-exposed containers cool and to protect men effecting the shut-off.

USUAL SHIPPING CONTAINERS: Steel pressure cylinders.

STORAGE: Protect against physical damage. Outside or detached storage is preferred, inside storage should be in a cool, well-ventilated, noncombustible location, away from all possible sources of ignition. Store away from nitric acid, strong oxidizing materials, corrosive liquids or gases, cylinders or other containers under high pressure, and possible sources of ignition. Protect against static electricity, direct sunlight and excessive heat.

REMARKS: Electrical installations in Class I hazardous locations, as defined in Article 500 of the National Electrical Code, should be in accordance with Article 501 of the Code; and electrical equipment should be suitable for use in atmospheres containing hydrogen sulfide gas. See Explosion Venting Guide

(NFPA No. 68), Static Electricity (NFPA No. 77), Fire-Hazard Properties of Flammable Liquids, Gases and Volatile Solids (NFPA No. 325M), and Chemical Safety Data Sheet SD-36 (Manufacturing Chemists' Association, Inc.).

ISOPRENE CH_2:$C(CH_3)CH$:CH_2

DESCRIPTION: Colorless, volatile liquid.

FIRE AND EXPLOSION HAZARDS: Flammable liquid. Flash point, minus 65° F. Flammable limit, lower, 1.6 (estimated). Boiling point, 93° F. Liquid is lighter than water (specific gravity, 0.7). Not soluble in water. Usually contains an inhibitor to prevent self-polymerization. At elevated temperatures, such as in fire conditions, polymerization may take place. If the polymerization takes place in a container, there is a possibility of violent rupture of the container.

LIFE HAZARD: Toxic by inhalation.

PERSONAL PROTECTION: Wear self-contained breathing apparatus.

FIRE FIGHTING PHASES: In advanced or massive fires, fire fighting should be done from an explosion-resistant location. Use dry chemical, foam, or carbon dioxide. Water may be ineffective (see Explanatory), but water should be used to keep fire-exposed containers cool. If a leak or spill has not ignited, use water spray to disperse the vapors. If it is necessary to stop a leak, use water spray to protect men attempting to do so. Water spray may be used to flush spills away from exposures.

USUAL SHIPPING CONTAINERS: Metal drums; tank trucks and tank cars.

STORAGE: Protect against physical damage. Outside or detached storage is preferred. Inside storage should be in a standard flammable liquids storage room or cabinet. Check at least weekly to determine inhibitor and polymer content. Separate from oxidizing materials.

REMARKS: Electrical installations in Class I hazardous locations, as defined in Article 500 of the National Electrical Code, should be in accordance with Article 501 of the Code. If explosionproof electrical equipment is necessary, it shall be suitable for use in Group C. See Flammable and Combustible Liquids Code (NFPA No. 30), National Electrical Code (NFPA No. 70), and Fire-Hazard Properties of Flammable Liquids, Gases and Volatile Solids (NFPA No. 325M).

ISOPROPYLAMINE $(CH_3)_2CH \cdot NH_2$

DESCRIPTION: Colorless liquid below 90° F (32° C), ammoniacal odor.

FIRE AND EXPLOSION HAZARDS: Flammable liquid. Flash point, below 0° F (minus 18° C). Reacts vigorously with oxidizing materials. Autoignition temperature, 756° F (402° C). Vapor is heavier than air (vapor density, 2.03) and may travel considerable distance to ignition source and flash back. Soluble in water.

LIFE HAZARD: Strong eye, skin and respiratory irritant.

PERSONAL PROTECTION: Wear full protective clothing.

FIRE FIGHTING PHASES: Use water spray, dry chemical, "alcohol" foam, or carbon dioxide. Use water to keep fire-exposed containers cool. Water spray may be ineffective as an extinguishing agent (see Explanatory). Direct hose streams from a protected location. If a leak or spill has not ignited, use water spray to disperse the vapors and to protect men attempting to stop a leak. Water spray may be used to flush spills away from exposures and to dilute spills to nonflammable mixtures.

USUAL SHIPPING CONTAINERS: Glass bottles, metal cans, drums.

STORAGE: Protect against physical damage. Separate from other storage. Outside or detached storage is preferred. Inside storage should be in a standard flammable liquids storage room.

LIQUEFIED NATURAL GAS

DESCRIPTION: Colorless, odorless liquid consisting primarily of methane (83-99%), with lesser amounts of ethane (1-13%), propane (0.1-3%), and butane (0.2-1.0%).

FIRE AND EXPLOSION HAZARDS: Flammable liquefied gas. Liquefied natural gas (LNG) is usually shipped and stored at a temperature slightly above its boiling point at atmospheric pressure. Boiling point, minus 255-263° F. Vapors form flammable mixtures with air. Flammable limits of methane at normal temperatures, 5 and 14%; near minus 260° F, 6 and 13%. Unlike natural gas at normal temperatures, the cold gas as it comes from the liquid is about 1½ times heavier than air and will not rise immediately but spread at ground level until it warms up. Fog formed when the cold gas comes in contact with the moisture of the air will indicate where the gas is spreading. However, flammable mixtures can exist outside the visible fog. If the flammable vapor-air mixture reaches an ignition source, flame is likely to propagate back to the liquid. After the gas warms to about minus 170° F, it becomes lighter than air and will start to rise.

LNG is lighter than water (specific gravity, about 0.45). When water contacts LNG, a white solid and ice forms. The solid exists only at low temperatures and evaporates rapidly. It may erupt violently if severely disturbed.

LIFE HAZARD: Nontoxic but the gas can cause asphyxiation by displacement of air in a closed space. The liquid can cause severe frostbite or "burn" to the skin or other body tissues. Prolonged contact can cause embrittlement of clothing.

PERSONAL PROTECTION: Wear special clothing designed to prevent LNG or the cold vapors from coming in contact with the body.

FIRE FIGHTING PHASES: If a spill has not ignited, use water spray to direct flammable gas-air mixtures away from ignition sources. If it is desirable to evaporate a spill quickly, water spray may be used to increase the rate of evaporation, if the increased vapor evolution can be controlled. Do not discharge solid streams into liquid.

If LNG has ignited, use water to keep fire-exposed containers and equipment cool and to protect men if an effort is to be made to stop the source of the leak. High expansion foam may be used to reduce the rate of burning of relatively small spills. Because of danger of reignition, liquefied natural gas fires normally should not be extinguished until the supply of gas has been shut off. If it is desirable to extinguish the fire, use dry chemical, carbon dioxide or halogenated extinguishing agent.

USUAL SHIPPING CONTAINERS: Insulated tank trucks, tank cars, tank ships and barges.

REMARKS: See NFPA No. 59A, Standard for the Production, Storage and Handling of Liquefied Natural Gas (LNG).

LITHIUM HYDRIDE LiH

DESCRIPTION: Off-white, translucent powder.

FIRE AND EXPLOSION HAZARDS: May ignite spontaneously in air. Violently reactive with strong oxidizers. On heating or in contact with moisture or acids an exothermic reaction results with the evolution of hydrogen. Exothermic heat is often sufficient for ignition. Can form airborne dust clouds which may explode on contact with flame, heat or oxidizing materials.

LIFE HAZARD: Slight toxicity. Exposure to dust will irritate mucous membranes of the eyes and upper respiratory tract. Ingestion of large amounts may cause dizziness and collapse. Lithium hydroxide formed from lithium hydride and water is very caustic.

PERSONAL PROTECTION: Wear full protective clothing.

FIRE FIGHTING PHASES: Do not use water, carbon dioxide, dry chemical or halogenated extinguishing agents such as carbon tetrachloride. Fires can be smothered by inverting a can over them. If fire is in drum, place cover loosely on drum. Dry graphite or ground dolomite can be used to smother fires in lithium hydride.

USUAL SHIPPING CONTAINERS: Metal cans packed inside wooden boxes. Metal barrels or drums.

STORAGE: Protect against physical damage. Store in isolated, well-ventilated, cool, dry area. Every means should be taken to keep water from entering storage area. Building must be well ventilated and so constructed as to eliminate pocketing of gases.

REMARKS: Open containers only in inert atmospheres or very low humidity rooms. If the use of lithium hydride can result in hazardous concentrations of hydrogen electrical installations should conform to the National Electrical Code requirements for Class I hazardous locations and electrical equipment should be suitable for Group B atmospheres. See National Electrical Code (NFPA No. 70).

METHYLAMINES (Mono-, di-, and trimethylamine) CH_3NH_2(mono-); $(CH_3)_2NH$(di-); $(CH_3)_3N$(tri-)

DESCRIPTION: Gases at ordinary temperatures with fish-like odor in low concentrations and ammonia-like odor in high concentrations. May be in water solutions (25% to 40%).

FIRE AND EXPLOSION HAZARDS: Methylamine gases are flammable. Flammable limits: monomethylamine, 4.9% and 20.7%; dimethylamine, 2.8% and 14.4%; trimethylamine, 2.0% and 11.6%. Ignition temperatures: monomethylamine, 806° F.; dimethylamine, 755° F. and trimethylamine, 374° F. All are heavier than air (vapor densities: mono-, 1.1; di-, 1.6; tri-, 2.0) and may travel considerable distance to a source of ignition and flash back. Liquid solutions are flammable. Vapor forms explosive mixtures with air. Contact with mercury can produce an explosive reaction.

LIFE HAZARD: Eye, skin and respiratory irritant. Direct or prolonged contact can cause burns and serious injury.

PERSONAL PROTECTION: Wear full protective clothing.

FIRE FIGHTING PHASES: Stop flow of gas. Use water to keep fire-exposed containers cool and to protect men effecting the shut-off. Water spray, carbon dioxide, dry chemical and "alcohol" foam can be used on fires involving water solutions of the methylamines.

USUAL SHIPPING CONTAINERS: For gases: Steel cylinders, tank cars and tank trucks. For solutions: Steel drums, tank cars, tank trucks, tank barges.

STORAGE: Protect against physical damage. Outside or detached storage is preferable. Inside storage of liquid solutions should be in a standard flammable liquids storage room or cabinet. Insure against accidental contact with mercury. Inside storage of gas should be in a cool, well-ventilated, noncombustible location, away from all possible sources of ignition. Separate from oxidizing materials.

REMARKS: Electrical installations in Class I hazardous locations, as defined in Article 500 of the National Electrical Code, should be in accordance with Article 501 of the Code; and electrical equipment should be suitable for use in atmospheres containing methylamine gas. See Flammable and Combustible Liquids Code (NFPA No. 30), National Electrical Code (NFPA No. 70), Static Electricity (NFPA No. 77), Lightning Protection Code (NFPA No. 78), Fire-Hazard Properties of Flammable Liquids, Gases and Volatile Solids (NFPA No. 325M) and Chemical Safety Data Sheet SD-57 (Manufacturing Chemists' Association, Inc.).

METHYL CHLORIDE CH₃Cl

DESCRIPTION: Colorless gas with faintly sweet ethereal odor.

FIRE AND EXPLOSION HAZARDS: Flammable gas. Forms flammable mixtures with air. Flammable limits, 10.7% and 17.4%. Ignition temperature, 1,170° F. Vapor heavier than air (vapor density, 1.8).

LIFE HAZARD: Inhalation of high concentrations of methyl chloride causes serious central nervous system damage, lingering illness and sometimes death. Because methyl chloride has so little odor and only a mild narcotic action, persons may be exposed to considerable concentrations without being aware of the danger. The onset of symptoms of poisoning, such as dizziness, headache, optical difficulties, nausea and vomiting, may be delayed for many hours.

PERSONAL PROTECTION: Wear self-contained breathing apparatus.

FIRE FIGHTING PHASES: Stop flow of gas. Use water to keep fire-exposed containers cool and to protect men effecting the shut-off.

USUAL SHIPPING CONTAINERS: Shipped as a liquid under pressure in cylinders and tank cars.

STORAGE: Protect against physical damage. Store in cool, well-ventilated area of noncombustible construction, away from sources of ignition. Outside or detached storage is preferred.

REMARKS: Electrical installations in Class I hazardous locations, as defined in Article 500 of the National Electrical Code, should be in accordance with Article 501 of the Code; and electrical equipment should be suitable for use in atmospheres containing methyl chloride gas. See National Electrical Code (NFPA No. 70) and Chemical Safety Data Sheet SD-40 (Manufacturing Chemists' Association, Inc.).

METHYL FORMATE CH₃OCHO

DESCRIPTION: Colorless liquid.

FIRE AND EXPLOSION HAZARDS: Flammable liquid. Vapor forms explosive mixtures with air. Flammable limits, 5% and 23%. Flash point, minus 2° F. Boiling point, 90° F. Ignition temperature, 853° F. Liquid is soluble in water. Vapor is heavier than air (vapor density, 2.1) and may travel considerable distance to a source of ignition and flash back. Can react vigorously with oxidizing materials.

LIFE HAZARD: Moderately toxic by inhalation or ingestion, also an eye irritant. Can emit toxic fumes on exposure to elevated temperatures.

PERSONAL PROTECTION: Wear self-contained breathing apparatus; wear goggles if eye protection not provided.

FIRE FIGHTING PHASES: Use dry chemical, "alcohol" foam, or carbon dioxide; water may be ineffective (see Explanatory), but water should be used to keep fire-exposed containers cool. If a leak or spill has not ignited, use water spray to disperse the vapors and to protect men attempting to stop a leak. Water spray may be used to flush spills away from exposures and to dilute spills to non-flammable mixtures.

USUAL SHIPPING CONTAINERS: Drums and tank cars.

STORAGE: Protect against physical damage. Store in cool, well-ventilated area, isolated from oxidizing materials. Keep containers closed. Outside or detached storage preferable. Inside storage should be in a standard flammable liquids storage room or cabinet.

REMARKS: Electrical installations in Class I hazardous locations, as defined in Article 500 of the National Electrical Code, should be in accordance with Article 501 of the Code; and electrical equipment should be suitable for use in atmospheres containing methyl formate vapors. See Flammable and Combustible Liquids Code (NFPA No. 30), National Electrical Code (NFPA No. 70), and Fire-Hazard Properties of Flammable Liquids, Gases and Volatile Solids (NFPA No. 325M).

MOTOR FUEL ANTIKNOCK COMPOUNDS
(Contain lead)

DESCRIPTION: Red, orange or blue (dyed) liquids with sweet musty odor. Comprise a range of mixtures of tetraethyl lead (TEL), tetramethyl lead (TML), methylethyl lead (MEL), ethylene dibromide, ethylene dichloride, solvent, antioxidant, dye and inerts.

FIRE AND EXPLOSION HAZARDS: Flammable or combustible liquids. Flash points range from 89° F (oc) (32° C) to 265° F (oc) (130° C). Specific gravity, greater than 1. Not soluble in water. Thermal decomposition may occur above 212° F (100° C). With TML compounds thermal decomposition is more likely to take the form of decomposition of vapors at the surface; with TEL compounds it is more likely to be in the form of homogeneous bulk decomposition. Both types of decomposition are considered hazardous and in either case rapid decomposition will cause container explosion.

LIFE HAZARD: Vapors are very toxic. Fatal lead poisoning may occur by ingestion, vapor inhalation or skin absorption.

PERSONAL PROTECTION: Wear full protective clothing.

FIRE FIGHTING PHASES: Fight fires from an explosion-resistant location. Use water from unmanned monitors and hose-holders to keep fire-exposed containers cool. On fires in which containers are not exposed, use water spray, dry chemical, foam, or carbon dioxide. If a leak or spill has not ignited, use water spray to disperse vapors. If it is necessary to stop a leak, use water spray to protect men attempting to do so. Water spray may be used to flush spills away from exposures.

USUAL SHIPPING CONTAINERS: Metal cans in wooden boxes, metal drums, cylinders, tanks, tank cars, tank trucks, tank barges.

STORAGE: Protect against physical damage. Store in a cool, isolated, well-ventilated area. Keep away from fire, heat and strong oxidizing agents. Tanks should be stored in a sprinklered area.

REMARKS: Electrical installations in Class I hazardous locations as defined in Article 500 of the National Electrical Code should be in accordance with Article 501 of the Code; and electrical equipment should be suitable for use in atmospheres containing vapors of the antiknock compound. See Flammable and Combustible Liquids Code (NFPA No. 30), National Electrical Code (NFPA No. 70), Static Electricity (NFPA No. 77), and Fire Hazard Properties of Flammable Liquids, Gases and Volatile Solids (NFPA No. 325M).

NITROGEN (Liquefied) N_2

DESCRIPTION: Colorless, odorless liquid.

FIRE AND EXPLOSION HAZARDS: Noncombustible liquefied gas. Liquefied nitrogen is usually shipped and stored at a temperature slightly above its boiling point at atmospheric pressure. Boiling point, minus 320° F. Vapor density at normal temperature approximately the same as air. Cold gas as it comes from the liquid is heavier than air. Liquefied nitrogen is lighter than water.

LIFE HAZARD: Nontoxic but the gas can cause asphyxiation by displacement of air from a closed space. The liquid can cause severe frostbite or "burns" to the skin or other body tissues.

PERSONAL PROTECTION: Wear special protective clothing designed to prevent liquefied nitrogen or the cold vapors from coming in contact with the body.

FIRE FIGHTING PHASES: Use water to cool fire-exposed containers. If it is desirable to evaporate a liquefied nitrogen spill quickly, water spray may be used to increase the rate of evaporation, if the increased vapor evolution can be controlled. Do not discharge solid streams into liquid.

USUAL SHIPPING CONTAINERS: Insulated containers, insulated tank trucks, tank cars.

STORAGE: Protect against physical damage. Inside storage should be in a well-ventilated area.

OXYGEN (LIQUID) O₂

DESCRIPTION: A blue liquid (blue solid when frozen).

FIRE AND EXPLOSION HAZARDS: Nonflammable itself, but promotes combustion in proportion to the concentration of oxygen present. Liquid oxygen forms explosive mixtures with organic and other readily oxidizable materials.

LIFE HAZARD: The liquid can cause severe "frost-bite" or "burn" to the skin or other bodily tissues. Gaseous oxygen from the liquid is absorbed readily in clothing and any source of ignition may cause flash burning. In gaseous form, no hazard, except that of irritation of mucous membranes is likely if 100% oxygen is inhaled continuously for more than a few hours. Smoking in oxygen enriched atmospheres is extremely hazardous.

PERSONAL PROTECTION: Wear special protective clothing that will not ignite on contact with liquid oxygen and that is designed to prevent liquid oxygen from coming in contact with the skin.

FIRE FIGHTING PHASES: Use water to cool tanks exposed to fire in order to prevent pressure rupture of the containers, which would disastrously magnify the fire. Avoid direct application of water on safety relief devices, since resultant cold exposure may cause ice to form, preventing proper function of safety relief devices.

When the fire results from a leak or flow of liquid oxygen onto wood, paper, waste or another similar material, the first thing to do is stop the flow if possible. For small spills, or after the leak or flow of liquid oxygen has been stopped, use enough water to put the fire out quickly. When the fire involves liquid oxygen and liquid fuels, control it as follows:

(a) When liquid oxygen leaks or flows into large quantities of fuel, shut off the flow of liquid oxygen, and put the remaining fuel fire out with extinguishing agents suitable for use on Class B fires. When fuel leaks or flows into large quantities of liquid oxygen, shut off the flow of fuel.

(b) When fuel and liquid oxygen are mixed or mixing but are not yet burning, isolate the area from sources of ignition and get out quickly, allowing the oxygen to evaporate. When large pools of water-soluble fuel are present, use water to dilute the fuel and to reduce the intensity of the fire. This method cannot be used with fuels which do not mix with water. Appropriate extinguishing agents may be used to put out fuel fires after the oxygen has evaporated.

USUAL SHIPPING CONTAINERS: Dewar flasks and tank trucks.

STORAGE: Protect against physical damage. Isolate from combustible gas installations and combustible materials by adequate distance or by gas-tight fire-resistive barriers. Protect against overheating. Outside storage of liquid oxygen tanks is recommended.

REMARKS: See Standard for Nonflammable Medical Gas Systems (NFPA No. 56F) and Standard for the Installation of Bulk Oxygen Systems at Consumer Sites (NFPA No. 50).

PROPYLAMINE (Normal) $CH_3(CH_2)_2NH_2$

DESCRIPTION: A colorless liquid with a pungent, irritating odor which is similar to ammonia when concentration is heavy.

FIRE AND EXPLOSION HAZARDS: Flammable liquid. Vapor forms explosive mixtures with air. Flammable limits, 2.0% and 10.4%. Flash point, minus 35° F. Ignition temperature, 604° F. Liquid is lighter than water (specific gravity, 0.7) but soluble in water in all proportions. Vapor is heavier than air (vapor-air density at 100° F., 1.7) and may travel a considerable distance to a source of ignition and flash back. Soluble in water.

LIFE HAZARD: A severe eye, skin, and respiratory irritant.

PERSONAL PROTECTION: Wear full protective clothing.

FIRE FIGHTING PHASES: Use dry chemical, "alcohol" foam, or carbon dioxide; water may be ineffective (see Explanatory), but water should be used to keep fire-exposed containers cool. If a leak or spill has not ignited, use water spray to disperse the vapors and to protect men attempting to stop a leak. Water spray may be used to flush spills away from exposures and to dilute spills to nonflammable mixtures.

USUAL SHIPPING CONTAINERS: Glass bottles, cans, steel drums, tank cars.

STORAGE: Protect against physical damage. Outside or detached storage is preferable. Inside storage should be in a standard flammable liquids storage room or cabinet. Separate from oxidizing materials.

REMARKS: Electrical installations in Class I hazardous locations, as defined in Article 500 of the National Electrical Code, should be in accordance with Article 501 of the Code; and electrical equipment should be suitable for use in atmospheres containing propylamine vapors. See Flammable and Combustible Liquids Code (NFPA No. 30), National Electrical Code (NFPA No. 70), Static Electricity (NFPA No. 77) and Fire-Hazard Properties of Flammable Liquids, Gases and Volatile Solids (NFPA No. 325M).

PROPYLENE OXIDE CH₃CHOCH₂

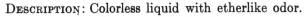

DESCRIPTION: Colorless liquid with etherlike odor.

FIRE AND EXPLOSION HAZARDS: Flammable liquid. Vapor forms explosive mixtures with air. Flammable limits, 2.8% and 37%. Flash point, minus 35° F. Boiling point, 95° F. Liquid is lighter than water (specific gravity, 0.9). Vapor is heavier than air (vapor density, 2.0), and may travel considerable distance to a source of ignition and flash back. May polymerize with evolution of heat when in contact with highly active catalytic surfaces such as anhydrous chlorides of iron, tin, and aluminum, peroxides of iron and aluminum, and alkali metal hydroxides. If the polymerization takes place in a container there is possibility of violent rupture of the container. Soluble in water.

LIFE HAZARD: Moderately toxic by inhalation; eye, skin, and respiratory irritant; prolonged contact with skin may result in delayed burns.

PERSONAL PROTECTION: Wear self-contained breathing apparatus.

FIRE FIGHTING PHASES: In advanced or massive fires, fire fighting should be done from a safe distance or from a protected location. Use dry chemical, "alcohol" foam, or carbon dioxide. Water spray may be ineffective as an extinguishing agent (see Explanatory), but water should be used to keep fire-exposed containers cool. If a leak or spill has not ignited, use water spray to disperse the vapors. If it is necessary to stop a leak, use water spray to protect men attempting to do so. Water spray may be used to flush spills away from exposures and to dilute spills to nonflammable mixtures.

USUAL SHIPPING CONTAINERS: Glass bottles, cans, metal drums, tank trucks, tank cars, tank barges, usually with nitrogen over the propylene oxide.

STORAGE: Protect against physical damage. Detached outside storage is preferred. Inside storage should be in a standard flammable liquids storage room or cabinet. Isolate from combustible materials and from oxidizing materials.

REMARKS: Electrical installations in Class I hazardous locations, as defined in Article 500 of the National Electrical Code, should be in accordance with Article 501 of the Code. If explosionproof electrical equipment is necessary, it shall be suitable for use in Group B, except that Class I, Group C equipment may be used if such equipment is isolated in accordance with Section 501–5(a) by sealing all conduits $\frac{1}{2}$-inch size or larger. See Flammable and Combustible Liquids Code (NFPA No. 30), National Electrical Code (NFPA No. 70), and Fire-Hazard Properties of Flammable Liquids, Gases and Volatile Solids (NFPA No. 325M).

SULFURIC ACID H_2SO_4

DESCRIPTION: Colorless (pure) to dark brown, oily, dense liquid.

FIRE AND EXPLOSION HAZARDS: Not flammable but highly reactive and capable of igniting finely divided combustible materials on contact. Reacts violently with water and organic materials with evolution of heat. Extremely hazardous in contact with many materials, particularly carbides, chlorates, fulminates, nitrates, picrates, powdered metals and other combustible materials. Attacks many metals, releasing hydrogen.

LIFE HAZARD: Causes severe, deep burns to tissue; very corrosive effect. Avoid any contact.

PERSONAL PROTECTION: Wear full protective clothing.

FIRE FIGHTING PHASES: Fires involving small amount of combustibles may be smothered with suitable dry chemical. Use water on combustibles burning in vicinity of this material but use care as water applied directly to this acid results in evolution of heat and causes splattering.

USUAL SHIPPING CONTAINERS: Glass bottles and carboys, metal barrels or drums with or without linings, tank trucks, tank cars, tank barges.

STORAGE: Protect against physical damage and water. Separate from carbides, chlorates, fulminates, nitrates, picrates, powdered metals and combustible materials.

REMARKS: See Chemical Safety Data Sheet SD-20 (Manufacturing Chemists' Association, Inc.).

TETRAFLUOROETHYLENE
MONOMER $CF_2 = CF_2$

Nonfire   Fire

DESCRIPTION: Colorless gas at normal temperatures.

FIRE AND EXPLOSION HAZARDS: Flammable limits, 10% and 50% for hot wire ignition in conventional apparatus. Upper limit increases to 100% when large volumes and powerful igniters are used. Ignition temperature for mixtures with air, 370° F. Boiling point, minus 105° F. Vapor is heavier than air (vapor density about 3) and may travel considerable distance to a source of ignition and flash back. Inhibited monomer can decompose explosively when exposed to materials with which it can react exothermally, or to low energy thermal sources sufficient to create local temperatures of about 900° F., or when containers are exposed to fire. Sensitivity to mechanical shock is minimal. Uninhibited monomer behavior is similar, with the additional hazard that uncontrolled polymerization may initiate explosive decomposition.

LIFE HAZARD: Tetrafluoroethylene is itself only moderately toxic. When burned in air, however, it forms carbonyl fluoride which hydrolyzes to hydrogen fluoride. Exposure of the vapor to high temperature in the absence of air leads to formation of other toxic compounds.

PERSONAL PROTECTION: Wear full protective clothing.

FIRE FIGHTING PHASES: Fire fighting should be done from an explosion-resistant location. Use water from unmanned monitors or hoseholders to keep fire-exposed containers cool.

USUAL SHIPPING CONTAINERS: High pressure compressed gas cylinders.

STORAGE: Protect against physical damage. Cylinders should be stored in a well-ventilated building. Outside or detached storage is preferred with protection from direct sunlight. Store separate from other flammable materials to reduce the possibility of exposure to fire. Provide a water spray system for cooling in the event of fire. Separate from oxidizing materials.

REMARKS: Tetrafluoroethylene, like carbon disulfide, does not meet any of the group classifications which are specified in Article 500–2 of the National Electrical Code. When tested in the UL explosion chamber, tetrafloroethylene-air mixtures produced explosions which propagated through the smallest clearance which could be set in the adjustable gap between the explosion chambers. Therefore, the concept of controlling clearances to prevent the propagation of explosions initiated within electrical equipment to the surrounding atmosphere is not valid for tetrafluoroethylene.

TRICHLOROSILANE HSiCl₃

DESCRIPTION: Colorless liquid below 89° F (32° C) with an acrid odor; fumes in air.

FIRE AND EXPLOSION HAZARDS: Flammable liquid. Vapor forms flammable mixtures in air. Flash point, 7° F (oc) (minus 14° C); flammable limits not reported; vapor is heavier than air (vapor density, 4.7). Reacts violently with water, yielding hydrochloric acid. (See Hydrogen Chloride.)

LIFE HAZARD: Vapor and liquid cause burns; toxic on inhalation. Reacts with water to form hydrochloric acid. (See Hydrogen Chloride.)

PERSONAL PROTECTION: Wear full protective clothing.

FIRE FIGHTING PHASES: Use dry chemical or carbon dioxide to extinguish small fires. Flooding with water may be necessary to prevent reignition. Water may be used if large amounts of combustible materials are involved and if fire fighters can protect themselves by distance or barriers from the violent trichlorosilane-water reaction. Water may be used to keep fire-exposed containers cool.

USUAL SHIPPING CONTAINERS: 55-gallon drums; 1-gallon glass bottles.

STORAGE: Protect against physical damage. Outside or detached storage is preferred. Inside storage should be in a standard flammable liquids storage room or cabinet. Separate from oxidizing materials.

REMARKS: Spills can be neutralized by flushing with large quantities of water followed by treatment with sodium bicarbonate. Provide adequate protection against generated hydrogen chloride. Do not allow water to get into the container since resulting pressure could cause the container to rupture. Electrical installations in Class I hazardous locations, as defined in Article 500 of the National Electrical

Code, should be in accordance with Article 501 of the Code, and electrical equipment should be suitable for use in atmospheres containing trichlorosilane vapors.   See National Electrical Code (NFPA No. 70), and Flammable and Combustible Liquids Code (NFPA No. 30).

Explanatory

Water may be ineffective in fighting fires in liquids with low flash points. This precautionary wording is used for materials having a flash point below 100° F. Obviously, the lower the flash point, the less effective water will be. However, water can be used on low flash point liquids when applied in the form of a spray to absorb much of the heat and to keep exposed materials from being damaged by the fire.

Much of the effectiveness of using water spray, particularly from hose lines, will depend on the method of application. With proper nozzles, even gasoline spill fires of some types have been extinguished when coordinated hose lines were used to sweep the flames off the surface of the liquid.

Water also has been used to extinguish fires in water-soluble flammable liquids by cooling, diluting and mixing the flammable liquid with water.

The inclusion of the phrase "water may be ineffective" is to indicate that although water can be used to cool and protect exposed material, water may not extinguish the fire unless used under favorable conditions by experienced fire fighters trained in fighting all types of flammable liquid fires.

§8.5   NFPA Fire Hazard Labeling System

The NFPA has established within its published standard, NFPA 704, Recommended System for the Identification of the Fire Hazards of Materials, a standard system for the labeling of various materials to easily identify the potential hazards of them. The system provides an easily recognizable method by which the inherent hazards of materials may be identified by anyone involved either in their use or in fire fighting. Because the principal thrust of the system is to assist those involved in fire fighting, it may seem to have only marginal interest for the litigator. However, given the prevalence of litigation concerned with labeling, this system is worthy of consideration in any case analysis.

The system is based on a diamond-shaped figure with three points of the diamond bearing numbers which correspond to a range of factors dealing with the health hazard, the flammability hazard, and the reactivity hazard of the material. Each of the three hazards is always designated by the same point on the diamond, and there is a standard coloring system for easy identification of the potential hazard. Figure 8-1 shows how the diamond is set up in practice.

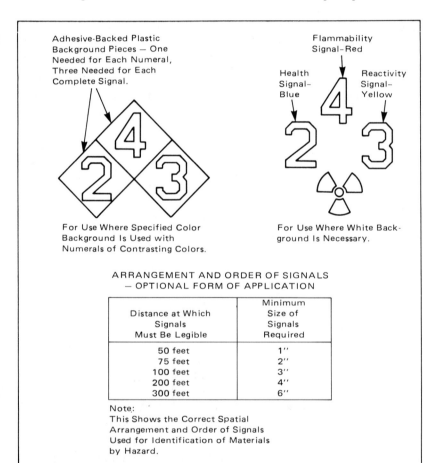

FIGURE 8-1.

The number in the upper point of the diamond designates the flammability hazard of the material. If the symbol is color coded, it will be colored red. The number in the right-hand point designates the reactivity hazard. If the symbol is color coded, it will be colored yellow. The number in the left-hand point designates the health hazard of the material. If the symbol is color coded, it will be colored blue.

Table 8-2 on the next page describes what each of the numbers represents, with regard to the particular hazard with which it is used.

The lower point of the diamond is reserved for particular properties of a material which may require special fire-fighting techniques or which may raise other special problems. The three symbols currently approved for use in this space are: W - for materials which have an unusual reactivity with water; OXY - for materials which possess oxidizing properties; and the standard radioactivity symbol for those materials which are radioactive. See Figure 8-2.

Note that the various numbers are not assigned based on definite quantitative factors. Rather, there is a range of relative factors which

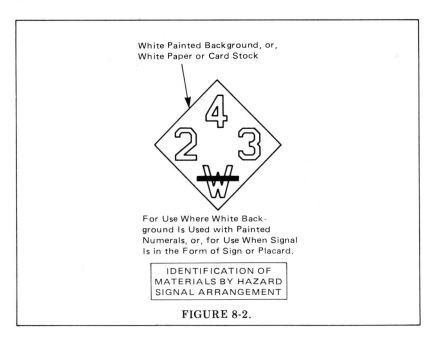

White Painted Background, or,
White Paper or Card Stock

For Use Where White Back-
ground Is Used with Painted
Numerals, or, for Use When Signal
Is in the Form of Sign or Placard.

IDENTIFICATION OF
MATERIALS BY HAZARD
SIGNAL ARRANGEMENT

FIGURE 8-2.

TABLE 8-2

	Identification of Health Hazard — Color Code: BLUE	Identification of Flammability — Color Code: RED	Identification of Reactivity (Stability) — Color Code: YELLOW
Signal	Type of Possible Injury	Susceptibility of Materials to Burning	Susceptibility to Release of Energy
4	Materials which on very short exposure could cause death or major residual injury even though prompt medical treatment were given.	Materials which will rapidly or completely vaporize at atmospheric pressure and normal ambient temperature, or which are readily dispersed in air and which will burn readily.	Materials which in themselves are readily capable of detonation or of explosive decomposition or reaction at normal temperatures and pressures.
3	Materials which on short exposure could cause serious temporary or residual injury even though prompt medical treatment were given.	Liquids and solids that can be ignited under almost all ambient temperature conditions.	Materials which in themselves are capable of detonation or explosive reaction but require a strong initiating source or which must be heated under confinement before initiation or which react explosively with water.
2	Materials which on intense or continued exposure could cause temporary incapacitation or possible residual injury unless prompt medical treatment is given.	Materials that must be moderately heated or exposed to relatively high ambient temperatures before ignition can occur.	Materials which in themselves are normally unstable and readily undergo violent chemical change but do not detonate. Also materials which may react violently with water or which may form potentially explosive mixtures with water.
1	Materials which on exposure would cause irritation but only minor residual injury even if no treatment is given.	Materials that must be preheated before ignition can occur.	Materials which in themselves are normally stable, but which can become unstable at elevated temperatures and pressures or which may react with water with some release of energy but not violently.
0	Materials which on exposure under fire conditions would offer no hazard beyond that of ordinary combustible material.	Materials that will not burn.	Materials which in themselves are normally stable, even under fire exposure conditions, and which are not reactive with water.

go into the assigning of a particular number to a particular material. The system is not intended to give precise scientific analysis of hazardous properties. Instead, it provides a quick, standardized reference expressing the relative hazard of a material in regard to the three important areas.

The presence or accuracy of labels may provide a subject for litigation in the case of stored liquids. However, there are a number of complex requirements for the storage of hazardous materials, particularly liquids. Requirements among the most important are found in NFPA 30, Flammable and Combustible Liquids Code, to which reference should be made for any case involving fire resulting from the storage or use of a flammable liquid.

CHAPTER 9

BUILDING STRUCTURES AND
FIRE TECHNOLOGY

§9.1 Introduction

Most fire deaths and injuries in this country occur in buildings. Property loss from fire almost always involves a building, its contents, or both. This interrelationship between fire damage and buildings makes it necessary for the litigator to have some understanding of building construction as it pertains to fire.

This chapter discusses design and structural characteristics of a building, how they can be set up to prevent fire, and to some extent how they will react to fire. By understanding some basic concepts, the litigator can better conduct an investigation and determine what could have been done to prevent a loss. The defense attorney, in a more reactive posture, will be no less aided by an understanding of building construction as it relates to fire. Although the litigator cannot be transformed into an architect, builder, or fire protection engineer, the attorney who achieves an understanding of the concepts and materials with which these professionals work, and the constraints under which they work, can be more effective in preparing and presenting a case.

While this chapter discusses the overall structure of a building, Chapter 11, Building Operations, looks inside the walls of a building and deals with the operating systems. This breakdown is not to suggest that the structure and its operating systems are not interrelated, rather, it is a convenient way of examining two factors present in every building and every building fire.

An important cautionary note must be included. The topics discussed in this chapter are the subjects of many volumes by specialists in the various facets of building construction. Also, the vast range of types and uses of buildings could expand this chapter to hundreds of pages if each type and use were to be individually discussed. The treatment here is necessarily brief, intended only to raise questions for further inquiry or to suggest areas for further study.

§9.2 Site and Design

Before the first foundation hole is dug for any structure, careful work should go into the site selection and design of the building for firesafety. While an industrial facility is likely to have a more extensive firesafety plan than a single-family house, living units should not be exempt from such planning. The majority of deaths related to fire occur in one- and two-family dwellings. "America Burning: The Report of the National Commission on Fire Prevention and Control[1]," went so far as to identify areas in which designers created unnecessary hazards for building occupants.

When a building is being designed, whether it be an industrial plant, a single-family home, or an institutional care facility, the designer must have an awareness of the basics of firesafety design. Generally, these basics revolve around three areas:

1. *Life Safety* - This will require attention to local and national fire and building codes and will center on alerting the occupants to a fire as well as allowing adequate time and paths for escape.

2. *Property Protection* - Specific items of property, areas of high value, or areas of high hazard concentrated within the building may require special consideration in the design stage.

3. *Continuity of Operations* - This applies generally to business uses. What part or parts of the building will house activities absolutely vital to maintenance of continuity? The absolutely essential areas may require differently designed protection systems to protect their vital operations.

[1] US Government Printing Office, 1973.

§9.3   Elements of Building Firesafety

Building firesafety may be achieved either by fire (ignition) prevention, or if a fire does occur, by managing its impact through building design.[2] Table 9-1 lists most of the firesafety factors which have to be considered from a prevention point of view. Naturally, any fire scene should be examined for the presence or consideration of each of the factors. Any failures should lead to further inquiry. This table should be regarded only as an overview.

TABLE 9-1.   Fire Prevention and Emergency Preparedness

1. Ignitors
    a. Equipment and devices
    b. Human accident
    c. Vandalism and arson

2. Ignitable Materials
    a. Fuel type and quantity
    b. Fuel distribution
    c. Housekeeping

3. Emergency Prepardness
    a. Awareness and understanding
    b. Plans for action
        —Evacuation or temporary refuge
        —Self help extinguishment
    c. Equipment
    d. Maintenance — operating manuals available

From a design point of view, there are numerous basic elements which should be considered, including the following:

a. *Fire Growth Hazard* - If a fire is to be properly contained, has

[2] McKinnon, Gordon, et al, *Fire Protection Handbook*, 15th ed., National Fire Protection Association, Quincy, MA, 1981, page 5-4.

proper consideration been given to methods of containment? Under this category, for each space there should be thorough consideration of the fuel load; the interior finish that is designated; the air supply which will be available; and the size and shape of the space. Much of the material discussed in this chapter will expand on these considerations.

b. *Automatic Suppression* - Sprinklers have historically been the most widely used and effective means of automatic fire suppression. Are these properly designed, installed, and maintained?

c. *Manual Suppression* - Has the building been designed to facilitate the efficient operations by the community fire department? Are heat and smoke detection systems in place, as well as methods by which the fire department can be notified? Is there adequate water pressure for fire extinguishment once the fire department has arrived?

d. *Barrier Design* - Are barriers properly designed and placed to assure not only containment of the fire but also their own integrity against penetration or collapse?

e. *Structural Collapse* - The potential for structural collapse must be considered. Fire severity and the resistance to collapse are the principal factors to examine.

f. *Smoke Movement* - Extreme life hazards are caused by smoke. Smoke must be contained or at least adequate time must be allowed for people to react to the fire before being overcome by products of combustion.

g. *Protection of Persons* - Have sufficient factors been considered to assure that people will have the maximum warning about a fire? Have alternate escape paths been provided? Has evacuation been preplanned and practiced?

h. *Protection of Property* - Given the nature of the property and activity in the building, was adequate consideration given to any special protection that might be necessary? Were highly sensitive, hazardous or valuable operations or materials considered and protected?

Designers are not without guidance in these matters as there are local and state building and fire codes which control design, as well as numerous NFPA and other codes and standards. In any fire case involving a building, the litigator should have complete familiarity with all applicable codes to construct the best possible case for the plaintiff or defense. Without this familiarity, an important aspect of the case could be overlooked.

§9.3.1   Site Planning

You have already been introduced to the necessities of site planning. Since site planning is at least simultaneous with, if not before, the design of the building, it can be broken down into two considerations:

a.   Has adequate attention been given to the site-related factors that will aid in the suppression of fire?

b.   Has the building been properly sited with regard to other buildings, leaving adequate separation or making other preparations to prevent the spread of fire?

Consideration *a* is similar in nature to those which were raised with regard to building design. There are those external factors which will assist in the manual suppression of a fire after it is started. There is no single, all-inclusive list of factors that must be considered but there are three basic factors.

The first factor in site planning for fire suppression is traffic and transportation. Fire apparatus must be able to reach a given building promptly. Of course, all buildings cannot be built near a fire station or be easily accessed, but when being sited, access should be considered. Busy streets may be overcrowded during certain times of the day so alternative routes should be available. Limited-access highways provide fast movement, but they may not allow easy traveling between the points of access.

The second factor is fire department access. Once at the scene, is it easy for the fire department to approach any point of the building? Blocked passageways and inaccessible walls will hamper efforts at fire suppression, either lengthening the time necessary to suppress the fire or making access for fire fighting impossible. Full accessibility may not always be possible in congested areas, but good firesafety planning should provide for the maximum access that is practical.

The third factor in site planning for fire suppression is the basic water supply. Adequate attention should be given to the number and location of hydrants and their capacity to provide water. The underground water mains that feed into them must be of sufficient volume and pressure to take care of fire fighter needs. Many local codes include requirements regarding the location and capacity of hydrants; a fertile field for litigation concerns the question of whether these codes are sufficient. Again, adequate design consideration should be given to this factor.

Consideration *b* is involved with the placement of and the protection around a particular building in regard to other buildings in the immediate vicinity. The primary danger which must be avoided is the exposure of the building to other buildings that may be on fire, or the exposure of other buildings when the particular building is on fire. This topic is covered by NFPA 80A, Recommended Practice for Protection of Buildings from Exterior Fire Exposures, which advises on means of protection as well as separation distances.

NFPA 80A also includes information on exposure severity. This is a measure of the radiation level developed per unit of window area by the exposing fire. NFPA 80A ranks severity of exposing fires as light, moderate, and severe, basing these classifications upon two factors: 1) average combustible load per unit of floor area, and 2) characteristics of an average flame-spread rating of the interior wall and ceiling finishes. While there are other factors that help determine the severity of a fire, these two factors are basic.

Exposure protection accomplished largely through adequate separation is a necessary consideration in the site planning of any building. Anticipated fire severity is a key factor in determining this necessary separation.

NFPA 80A recognizes that the ideal separation distances will often be precluded by cost and existing site conditions. The document includes a list of separation distances with ways that a designer can allow for a decrease in the initially required distances. Among the factors which allow for the decrease are construction of free-standing barrier walls, total automatic sprinkler protection, extension of exterior masonry walls to form parapets or wings, and the inclusion of extensive automatic sprinklering and closing devices. The inclusion of any or a combination of these factors allows a decrease in the initially recommended separations.

266

§9.3.2   Classifications of Building Construction

The actual buildings themselves must come under careful scrutiny for firesafety considerations. When entering the discovery stage of litigation, it may be necessary to determine the extent to which these firesafety principles are actually used in the construction of specific buildings.

Five numerical classifications of buildings, Standard Types I through V, are currently in use under NFPA 220, Standard on Types of Building Construction. In the past, the five different types of construction have been labeled with more descriptive names such as (a) fire resistive, (b) noncombustible/limited combustible, (c) heavy timber, (d) ordinary, and (e) wood frame. In a general way the five numerical classifications can be correlated to the descriptive names. The model building codes also use numerical classification which will likewise have a general correlation to the five numerical classifications. One of the advantages of the numerical system is that it allows the use of additional numbers which will indicate the required fire-resistance rating of three major building components. In this system the Roman numeral is followed by three Arabic numbers. The first of the Arabic numbers designates the required fire-resistance rating of the exterior bearing walls. The second designates the required fire-resistance rating of the structural frame or columns and girders, supporting loads for more than one floor. The third designates the fire-resistance rating of the floor construction. Thus, a classification such as I-443 not only designates a particular type of construction, but also denotes fire resistance ratings required of the three mentioned structural members.

Whether the numerical system or the more descriptive system of building classification is used, the basic concept is to classify these buildings on the dual basis of:

a.   The potential contribution of fuel from the building structural systems and enclosing envelope (exterior walls and roof) to fire development.

b.   The ability of the building's structural systems, floor systems, and bearing walls to fulfill both their load-bearing and fire-barrier functions should a full, developed fire occur within the structure.[3]

[3] McKinnon, Gordon, et al, *Fire Protection Handbook*, 15th ed., National Fire Protection Association, Quincy, MA, 1981, page 5-22.

The term "fire resistance," which is basic to this system of classification, refers to the time in minutes or hours that the materials or assemblies have withstood a fire exposure as established by a standardized test, and the test method is described/defined in detail in NFPA 251, Standard Methods of Fire Tests of Building Construction and Materials. (See also ASTM E119 test and UL 263.) The test is accepted throughout the construction industry and used by most testing laboratories.

§9.3.3   Classification Descriptions

It is worthwhile to describe in some detail the five previously mentioned classifications as designated by their descriptive titles. The numbered classifications, while correlating generally, are based more directly to the requirements of the standard, to which reference should be made in a specific application.

§9.3.3.1   Type I   This classification represents the superior type of fire rated construction. Traditionally, this type of construction refers to a building with structural members fabricated of noncombustible materials of such quality and so protected that they will resist the maximum severity of fire expected within the building without collapse.[4] In modern classification, such a structure can be expected to maintain shape and form even if the contents burn out. It is expected that such a type of construction will not contribute fuel to exacerbate an existing fire. Such construction, however, is not a guarantee of safety for life and property. Numerous other factors are present in a fire which can cause death and destruction other than collapsing walls.

§9.3.3.2   Type II   The key element of a building which falls into this class is that its materials of construction do not contribute to the development or spread of fire. Unlike the fire resistive construction, however, a building of this type is not expected to survive a burnout of

---

[4] McKinnon, Gordon, et al, *Fire Protection Handbook*, 15th ed., National Fire Protection Association, Quincy, MA, 1981, page 5-22.

the contents, and there is a potential for structural collapse. Many metal frame and metal clad buildings would be in this category, as well as many other nonwooden buildings constructed around an unprotected metal hoist system. This type of construction is popular with metal buildings erected in the field. Caution should be taken that the noncombustible nature is not compromised by the use of combustible materials for roofing, weather protection or vapor barriers.

§9.3.3.3 Type III This type of construction is, perhaps, the most difficult to define due to a combination of noncombustible exterior walls and requirements for sizes of the interior floor and beam construction, which is generally of heavy timber with a comparatively high fire resistance. Heavy timber construction requires a coming together of numerous characteristics involving plank width, construction techniques, and framing member widths. This type of construction first gained importance with large mills in New England, and even became known as "mill construction." It fell out of favor with the difficulty in obtaining the necessary large planks. However, recent techniques of creating the necessary planks by gluing together numerous layers of wood, making a timber equivalent to nature in fire resistance, has made it once again a feasible building method.

§9.3.3.4 Type IV This type of construction has wide use. It is typically defined as having exterior bearing walls of noncombustible, or limited combustible, materials with minimum hourly fire resistance ratings and stability under fire conditions. Roofs, floors, and interior framing are wholly or partly of wood and of smaller dimensions than required for heavy timber construction. In this type of construction, unlike fire resistive and noncombustible/limited combustible buildings, there will usually be concealed walls and ceiling spaces which may contain combustible materials.

Ordinary construction was, at one time, the most common style for commercial buildings, larger occupancy buildings, and institutional occupancies. It may still be found among many buildings in congested areas, particularly in cities. There are two key problems with such types of buildings. They have a propensity to spread fire due to their combustible interior walls and framing, and they have concealed spaces in which fire may be nurtured and spread.

As in heavy timber construction, a full description of the characteristics would consider the beam and plank size, as well as methods of attaching the beams to the planks through joists.

§9.3.3.5 *Type V* Although the most widely used method of construction, particularly for residential properties where most fire deaths and injuries occur, this type of construction offers the least fire resistance and contributes the most to the growth and spread of a fire. A building is considered to be of wood frame construction when the exterior walls and their supports, as well as the floors and roof, are all made of wood or other combustible material, but the construction does not qualify as either heavy timber or ordinary construction. The principal difference between wood frame and ordinary construction lies in the wooden exterior walls. Because of the higher potential of fire spread in this type of construction, the building codes will, in most cases, require firestopping in all joists or other areas where fire might spread through walls. A firestop is a piece of material, such as a two by four, gypsum board, or cement asbestos board, placed in a natural opening in any wall or other partition so that fire spread can be prevented or inhibited. Strategic placement of such stopping will aid in maintaining the fire resistance of the particular wall or ceiling construction which is being protected. The natural openings in any wall or other construction will greatly impair the ability of that wall to resist the spread of fire.

§9.3.4 Summary of Classifications

The five types of construction are based on two factors of resistance to fire spread and integrity of the structure itself during fire. While the treatment here is necessarily brief, the litigator at the outset of any investigation should become aware of what basic type of construction was involved in the structure. This may not be obvious, or in a specific situation may not be important. Nevertheless, given the differing requirements for differing types of structures, each of these requirements should be thoroughly researched to provide full development of theories and testing of all hypotheses. Furthermore, the type of construction may or may not have been appropriate for the location or use of the structure in question.

§9.4   Particular Construction Components

Classification of the different types of construction and their requirements is based on overall exterior and framing considerations. Of perhaps more interest in a specific loss situation is an awareness of the code or standard requirements and firesafety considerations of some specific building components, such as the particular assemblies and factors that go into a building. This section of this chapter discusses these particular components of construction, along with how they can and should be used for firesafety. The purpose of the discussion is to give the nonengineering-oriented attorney some background in the proper design, construction and installation of these factors. The attorney can then use this information to work with a qualified expert and/or for the initial development and proof of theories of liability or defense.

§9.4.1   Structural Frame

The great majority of buildings, regardless of their construction method, have an internal structural frame. As mentioned in the descriptions of the various building classifications, the nature of the frame is an important consideration in deciding into which classification a particular building falls. The frame is the skeleton of the building. For firesafety considerations, the foundation is generally not of great significance.

The principal firesafety considerations for framing are fire, firespread, and structural integrity. These topics, already covered in the previous sections, are also applicable to the frame because it plays an important role in preventing structural collapse.

The collapse of walls, floors or ceilings is one of the most serious dangers of fire. In discussing other functions of the structure's components, such as load bearing or nonload bearing, the firesafety importance of the distinction is the harm and damage which can occur as a result of collapse. In the case of the frame, however, its very purpose is to support and bear the load of the structure. Therefore, in every case including structural collapse, the frame should be throroughly investigated. Furthermore, the possibility of progressive collapse of the whole, or at least of large segments, of a structure may be traceable to one or two girders being weakened. If a fire causes weakening of columns to the point of collapse, it is likely that a major wall or roof/ceiling/floor area will collapse.

It should also be remembered that any weakening or collapse will not only be dangerous in itself, but will also compromise or destroy the integrity of any design for the retardation of flame spread or of the fire suppression system. This will be mentioned later when discussing walls and ceilings in detail.

The specific requirements of size, weight, spacing and other attributes of the framing members will be a function of the local building code, which should be scrutinized carefully. In all cases where the framing is of importance, the litigator should be conversant with the required fire resistance ratings of that assembly and the properties of the particular materials used, such as wood, steel or concrete; the dimensions of any girders, beams, or columns; their spacing from one another; the methods by which they are fastened to one another; and the particular load which they are intended to support. A full description of each of these can be found in a building code or in the listing of a specific assembly.

Two particular points raise issues which must be checked. The first of these, the presence of firestopping, was mentioned in a previous section. Particularly with wood frame construction, the presence of adequate firestopping to control fire spread and protect the integrity of partitions is a necessary requirement. A failure to have adequate provision for firestopping might well be considered a failure to provide full protection against a very foreseeable danger.

The second point which should be checked concerns the presence and quality of protective coverings on the framing members. Fire resistive protective coverings are often sprayed on or attached to the beams, girders and columns used to support the structure. Because these structural elements are often contained within concealed spaces, it may take considerable investigation to unearth the state of the protective covering, but it could prove to be an important factor in any fire theory.

Girders, beams and columns are so important to the integrity of the structure that they are usually required by the building code to have a certain fire resistance rating. This rating, usually expressed in minutes or hours, will indicate how long a particular assembly withstood the fire test set out in NFPA 251, Standard Methods of Fire Tests of Building Construction and Materials. Protective coverings are usually needed to bring the assembly up to required standards. Failure of these coverings may help explain fire spread or building collapse in a given case. Charts exist to show the specific thickness of covering that will be needed to bring particular types and sizes of framing members up to various fire ratings. These charts can be found in NFPA's *Fire Protection Handbook* as well as many building codes and manuals.

In addition, design criteria or fire resistance ratings for specific building construction assemblies can be found in indexes, or in directories as published by major testing laboratories. As an example, references frequently are made to the UL Building Materials Directory or the Factory Mutual Approval Guide.

§9.4.2   Walls

The walls are probably the most important building assembly for firesafety purposes. While almost every aspect of firesafety design and construction focuses on wall construction and placement, it is significant that walls prevent firespread and structural collapse.

Walls are either interior or exterior to the structure. As exterior walls, they not only serve as general protection against the elements, but they are also most commonly load bearing. In some cases, however, the skeletal frame will be designed to carry the full weight of the structure. Thus the exterior wall is nonload bearing and may even be what is known as a curtain wall. From a structural integrity point of view, information concerning whether a wall is or is not load bearing is important when reviewing its construction. Load-bearing exterior walls are most often constructed of stone, brick, concrete block, or some combination of these.

In wood frame construction, such as that used in most single-family homes, the exterior wall is of wood and will be load bearing. The frame is designed to use the wall to bear the weight of the structure, though the particular sheathing on the wall will usually be nonload bearing and principally serve for protection.

Curtain walls are generally prefabricated, by definition are nonload bearing and usually are made of aluminum, stainless steel, other metal material, or of glass. The walls are set within the frame for protection purposes. For curtain walls in particular, the method of attachment to the frame is of great importance. If the structure is more than one story in height, an improperly attached curtain wall can leave open spaces between the frame and the wall. These spaces can provide a natural opening that will allow the vertical spread of fire. Once again, strategically placed firestopping is necessary where such openings exist.

Other wall construction types designed for specific use include enclosure wall, party wall, faced wall, hollow wall, or veneered wall. These types of walls are defined in the glossary of this book and in more complete texts on building construction.

Interior walls comprise the other general classification of walls. While some interior walls may be load bearing, they usually are not. Most new construction, particularly in office or commercial buildings, has a minimum of bearing walls to allow for maximum flexibility in arranging the interior spaces. Because most interior walls have to bear only their own weight, they are usually constructed of plaster or wallboard attached to wood or metal. These walls may or may not be part of the frame. In modern office building construction, the studs generally are not load bearing. In most older buildings, however, some stud assemblies will be either fully load bearing or will at least assist in bearing a load so that their collapse would greatly weaken the entire structure. From a firesafety point of view, the importance of interior walls is their role in stopping firespread more than their role in supporting the structure.

When considering the firespread prevention of walls, the focus is more on the interior than the exterior walls. However, exterior walls are important to protect surrounding structures. Since surrounding structures may be affected by radiated heat and exposure to flames or burning projectiles, the exterior walls must contain the fire in order to protect other structures in the area.

Exterior walls also provide protection against fire in surrounding structures. For this function, the key factors are the height of the building being exposed to fire and the fire resistance rating of the walls which protect against radiated heat. NFPA 80A, Recommended Practice for Protection of Buildings from Exterior Fire Exposures, sets recommendations for fire resistance ratings of walls and describes the makeup of the exterior walls necessary for proper fire protection.

As stated above, interior walls bear the maximum responsibility in preventing the spread of fire. This prevention of fire spread is done through the placement of walls, their internal integrity and fire resistance, and protection of wall openings by fire doors and dampers.

Placement of walls is largely a design factor. From a firesafety point of view, the goal should be to compartmentalize as much as possible. To fully describe the necessity for and the theory of compartmentalization is a function of the *Life Safety Code®* along with a construction or engineering text. Thus if a fire and the products of combustion can be confined to a small area, not only will there be less property damage, but there will most likely be less opportunity for bodily injury, and the fire will be more quickly extinguished. Therefore, on any fire loss the litigator should become familiar with the floor plan. Was every method of construction reasonably used to keep a fire contained in a small area? Were walls designed to control potential fire spread or placed for decorative purposes only?

In order to successfully contain any fire and in order to successfully implement a plan of compartmentalization, the fire resistance of the wall must be known. The fire resistance must meet minimum standards which will usually be set by the locally adopted building codes. Fire resistance as previously mentioned is based on a test described in NFPA 251, Standard Methods of Fire Tests of Building Construction and Materials. The fact that a wall assembly has a 1-, 2-, or 3-hour rating does not necessarily mean that under actual conditions it will withstand the full force of a fire for this specific period of time. This can best mean that the most highly rated assembly is likely to withstand collapse longer than the lower rated assembly. The fact that any given assembly may resist collapse does not mean that it will likely resist the spread of fire. For example, a steel wall may resist failure for a long period; meanwhile, it may heat up to the point that it is radiating significant heat to ignite fires in adjoining rooms. Therefore, the fire resistance rating of all walls should be known.

When wall assemblies are tested, the test is done without concern for openings such as doorways, windows, pipes or beams. Since the basic purpose of the test is to measure the ability of the wall to prevent fire spread, such openings can compromise the entire inhibition value of the wall. Adequate firestopping should be provided to protect the fire resistance rating of the wall. Windows or other openings may be part of the design, and of themselves may be adequate as long as it is recognized that they may destroy any required fire rating.

It is the material of which the wall is made and the means by which it is put together that will largely determine the fire resistance rating. Various materials have been rated for their fire resistance and, like framing members, there are charts available to indicate the fire resistance ratings of materials. Publications of the National Bureau of Standards and listings from the Underwriters Laboratories will assist in figuring the fire resistance rating of a particular wall assembly. Engineers and architects also have developed mathematical formulae which by interpolation can estimate the resistance rating of a wall. Finish material will have a significant bearing on the fire resistance of a wall, as well as on the general fire propensity and severity of any room.

§9.4.3   Ceiling Assemblies

For fire protection purposes, ceilings are often thought of along with floors due to their natural relationship from one story to the next. A fire resistance rating is usually assigned to the floor/ceiling assembly, not the ceiling itself. Since ceilings by themselves are not load bearing, their relationship to framing and to walls can be put aside. What is of key importance is the ability of the ceiling to contain the fire and to adhere to the frame without collapse. The same considerations concerning fire resistance ratings for walls are important for ceilings as well.

The ability to suppress a fire is greatly diminished once it has spread vertically to another story. Also, since the natural tendency of heated air and gases is to rise, the integrity of the ceiling assembly must be maintained. Ceiling assemblies are largely of the same material as walls and are tested and rated in the same manner. Likewise, the same considerations regarding pokethrough holes for columns, electrical wires, air ducts or other material must be met.

The method by which a ceiling is attached to the frame is important. Usually, a gypsum wallboard or lath is nailed to the framing members. The nails should not be driven in so that they crush the gypsum core, which would reduce its fire resistance. Also, the type of nails or other attachment method should be checked. Longer, thinner nails will conduct less heat to their surrounding wood than will common wire nails; likewise, nails with protective coatings are generally less likely to conduct heat. Screws which are made particularly for the attachment of gypsum board generally provide greater holding power, thus making collapse less likely.

The fact that many ceilings today are suspended does not substantially change their importance or character in firesafety. The material which comprises the suspended ceiling tiles or slabs should likewise be fire rated, and the holding assembly should not be allowed to be a source of weakening of the fire spread barrier function of the ceiling.

Finally, ceilings often house recessed light fixtures which can become a source of heat. The National Electrical Code® has specific requirements for such assemblies so that they do not become concealed sources of fire ignition. Nonrecessed lighting fixtures and other materials or assemblies which lessen the integrity of the ceiling should also be checked. They must not be a means by which fire travels to higher floors in spite of an acceptable fire rating.

§9.4.4 Floors

Floors are, by their nature, load bearing. The building may be framed so that floors do not bear any weight from the walls, though they do clearly come in contact with and intersect at least one wall.

The floor, however, must bear the weight of any interior furnishings without being significantly weakened. Also, in case a fire is being fought, the floor may have to support not only the fire fighters who enter the building, but also a great weight of water used to fight the fire. Ideally, a floor should be watertight to prevent heavy water damage to lower floors. Given the weight of water and the amount that can be used, this is a significant consideration. Unfortunately, it is not always possible to make floors as watertight as might be liked. Wood, one of the most popular flooring materials, is almost impossible to make watertight.

From the standpoint of fire litigation, the watertightness of a floor may be of little significance. However, since water damage to lower floors is often a factor in considerable property loss, the tightness of the upper floor construction or design should be considered.

Inasmuch as a floor, like a wall or ceiling, is an important factor in preventing fire spread, considerations of fire resistance which have already been examined for those assemblies should prevail. As was originally pointed out for ceilings, the floor/ceiling assembly is often considered as one for the purpose of setting a fire resistance rating. What is of great importance for a floor is its composition and surfacing as well as the covering which may be laid on it. The danger to be considered is the horizontal spread of fire across the room of origin should the fire originate on the floor, or should embers or other ignited material drop to the floor. Of course, concrete will have less propensity to spread fire than wood. It will be necessary to investigate any tile, asphalt or membrane surfacing which may be present. Besides the material itself, the glue, cement or other method by which it is bonded to the floor should be checked. Publications from the National Bureau of Standards and Underwriters Laboratories, among other sources, describe the various materials and their fire spread ratings. NFPA 101,® Life Safety Code®, sets standards for the class of floor finish that is allowed.

Unbonded floor coverings, principally rugs, are considered part of the furnishings, and thus are not part of the structure. There are specific standards which such coverings must meet under NFPA 253, Standard Method of Test for Critical Radiant Flux of Floor Covering Systems Using a Radiant Heat Energy Source.

§9.4.5   Roof Systems

The roof system of any building is made up of two components: the roof deck and the roof covering. From a firesafety point of view, the roof deck has much the same considerations and characteristics as were set out for frames, walls, and ceilings. In many buildings, particularly flat-roofed industrial buildings, the roof deck is the ceiling of the top floor. This may not be the case in pitched roof buildings, but the roof deck will usually be part of the frame covered by a plywood or metal exterior. While the deck is not load bearing by itself, it will often be used as a point of entry in case of fire and must be designed to bear at least this easily foreseen weight. In addition, a roof deck might well be immediately adjacent to a concealed attic space where fire could have an opportunity to develop and weaken the underside of the roof even before any load is placed upon it. There also may be architectural considerations which require that roof framing spans be larger than those employed for the floor.

A necessary corollary between roof deck considerations and those for ceilings and walls is that fire spread characteristics of roof decks should be no less than those for ceilings and walls. Any litigation involving a fire spread or roof collapse should include a careful examination of how the roof was designed to reduce or contain the spread of fire, or to bear the load that would foreseeably be put upon it. NFPA 203M, Manual on Roof Coverings and Roof Deck Constructions, which deals more directly with roof coverings, contains a discussion of the dangers which are prevalent in this construction and which can be avoided by proper design, materials, and construction.

Roof coverings, whose principal function is protection against the elements, may have great impact for firesafety purposes. The roof covering usually will form the initial point of contact for burning material falling from another structure, or for heat radiated from fire at a higher level of an adjacent building. Also, the roof covering could be exposed to flame and heat from an internal fire which can build up significantly in a concealed area. Because many roof coverings, whether for flat industrial-type roofs or shingles for pitched-roof residences, have historically been made of various types of asphalt or, in some instances, wood, there have been numerous cases of the roof covering acting to exacerbate or spread an existing fire, or to serve as a point of ignition.

NFPA 203M, Manual on Roof Coverings and Roof Deck Constructions, deals with roof coverings. It covers four general types of roof coverings:

a.   Composition Built-Up, which is made of alternate layers of felt and bitumen built up into a weatherproof membrane. Gravel or slag may be imbedded in the surface to reflect heat, prevent flow and cracking of the bitumen, and improve the fire performance.[5]

b.   Prepared, which is factory-produced and ready for attachment to the deck. Prepared roofing can be made of numerous materials and will usually be mixed with asphalt and surfaced with granules. This is the most common type of roofing for one- and two-family houses with pitched roofs for drainage.

c.   Wood shingles and shakes are sawed or split wood attached to the roof so a portion of each shingle is exposed. Attached in an overlapping fashion, shingles and shakes are the most combustible of the basic roofing types and should not be used unless treated to be flame retardant. If used, they should be treated by a pressure impregnation process, classified in accordance with NFPA 256, Standard Methods of Fire Tests of Roof Coverings.

d.   Elastomers are essentially one-ply coats of membranelike material having elastic properties. Elastomers are applied to the roof deck either in the form of sheets cemented to the deck, or brushed, rolled or sprayed onto the deck if supplied in a liquid fashion. Firesafety considerations of these roofs are based on the cements and liquids used because they may be flammable. No open flames should be permitted near the areas of application.[6]

Beyond simply the design and nature of the roof coverings, products in each of these categories are classified as A, B, or C on the basis of five different tests. The tests and test methods which are discussed in NFPA 256, Standard Methods of Fire Tests of Roof Coverings, are:

---

[5] NFPA 203M, *Manual on Roof Coverings and Roof Deck Constructions*, National Fire Protection Association, Quincy, MA, 1980.

[6] *Id.*

(a) Intermittent Flame Exposure Test, (b) Spread of Flame Test, (c) Burning Brand Test, (d) Flying Brand Test, and (e) Rain Test. Each of these tests has a detailed procedure and specific criteria for determining the A, B, or C classification.

In any case where roof covering plays an important part either in fire ignition or spread, detailed information concerning classification of the particular covering involved should be studied. The tests described in NFPA 256 should be thoroughly understood, as this may provide a fertile ground for testing an expert's knowledge of roofing materials and procedures.

Requirements have been developed to classify various roofing materials on the basis of the degree of the roof incline, the material of which the roofing is made, and the deck surface material used. These requirements are set out in Tables 9-2 and 9-3.

§9.4.6 Interior

Three principal factors which determine the fire hazard of a building include fire resistance of the structure, the contents or process enclosed by the structure, and the characteristics of the interior finish of the structure. A fourth principal factor, building services, is discussed in Chapter 11. Some facets of the structure itself have already been discussed as to their importance in a fire and, thereby, their impact on litigation. The contents or process presents a set of variables too numerous to deal with here. In this category, there are many different types, styles, and uses of furniture or machinery made out of an endless variety of materials and used for a vast number of purposes. To begin an examination of all such interior furniture or machinery would require a more comprehensive treatment than is possible here. If, in a particular case, some part of the interior furniture or machinery is important, then general principles that apply to fire as discussed in previous chapters should be reviewed. Also, materials should be analyzed at a laboratory for their fire spread characteristics, and material from the Consumer Product Safety Commission should be reviewed. The Flammable Fabrics Act could be helpful in cases where the particular material involved may be listed.

TABLE 9-2 Typical Prepared Roof Coverings*

Description	Minimum Incline In. to Ft	CLASS A	CLASS B	CLASS C
Brick		Brick, 2¼ in. thick. Reinforced portland cement, 1 in. thick.		
Concrete		Concrete or clay floor or deck tile, 1 in. thick.		
Tile		Flat or French-type clay or concrete tile, ⅜ in. thick with 1½ in. or more end lap and head lock, spacing body of tile ½ in. or more above roof sheathing, with underlay of one layer of Type 15 asphalt-saturated asbestos felt or one layer of Type 30 or two layers of Type 15 asphalt-saturated organic felt. Clay or concrete roof tile, Spanish or Mission pattern, 7/16 in. thick, 3-in. end lap, same underlay as above.		
Slate		Slate 3/16 in. thick, laid American method.		

TABLE 9-2 Typical Prepared Roof Coverings (Cont.)*

Metal Roofing	12	Sheet roofing of 16-oz copper or of 30-gage steel or iron protected against corrosion Limited to non-combustible roof decks or noncombustible roof supports when no separate roof deck is provided.	Sheet roofing of 16-oz copper or of 30-gage steel or iron protected against corrision or shingle-pattern roofings with underlay of one layer of Type 15 saturated asbestos-felt, or one layer of Type 30 or two layers of Type 15 asphalt-saturated organic felt.	Sheet roofing of 16-oz copper or of 30-gage steel or iron, protected against corrosion, or shingle-pattern roofings, either without underlay or with underlay of rosin-sized paper.
				Zinc sheets or shingle roofings with an underlay of one layer of Type 30 or two layers of Type 15 asphalt-saturated organic-felt or one layer of 14 lbs unsaturated or one layer of Type 15 asphalt-saturated asbestos felt.
Cement-Asbestos Shingles	Exceeding 4	Laid to provide two or more thicknesses over one layer of Type 15 asphalt-saturated asbestos felt.	Laid to provide one or more thicknesses over one layer of Type 15 asphalt-saturated asbestos felt.	

TABLE 9-2 Typical Prepared Roof Coverings (Cont.)*

Description	Minimum Incline In. to Ft	CLASS A	CLASS B	CLASS C
Asphalt-Asbestos Felt Sheet Coverings	Not Exceeding 12	Factory-assembled sheets of 4-ply asphalt and asbestos material.	Factory-assembled sheets of 3-ply asphalt and asbestos material or sheet coverings of single thickness with a grit surface.	Single thickness smooth surfaced.
Asphalt-Asbestos Felt Shingle Coverings	Exceeding 4		Asphalt-asbestos felt grit surfaced.	
Organic-Felt (previously referred to as rag felt) Shingle Coverings	Exceeding 4			Sheet coverings of asphalt organic felt either grit surfaced or aluminum surfaced.
Organic-Felt (previously referred to as rag felt) Shingle Coverings, with special coating	Sufficient to permit drainage	Grit surfaced, two or more thicknesses.	Grit surfaced, two or more thicknesses.	
Organic-Felt (previously referred to as rag felt) Shingle Coverings	Sufficient to permit drainage	Grit surfaced, two or more thicknesses.	Grit surfaced, two or more thicknesses.	Grit surfaced shingles, one or more thicknesses.

TABLE 9-2 Typical Prepared Roof Coverings (Cont.)*

Asphalt Glass Fiber Mat Shingle Coverings	Sufficient to permit drainage	Grit surfaced, two or more thicknesses.	Grit surfaced, one or more thicknesses.	Grit surfaced shingles, one or more thicknesses.
Asphalt Glass Mat Sheet Covering	Sufficient to permit drainage			Grit surfaced.
Fire-retardant treated red cedar wood shingles and shakes	Sufficient to permit drainage			Treated shingles or shakes, one or more thicknesses; shakes require at least one layer of Type 15 felt underlayment.

* Prepared roof coverings are classified as applied over square-edge wood sheathing of 1-in. nominal thickness, or the equivalent, unless otherwise specified. Laid in accordance with instruction sheets accompanying package. Limited to decks capable of receiving and retaining nails.

Where organic-felt is indicated, asbestos felt of equivalent weight can be substituted.

By end lap is meant the overlapping length of the two units, one placed over the other. Head lap in shingle-type roofs is the distance a shingle in any course overlaps a shingle in the second course below it. However, with shingles laid by the Dutch-lap method, where no shingle overlaps a shingle in the second course below, the head lap is taken as the distance a shingle overlaps one in the next course below.

Prepared roofings are labeled by Underwriters' Laboratories which indicate the classification when applied in accordance with direction for application included in packages.

TABLE 9-3   Built-up Roof Coverings*

Description	Minimum Incline In. to Ft	CLASS A	CLASS B	CLASS C
Asphalt organic-felt, bonded with asphalt and surfaced with 400 lbs of roofing gravel or crushed stone, or 300 lbs of crushed slag per 100 sq ft of roof surface, on coating of hot mopping asphalt.	3	4 (plain) or 5 (perforated) layers of Type 15 felt. 1 layer of Type 30 felt and 2 layers of Type 15 felt. 1 layer of Type 15 felt and 2 layers of Type 15 or 30 cap or base sheets. 3 layers of Type 15 or 30 cap or base sheets. 3 layers of Type 15 felt. Limited to noncombustible decks.	4 layers of perforated Type 15 felt. 3 layers of Type 15 felt. 2 layers of Type 15 or 30 cap or base sheets.	
Tar-asbestos-felt or organic-felt bonded with tar and surfaced with 400 lbs of roofing gravel or crushed stone, or 300 lbs of crushed slag per 100 sq ft of roof surface on a coating of hot mopping tar.	3	4 layers of 14-lb asbestos-felt or Type 15 organic-felt. 3 layers of 14-lb asbestos-felt or Type 15 organic-felt.	3 layers of Type 14-lb asbestos-felt or Type 15 organic-felt.	

TABLE 9-3 Built-up Roof Coverings (Cont.)*

Steep tar organic-felt	5	4 layers of Type 15 tar-saturated organic-felt, bonded with steep coal-tar pitch, surfaced with 275 lbs of 5/8-in. crushed slag per 100 sq ft of roof surface on steep coal-tar pitch.
Asphalt organic-felt, plain or perforated bonded and surfaced with a cold application coating.	12	3 layers of Type 15 felt. 1 layer of Type 30 felt and 1 layer of Type 15 felt. 2 layers of Type 15 or 30 cap or base sheets. 2 layers of Type 15 felt and 1 layer of Type 15 or 30 cap or base sheets.

* Built-up roof coverings are classified as applied over square-edge wood sheathing of 1-in. nominal thickness, or the equivalent, unless otherwise specified.

From the standpoint of relative effectiveness of the different types of wood roof sheathing, the tongue-and-groove boards and 3/4-in. moisture-resistant plywood give better results in the brand and flame tests than square-edge sheathing with boards spaced about 1/4-in. apart. For classifications based on square-edge sheathing, tongue-and-groove or plywood sheathing can be substituted. Square-edge sheathing boards should be butted together as closely as possible. Reference to 1/4-in. spacing is to indicate fire test procedure intended to simulate actual conditions after shrinkage of boards due to age or other reasons.

The minimum weight of cementing material between separate layers of felt is considered to be 25 lbs per 100 sq ft of roof surface. Types 15 and 30 felts are defined as saturated felts weighing a minimum of 14 lbs and 28 lbs per 100 sq ft of the finished materials, respectively. Where saturated felts are referred to by weight, the weight is minimum and is expressed in pounds per 100 sq ft of the finished material.

Materials intended for built-up roof coverings are labeled by Underwriters' Laboratories. The classifications indicated are of generally accepted combinations.

The interior finish is the third factor controlling the determination of a fire hazard. Because of the part such finish can play in fire, it is a factor which must be considered in every fire case.

What is interior finish? It is generally considered to consist of those materials or a combination of materials that form the exposed interior surface of walls and ceilings. When the term "finish" is used by most building codes, it will not include trim and incidental finish in the requirements for wall and ceiling finish. A distinction is often made in building codes as well as in the NFPA *Life Safety Code* between interior finish and interior floor finish. Because most interior floor finish consists of rugs or other textile materials, it will usually have separate building code requirements which will often require testing of such covering under NFPA 253, Standard Method of Test for Critical Radiant Flux of Floor Covering Systems Using a Radiant Heat Energy Source. Previously, the tunnel test, described by Underwriters Laboratories, was often used to test floor coverings (UL 723; ASTM E-84; NFPA 255).

The general construction uses of interior finish are for aesthetic, acoustical, and insulating purposes. The materials which are most commonly employed as interior finish are synthetics, plastics, wood and plywood. Depending on the way it is used, plaster or gypsum wallboard is generally considered part of the interior finish and is required to meet flame spread tests as set out by the local building code or the *Life Safety Code*. Previously, in Section 4.2 of this chapter, plaster and gypsum wallboard were treated as part of wall construction; even though it may be part of the wall it may also be part of the interior finish, giving it a dual use which the litigator may find interesting.

Highly decorative finish, such as the type often found in well-adorned places of public accommodation, has often been made of plastics, tiles or other such materials, and should be carefully considered as an ignition or spread factor in any fire case arising from such an area. Paint and wallpaper less than $1/28$ inch in thickness are not generally considered as interior finish material. Thick decorative wallpaper greater than $1/28$ inch in thickness may be considered as interior finish in the view of the local building inspector or local building code. Plastics particularly should be investigated in any fire scenario. There have been incidents of rapid fire spread as a result of flammable plastic finish and decoration. Most plastics in use today will be impregnated with fire retardants to meet building code requirements. NFPA 101, *Life Safety Code*, indicates that in the absence of certain conditions, cellular or foam plastics materials shall not be used as interior finish.

In any situation involving interior finish, its method of application and the method by which it is held to a wall should be given serious consideration. Many types of wall cement or adhesives are flammable or at least have a chemical composition which will not deter fire. In a large room or whole building, this could become a serious factor. Also, from another flame spread point of view, the integrity of the wall will be weakened if the finish fails to adhere properly. In this factor, nailed-on finish also should be scrutinized in case the method that was used did not provide sufficient strength to hold up under fire. Falling finish materials will not only compromise wall integrity, but they also will provide more surface area of combustible material in the form of the finish itself, and they will expose adhesives which may be made of flammable material.

§9.4.6.1 Fire Reaction of Interior Finish   To understand the full importance of interior finish to a fire, the four ways in which interior finish relates to a fire should be understood. In any case where the attorney feels interior finish is involved, it would be helpful to relate the action or reaction of the finish through one of more of these ways.

a. Interior finish affects the rate of fire buildup to a flashover condition. Flashover is the point in a fire when a room's entire contents catch fire simultaneously. Prior to this point, the fire had been localized in one part of the room; after this point, the destruction is throughout the room. Not long after flashover, the entire room is destroyed. Interior finish is important because flashover occurs when heat radiated from walls and ceilings raises the entire contents of the room to ignition temperature, causing the fire to simultaneously flash over the entire room. If the finish is such that it will absorb and conduct heat rather then redirect it back into the room, the flashover will be retarded.

b.   Interior finish contributes to fire extension by flamespread over its surface. Finish is often made of flammable materials which exacerbate the spread of fire.

c.   Interior finish adds to the intensity of a fire by contributing additional fuel. Total fuel load in the room is an important determination of fire severity. Flammable finish may add significantly to the total load that is available.

d.   Interior finish produces smoke and toxic gases that can contribute to life hazard and property damage.[7]

Because of its universal prevalence and its often flammable nature, interior finish should be considered in any fire loss situation. The materials of which it was made, the way it was designed to be used, the way it was applied and held to a wall or other surface, and the way in which it contributed to the full fire loss should all be examined carefully.

§9.4.7   Various Common Products

Before leaving the general field of structural fire technology, it is worthwhile to briefly examine a few of the more common materials involved in buildings for their firesafety characteristics.

1.   *Glass.* There are three common uses of glass in building construction: in windows (in many cases the window may be the entire wall) and doors, in fiberglass insulation, and as a reinforcement for fiberglass-reinforced plastic building products.

Glass used in windows and doors is, of course, the most obvious. In this use, the glass has little resistance to fire, although wire reinforcement will improve the ability of the glass to maintain its integrity. However, glass is not by itself a material which can be readily relied upon in a fire. Given its prevalent use, it may well be a factor to consider as one which failed to retard the growth of a fire. Building codes will often regulate the percentage of a wall area which can be a glass window.

The other two uses rely on glass in a fiberglass form. By itself, the

---

[7] Christian, W. J., "The Effect of Structural Characteristics on Dwelling Fire Fatalities," *Fire Journal*, Vol. 68, No. 1, January 1974.

fiberglass will not burn; however, in each case the fiberglass is used with resins which are usually combustible. Therefore, fiberglass insulation and many translucent window panels, siding materials, and prefabricated bathroom units are each flammable as a result of other materials within them.

2.   *Gypsum.* One of the most prevalent building products due to its use in wallboard, gypsum is also an excellent fire protection material. Within the gypsum is a large amount of chemically-combined water which will absorb a great deal of heat as it evaporates. It is a valuable, relatively inexpensive material.

3.   *Asbestos.* This is one of the best and most widely used materials for fire protection of structural members and other parts of the building. However, because asbestos fibers can cause a health hazard, their use has come under serious question and has been drastically curtailed. When mixed with cement, asbestos can be used in such products as asbestos insulation board and asbestos wood. These products have excellent fire resistance and protection characteristics. When mixed with portland cement, asbestos forms excellent fire resistance products which, unfortunately, have been known to shatter during the temperature rise of a fire, thus reducing their effectiveness.

4.   *Masonry.* Brick and concrete masonry products behave well when subjected to the elevated temperatures of a fire. Hollow concrete blocks may crack from the heat, but they generally retain their integrity. Brick can withstand high temperatures without severe damage.[8]

5.   *Plastics.* Clearly, plastics are among the most widely used products in construction. Because of their many different formulations and uses, as well as the many different shapes they can assume, plastics are likely to continue to grow in their application. From a firesafety point of view, however, all plastics must be considered combustible. While various chemical additives can retard their combustible nature, there is presently no known additive that will make plastics noncombustible. In addition, the toxicity of the products of combustion from certain kinds of plastics during a fire has recently

---

[8] McKinnon, Gordon, et al, *Fire Protection Handbook*, 15th ed., National Fire Protection Association, Quincy, MA, 1981, page 5-85.

come under great scrutiny. However, there is currently no universally acceptable test for toxicity, so it is impossible at this time to quantify the toxic nature of plastic building materials. In any event, because the plastic material is combustible, it will contribute to toxic smoke and gases, even if the precise nature and quantity of that contribution cannot definitely be determined.

6. *Lightweight Concrete.* An extremely common material in commercial and industrial structures, lightweight concrete has excellent fire resistance characteristics when made with noncombustible aggregates. The most common of these aggregates are vermiculite and perlite — types of crushed rock which, when mixed with the other material making up the concrete, give it the ability to resist high temperature without degradation.

§9.5 Summary

Almost every litigation involving a fire will involve a building. In order to fully handle that case, the attorney should examine every aspect of the building from its placement and design to the materials that make up its decorations. State and local building codes will control many aspects of the basic structure and should be examined thoroughly. Specific fire codes, especially those of the National Fire Protection Association, may be important whether or not they are actually written into local codes or ordinances.

Different types of construction have different requirements. Therefore, different standards must be applied to determine the presence or lack of negligence. Because of the detailed requirements of each construction type, structural engineers or other qualified experts may be necessary for testimony or pre-trial consultation on what should have been present on a particular occasion.

Each facet of building construction has a role to play in the firesafety of a building. Under a particular set of circumstances, the attorney must ask if that role was fulfilled. Standards for fire resistance and flame spread often exist; they must not only be discovered, but must also be fully understood. How particular fire ratings were established may prove fertile ground for exploration with a qualified expert.

Materials used in and around the point of fire origin and development should be thoroughly examined. Were the materials capable of retarding the spread of fire or did they enhance it? Was the use of a particular material an operative cause of ignition or spread? If so, was it negligent to have included that material in that area in the way it was designed?

The principles of fire ignition, development, spread, and extinguishment are the basis for any understanding of structural reaction to fire. A structure can be designed so as to minimize fire hazards. Whether the structure involved in a particular case was so designed should be an underlying consideration of fire litigation. Qualified experts may be necessary to provide knowledge about this question sufficient for full presentation, but most of the basic considerations have been considered in this chapter.

CHAPTER 10

LIFE SAFETY CONSIDERATIONS

§10.1 Introduction to the Life Safety Code®

For well over a half century, the National Fire Protection Association has been the publisher of the Life Safety Code. Formerly known as the Building Exits Code, the Code is prepared by the NFPA Committee on Safety to Life, one of the more than 175 committees operating within the framework of the NFPA standards-making activities. The Safety to Life Committee is made up of a group of well-qualified individuals who have demonstrated competence in the design and construction of buildings and structures, in the manufacturing and testing of a variety of building components and accessories, and in the enforcement of regulations pertaining to life safety from fire and other perils encountered in buildings and structures.

The Life Safety Code is a unique document; its contents address themselves specifically to requirements that have a direct influence on safety to life in both new and existing structures. Observance of Code requirements provides ancillary benefit to property construction through strengthened protection systems that provide faster response time during emergencies and reduce the spread of smoke and fire.

What impact application of the Code may have on saving lives is difficult to measure; however, it is reasonable to assume that its influence is significant. For example, of the many fatal fires investigated by the NFPA, invariably, one or more of the features contributing to loss of life from fire was in violation of the requirements of the Code.

The Code has gone through many revisions over the years. As the Building Exits Code, first issued in 1927, new editions were issued in 1929, 1934, 1936, 1938, 1939, 1942, and 1946 to incorporate amendments.

The Cocoanut Grove Nightclub fire in Boston in 1942 in which 492 lives were lost focused national attention upon the importance of adequate exits and related firesafety features. Public attention to exit matters was further stimulated by the series of hotel fires in 1946 (the LaSalle, Chicago — 61 dead; the Canfield, Dubuque, Iowa — 19 dead; and the Winecoff, Atlanta — 119 dead).

The Building Exits Code thereafter was used to an increasing extent for legal regulatory purposes. However, the Code was not in suitable form for adoption into law, as it had been drafted as a reference document containing many advisory provisions useful to designers of buildings, but not appropriate for legal use. This led to a decision by the Committee to re-edit the entire Code, limiting the body of the text to requirements suitable for mandatory application and placing advisory and explanatory material in notes.

This re-editing also involved adding provisions on many features to the Code in order to produce a complete document. Preliminary work was carried on concurrently with development of the 1948, 1949, 1951, and 1952 editions. The results were incorporated in the 1956 edition, and further refined in subsequent editions dated 1957, 1958, 1959, 1960, 1961, and 1963.

In 1955, separate documents, NFPA 101B and NFPA 101C, were published on nursing homes and interior finish, respectively. NFPA 101C was revised in 1956. These publications have since been withdrawn and incorporated into NFPA 101, Life Safety Code.

In 1963, the Safety to Life Committee was reconstructed. The Committee was decreased in size to include only those having very broad knowledge in fire matters and representing all interested factions. The Committee served as a review and correlating committee for seven Sectional Committees whose personnel included members having special knowledge and interest in various portions of the Code.

Under this revised structure, the Sectional Committees through the Safety to Life Committee prepared the 1966 edition of the Code which was a complete revision of the 1963 edition. The Code title was changed from Building Exits Code to the Code for Life Safety from Fire in Buildings and Structures, the text was put in "code language" and all explanatory notes were placed in an appendix. The contents of the Code were arranged in the same general order as contents of model building codes because the Code is used primarily as a supplement to building codes.

The Code was then placed on a three-year revision schedule, with new editions adopted in 1967, 1970, 1973, and 1976.

In 1977, the Committee on Safety to Life was reorganized as a Technical Committee with an Executive Committee and eleven standing Subcommittees responsible for various chapters and sections.

The 1981 edition contained major editorial changes including reorganization within the occupancy chapters to make them parallel to each other and the splitting of requirements for new and existing buildings into separate chapters. New chapters on detention and correctional facilities were added as well as new requirements for atriums, apartments for the elderly, and ambulatory health care centers.

For any revisions, groups particularly concerned in developing the various sections of the Code are consulted. All Public Proposals are reviewed and these proposals, along with Committee Proposals, and the Committee's response to all proposals, are published by the NFPA for review by all concerned. Public Comments received on these proposals are discussed and adopted either by the Committee or at an association meeting where NFPA members have the opportunity to vote to amend the Code.

The next edition of the Code will be considered at the 1984 Fall Meeting of the NFPA, and published in 1985. The Code is presently revised and published every four years. Code revision is prompted by recent events and input from Code users and other interested parties.

The Committee on Safety to Life is broken down into many subdivisions. Under the Committee are the Executive Committee and Subcommittee on Administration plus the following Subcommittees: Assembly and Educational Occupancies; Communications and Extinguishing Systems; Detention and Correctional Occupancies; Fire Protection Features; Health Care Occupancies; Industrial, Storage and Miscellaneous Occupancies; Interior Finish, Furnishings and Decorations; Means of Egress; Mercantile and Business Occupancies; Residential Occupancies; and Tents, Grandstands and Air-Supported Structures.

§10.2   *Revision Process*

It is possible that in 1985, for example, recent hotel fires may have a significant effect on revisions for the next edition of the Code. The purpose of the Code is to establish minimum requirements that will provide a reasonable degree of safety from fire in buildings and structures. The Code endeavors to avoid requirements that might involve unreasonable hardships or unnecessary inconvenience or interference with the normal use and occupancy of a building, but insists upon compliance with a minimum standard for firesafety consistent with the public interest.

However, the Code does contain several fundamental requirements:

1.   To provide for adequate exits without dependence on any one single safeguard.

2   To ensure that construction is sufficient to provide structural integrity during a fire while occupants are exiting.

3.   To provide exits that have been designed to the size, shape, and nature of the occupancy.

4.   To ensure that the exits are clear, unobstructed, and unlocked.

5.   To ensure that the exits and routes of escape are clearly marked so that there is no confusion in reaching an exit.

6.   To provide adequate lighting.

7.   To ensure early warning of fire.

8.   To provide for back-up or redundant exit arrangements.

9.   To ensure the suitable enclosure of vertical openings.

10.   To make allowances for those design criteria that go beyond the Code provisions and are tailored to the normal use and needs of the occupancy in question.

In considering these requirements, the Code addresses both new and existing structures. The Committee is very careful when making requirements that might necessitate revisions to existing buildings, due in part to the cost of retrofitting the buildings. Any changes in the Code, including those affecting existing buildings, must be technically substantiated before they are approved.

If changes are made in the Code, updating is required in structures located in jurisdictions that have adopted the Code. This is different from a building code which will allow an existing building to comply just to the Code in effect at the time of construction. Thus, the "once in compliance, always in compliance" rule is not effective for this Code.

§10.3 Organization of the Life Safety Code

To understand and use the Code, it is important to see how the printed Code is organized.

The first seven chapters of the Code are known as the core chapters. They contain administrative definitions and fundamental requirements establishing minimum acceptable criteria for all types of occupancies. Chapters 8-30 are the occupancy chapters. Chapter 31 serves a unique purpose in the Life Safety Code. It specifies activities to complement the structural features mandated by the Code to ensure a minimum acceptable level of safety. The chapter also focuses on the things that people (occupants, owners, tenants and maintenance personnel) can do to assist the Code in achieving life safety. This allows for involvement of people in providing for their own safety and security.

Each chapter is broken down into sections. Each section number corresponds to the identical number in other chapters. For instance, to find out the specifications for the number of exits in an occupancy, you can check the 2.4 section of each chapter.

If there are special occupancy provisions to be considered, these are included in the chapter. For example, under health care the litigator will find that for ambulatory health care facilities, 12-6 (13-6 for an existing facility) is the correct section for special occupancies.

The Code has three appendices with which the litigator needs to be familiar. Appendix A is used for additional explanations of certain sections of the Code. When an asterisk (*) is used in the body of the Code, it symbolizes an entry in Appendix A. Presently Appendix B lists referenced publications that coincide with particular sections of the Code, but for the 1985 edition, a new Chapter 32 will list mandatory referenced publications. Appendix C deals only with health care occupancies. The appendix helps to assess the facility to see if it provides an overall level of life safety equivalent to that intended by the Code. This assessment is done with the knowledge that the facility is not in strict compliance with the Code. The forms and evaluations included help the facilities to determine if they have equivalent life safety when not in compliance with the Code.

The litigator should note that these appendices are *advisory* only. The only mandatory wording is within the Code itself. The Code also provides a cross-reference table that helps to compare editions and changes.

As previously mentioned, it is important to understand that the Code applies to both new construction and existing buildings. This concept is essential to understanding the intent of the Code, which is to achieve at least a minimal level of life safety in all structures and in all occupancies. To this end, there are provisions throughout the document that either specifically apply to existing buildings or that are specifically modified for existing buildings. This is an attempt to limit the retroactive provisions of the Code to existing buildings. If no special provisions are made for existing structures, then the common provisions for new and existing construction apply, an approach quite different from that of a building code. This applies to the entire text; where life safety is involved, there is no reason for an existing structure not to conform to requirements for existing structures found in the Code.

Since only the individuals involved can properly assess extenuating circumstances, should they exist, the authority having jurisdiction is recognized as the best judge of determining to what extent the Code may be further modified. However, the Code will establish the bottom limit so that reasonable life safety against the hazards of fire, explosion, and panic is provided and maintained.

Because the Code can only reflect what is known and commonly practiced in fire protection design, there must be a device to recognize major breakthroughs or even simple changes in existing hardware that create new design capabilities.

This recognition of the limitations of the Code permits the use of "systems, methods, or devices of equivalent or superior quality, strength, fire resistance, effectiveness, durability, and safety." The stipulation is that the technical documentation be submitted to the authority having jurisdiction, justifying the use of the new approach.

This equivalency concept is important because it gives the Code a broader perspective than that of a specification code which only concerns itself with the construction details of individual items.

Chapter 3 of the Code includes any definitions with special meanings used in the Code. It is important for the litigator to understand what the terms of the Code mean in order to apply the Code correctly.

If a question arises concerning a word not defined in Chapter 3 of the Code, the litigator should apply its ordinarily accepted dictionary meaning.

One particular phrase used throughout the Code is "authority having jurisdiction." This phrase is used in NFPA documents in a broad manner since jurisdictions and "approval" agencies vary as do their responsibilities. Where public safety is primary, the authority having jurisdiction may be a federal, state, local, or other regional department or individual such as a fire chief, fire marshal, chief of a fire prevention bureau, labor department, health department, building official, electrical inspector, or others having statutory authority.

There may be more than one authority having jurisdiction for a specific occupancy. For instance, a health care occupancy may have four or five different authorities having jurisdiction, including the State Health Board, the municipal building authority, the Joint Commission on Accreditation of Hospitals, the U.S. Department of Health and Human Services, and the building insurer.

§10.4 Means of Egress

The real substance of the Code emerges in Chapter 5, Means of Egress. This part of the Code reverts to the origins of the Building Exits Code.

A means of egress is defined as a continuous, unobstructed way of exit travel from any building area to a public way. The public way is an important end to the means of egress since the Code requires complete exit to a street, alley, or parcel of land, open to the outside air, deeded to the public, with clear width and height of at least 10 feet.

A means of egress comprises the vertical and horizontal ways of travel and includes intervening room spaces, doorways, hallways, corridors, passageways, balconies, ramps, stairs, enclosures, lobbies, escalators, horizontal exits, courts, and yards.

From every location in a building, there must be a means of egress or path of travel over which a person can move to gain access to the outside, or gain access to a place of safety and refuge should the need arise.

Any person who gains entrance to a building (means of entry) has available that same route by which to exit. One important consideration makes exiting something more than just reversing one's route of entry, especially if emergency conditions exist. This reverse route, or any other route chosen for exiting, may present to the person leaving the building features which, though they were not obstacles upon entrance, prove to be such upon exit. For example, a door hinged to swing in the direction of entry becomes an obstacle when one attempts to leave the building in the opposite direction. The door swings against the flow of traffic — a flow that in an emergency situation is greatly increased as compared with the leisurely flow of people entering a building. In such an instance, time is limited and the way of entry may become blocked by fire. The path of travel must be one easily traversable and recognizable.

There are three components to a means of egress: the exit access, the exit, and the exit discharge.

The *exit access* is that portion of the means of egress which leads to an entrance to an exit. The exit access can include the room or space in a building in which a person is located and the aisles, ramps, passageways, corridors, and doors that must be traversed on the way to an exit. In general, no special protection is required for the exit access beyond that which is required for corridor separation and that which is normally required by building regulations for compartmentation within the building. An example of an exit access is shown in Figure 10-1.

The *exit* is that portion of a means of egress which is separated from all other spaces of the building or structure by construction or

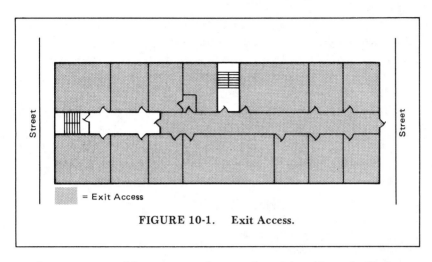

FIGURE 10-1.   Exit Access.

equipment to provide a protected way of travel to the exit discharge. The exit may include parts of corridors, stairs, smokeproof towers, escalators, outside balconies, ramps, and doors. In each case, the exit component must conform to the specifications for fire protection and the maximum or minimum dimensions established by the Code. In its simplest form, the exit is simply a door leading to the outside. An example of this is found in a schoolroom having a door or doors opening directly to the outside.

An example of an exit is shown in Figure 10-2.

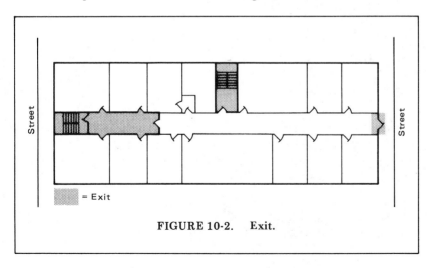

FIGURE 10-2.   Exit.

The *exit discharge* is that portion of a means of egress between the termination of an exit and a public way. An example of an exit discharge is out an exit door, traveling along a paved walk parallel to the building, then to a public sidewalk or street.

Where an exit stair opens onto an alley, court or yard of size insufficient to accommodate the people using the exit, a safe passageway must be provided to a public way or some other area adequate in size. Such a passageway would constitute a part of the exit discharge.

An example of an exit discharge is shown in Figure 10-3.

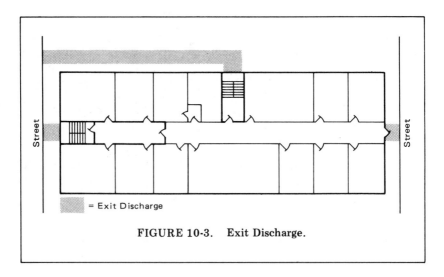

FIGURE 10-3.   Exit Discharge.

§10.4.1   Capacity of Means of Egress

The normal occupancy load is not necessarily a suitable criterion for determining the size of the means of egress, as the greatest hazard may occur when an unusually large crowd is present. This is a condition often difficult for the authority having jurisdiction to control by regulatory measures. The principle of this Code is to provide exits for the maximum probable number of occupants, rather than attempt to limit the number of occupants to a figure commensurate with available exits. There are, however, limits of occupancy specified in certain special cases for other reasons.

The number of occupants for which the egress system must be designed differs on a square foot basis with each occupancy. For example, a restaurant would use 15 sq ft per person while an office setting would use 100 sq ft per person.

The geometry of a building, its occupancy and related occupant load, and the travel distance to exits dictate in large measure the appropriate locations of exits, the number of exits, and the access thereto.

As a consequence, the exits themselves profoundly influence the plan and layout of the entire system of the means of egress. The ability of the means of egress to accommodate certain volumes of people is proportionately related to the ability of each component within it to accommodate a certain volume of people.

The number of people for which the means of egress will provide a path of travel must be determined first. This number should be based on the maximum number of occupants that can be anticipated to be in the building rooms or spaces at any one given time, under all occasions or unusual circumstances; it must not be based only upon the usual number for normal occupancy.

§10.4.2   Exit Widths

Means of egress is measured in units of exit width of 22 inches (55.88 cm). The measurement of exit width in terms of units representing the width occupied by one person, rather than measurement in feet and inches, is an important concept of the Code.

Each unit of exit width is given credit for being able to accommodate a designated number of people, depending on whether the travel surface is level or inclined. The exit is considered to be independent from the exit access and the exit discharge, because it is at this point — the entrance to the exit — that the known occupant load of a floor first comes together. Up to that point, the occupants have come from different directions, using different paths of travel, and have been increasing in number as they move along the path;

thus, the character of the access is not the same from end to end as for an exit. The same can be said of the exit discharge.

Width of the means of egress is measured in the clear at the narrowest point of the exit component.

Any projection, radiator, pipe or other object that extends into a corridor, irrespective of width, is undesirable, particularly where large crowds must be accommodated. In the case of a narrow stairway or passage, this feeling of restricted space might be conducive to panic under fire conditions.

The capacity of the means of egress differs with the various widths of the components.

Each unit of exit width is credited with being able to accommodate a designated number of reasonably alert and healthy people, the number being dependent on whether they are moving along a level path, a ramp, or on the steps of a stairway.

Each building has its own particular design needs which must be given consideration. Consequently, different solutions are needed to provide a safe and workable means of egress.

Every effort is made to write provisions that are flexible and, as a consequence, the Code provides unlimited possibilities to challenge the initiative of the designer in seeking economical exiting consistent with the safety principles contained therein.

§10.4.3   Travel Distances

Building codes generally, and the Life Safety Code in particular, specify the maximum distance that a person should have to travel from that person's position in a building to the nearest exit.

Travel distance is measured on the floor or other walking surface along the center line of the natural path of travel, starting one foot from the most remote point, curving around any corners or obstructions with a one-foot clearance, and ending at the center of the doorway or other point at which the exit begins.

There is no formula by which this distance can be established. However, there are factors on which the Committee on Safety to Life bases maximum travel distances.

1. The number, age, and physical condition of building occupants and the rate at which they can be expected to move.

2. The type and number of obstructions — display cases, seating, heavy machinery, etc. — that must be circumvented.

3. The number of people in any room or space and the distance from the farthest point in that room to the door.

4. The amount and nature of combustibles expected in a particular occupancy.

5. The rapidity with which a fire might spread — a function of the type of construction, the materials used, the degree of compartmentation, and the presence or absence of automatic fire detection and extinguishing systems.

While moving along any means of egress, many different components of the building or structure are encountered — components that go to make up the features of the means of egress. These components are items such as the doors, stairs, ramps, horizontal exits, exit passageways, hardware, handrails, balconies, etc. Their composition, properties, use, limits, and bearing relative to the total means of egress should be understood.

Portable ladders, rope ladders, and similar devices are not recognized by the Code as providing any portion of the required capacity of a means of egress. Neither should they be considered as in any way upgrading an inadequate means of egress in an existing building where the means of egress system is below minimum. While on occasion such devices have been used and have provided a way to safety, they most definitely should not be relied upon and may only lead to a false sense of security. The likelihood of their being unusable by small children, older people, and the physically handicapped, or those who simply have not used them before, is a possibility.

The components making up a means of egress must meet certain standards, must be built in a prescribed manner, and must perform at a level specified by the Code, all this depending upon whether they are part of the access to an exit, the exit itself, or the exit discharge. In some instances, the requirements will be the same throughout the means of egress.

Doors are probably the most common exit component. Doors serve three purposes related to the comfort and safety of building occupants.

They provide protection from weather, drafts, noise, and disturbance from adjoining areas; deter trespassing by unauthorized persons; and provide protection from as well as control and restriction of fire and smoke, the concern of this Code.

The term "door" within the use of the Code means door assembly — including the doorway, frame, door, and any necessary hardware.

Ideally, all doors in a means of egress should swing in the direction of exit travel. The Code requires that those doors which enter into the exit itself or are themselves an exit, such as a door leading directly to the outside, must swing in the direction of exit travel.

The importance of this requirement comes to light when one considers the potential tragedy in a nightclub or other place of assembly if the doors did not swing in the exit direction. People would pile up against the doors, unable to escape.

In 1981, the Committee addressed the joint need for security and life safety by recognizing a delay-release lock. The lock is used only where specifically permitted by the occupancy chapters. When in use, under nonemergency conditions, a person could come up against a door for 15 to 30 seconds before the door opened. If a building alarm sounds, the lock releases, allowing easy access to the exit.

§10.4.4   Stairs and Other Egress Components

Stairs, whether interior or exterior, serve three functions: they are a means of communication between the floors and different levels of a building; they serve as an emergency exit in case of fire; and they are essential for the rescue and fire control operations conducted by fire fighters.

The Code recognizes fire escape stairs, but only in certain circumstances and then in existing buildings only.

A horizontal exit is a way of passage from one building to an area of refuge in another building on approximately the same level. It may also be a way of passage through or around a wall or partition to an area of refuge on approximately the same level in the same building that affords safety from fire or smoke from the area of incidence and areas communicating therewith.

Horizontal exits should not be confused with egress through doors in smoke partitions. Doors in smoke partitions are designed only for temporary protection against smoke, whereas horizontal exits provide protection against serious fire for a relatively long period of time. Horizontal exits are used to a great extent in hospital settings where patients can be put into an area of refuge away from fire.

Although horizontal exits can often provide as fast and as safe a means of reaching an area of refuge as any other exit, they cannot be given credit for providing more than one half the required exit capacity of the building or buildings connected, except in health care occupancies — where they are permitted to provide two thirds of the exit capacity — and in detention and correctional occupancies.

Ramps are permitted as a part of a means of egress and, in fact, under some circumstances are to be preferred over stairs. The Code issues special requirements for ramps.

Exit passageways also serve as a horizontal means of exit travel and are protected from fire in a manner similar to an enclosed interior exit stair where it is desired to offset exit stairs in a multi-story building. An exit passageway can be used to preserve the continuity of the protected exit by connecting the bottom of one stair to the top of another stair that will continue to the exit discharge.

Probably the most important use of an exit passageway is to satisfy the requirement that exit stairs discharge directly to the outside. Thus, if it is impractical to locate the stairs on an exterior wall, an exit passageway can be connected to the bottom of the stairs to lead the occupants safely to an outside exit door. In buildings with an extremely large area, such as shopping malls and some factories, the exit passageway can be used to advantage where the travel distance to reach an exit would otherwise be excessive.

Exit passageways are different from access aisles, corridors, and hallways in that the exit passageway is required to be protected by a fire resistant enclosure with a rating equal to that required for exit stairwells.

The fact that the word "exit" is used in the expression "exit passageway" signals that it is not just any passageway; it is an area providing the same level of protection and safety that is required of any exit. Protection requirements for exit passageways are the same as those for stairs.

§10.4.5   Illumination of Means of Egress

When fire occurs in a building, the degree of visibility in corridors, stairs, and passageways may mean the difference between orderly evacuation and panic, and possibly between life and death. A brief glance at the history of fires reveals several significant fires in which the failure of normal or emergency lighting was a major factor in the casualties that were incurred.

These include:

Iroquois Theater, Chicago, 1903 ... 602 died

Cocoanut Grove Nightclub, Boston, 1942 ... 492 died

Arundel Park Hall, Baltimore, 1956 ... 11 died

Apartment House, Boston, 1971 ... 8 died

Illumination of the means of egress is provided for every building and structure when required in the occupancy chapters. Emergency lighting facilities must also comply with the occupancy chapters. This system is arranged so as to provide the required illumination automatically in the event of any interruption of normal lighting. Any type of illumination should be designed in accordance with the provisions of NFPA 70, National Electrical Code.

§10.4.6   Exit Marking

Where required by the provisions of the occupancy chapters, exits should be marked by an approved sign readily visible from any direction of exit access.

In the case of a main entrance also serving as an exit, the exit location will usually be sufficiently obvious to occupants and no exit sign is needed.

The character of the occupancy has a practical effect upon the need for signs. In any assembly occupancy, hotel, department store, or other building subject to transient occupancy, the need for signs will be greater than in a building subject to permanent or semipermanent occupancy.

There are many situations where the actual need for signs may be debatable. In case of doubt, however, it is best to provide signs; putting up the signs does not ordinarily involve major expense or inconvenience.

While the Code specifically does not mention color, red is the traditional color for exit signs and is required by law in many jurisdictions. The location of exit signs is not specified. Usually they are placed over the exit entrance or near the ceiling. There are those who argue, with reason, that smoke builds up more rapidly at higher levels and exit signs a foot or so above the floor would be visible for a much longer period in a fire situation. However, when several people are moving toward an exit, those in the back might not be able to see signs located at a low level. Also, in the absence of careful housekeeping, such signs might be damaged or blocked. Thus, the Code simply states that exit signs be located where they are readily visible and provide contrast with the surroundings.

§10.5   Features of Fire Protection

The Code in chapter 6 establishes basic requirements for features of fire protection which include such items as protection of vertical openings, fire barriers, smoke barriers, construction, protection of concealed spaces, protection from hazards, and interior finish. For the most part, chapter 6 sets requirements which are then referenced by the applicable Occupancy chapters. However, some requirements apply to all occupancies.

Lack of compartmentation and rapid fire development have been found to be primary factors in numerous multiple fatality fires, especially in residential occupancies. The Code attempts to preserve these compartments or "safe areas" through the use of wall, ceiling, and floor construction that will be reasonably free from penetration by fire and smoke for a safe period. This construction incorporates fire barriers that prevent the passage of heat and flame for a designated period of time.

These fire barriers, classified by their fire resistance ratings, are used to provide enclosure of floor openings or are used for subdivision of stories. These ratings are determined by the test method described in NFPA 251, Standard Methods of Fire Tests of Building Construction and Materials.

Floor openings, such as stairways and shaftways used for elevators,

311

lights, ventilation, or building services, shall be enclosed with fire barriers (vertical) such as wall or partition assemblies. These enclosures should be continuous from floor to floor and openings shall be protected as appropriate for the fire resistance rating of the fire barrier.

Protection of vertical openings is of extreme importance in order to reduce fire casualties. In an NFPA analysis of 500 fires involving fatalities, the vertical spread of fire in buildings is listed as the responsible factor in 250 incidents (50 percent of the cases).

Due to a growing architectural trend of the use of atriums, the 1981 Code first incorporated a section on the fire protection features necessary to provide a degree of safety to occupants of an atrium building. Because the atrium is a large, open space where smoke can dissipate, there exists a great need to ensure that dangerous concentrations of smoke are promptly removed from the atrium.

§10.5.1   Smoke Barriers

In the use of the Code, smoke barriers are provided to subdivide building spaces for the purpose of restricting the movement of smoke. These barriers are continuous from outside wall to outside wall, from a fire barrier to a fire barrier, from a floor to a floor, from a smoke barrier to a smoke barrier, or a combination thereof, including continuity through all the concealed spaces such as those found above a ceiling, including interstitial spaces.

Doors in smoke partitions, while not the equivalent of fire doors and not completely smoketight, are effective in restricting the spread of smoke and reducing drafts which might otherwise spread fire rapidly. It is important to maintain the integrity of smoke barriers over the life of a building.

§10.5.2   Interior Finish

The faster a fire develops, the greater the threat it represents to the occupants of a building and the more difficult it is to control. The wall and ceiling surfaces of a building have a major influence on how fast a fire develops, and it is important that these interior finishes meet certain standards.

In establishing these restrictions for the use of interior finish materials, the intention is to slow the early growth of fire and to limit the speed at which flame will travel across the interior surfaces of a building. These restrictions are also imposed on interior finish materials to minimize the contribution of "fuel" to the early growth of a fire and to limit the generation of smoke.

While the Code stipulates several forms of interior finish, the authority having jurisdiction is allowed to exercise judgment in determining what constitutes interior finish. For example, a tapestry would not normally be construed as interior finish, but one which is secured to and covers a major portion of a wall could promote the rapid growth of fire and may warrant regulation.

There are various tests that can be applied on interior finishes under the Code, and these should be checked when questioning the use and rating of the particular materials. NFPA 255, Test of Surface Burning Characteristics of Building Materials, is one such test to be applied.

§10.6 Building Service and Fire Protection Equipment

The Code is designed to be used with other codes and standards that provide design guidance for building service equipment. The provisions of these various codes and standards must be met in order for the facility to comply with the Life Safety Code. Any references made to these standards or codes in the main body of the Code is meant to reinforce the fact that compliance with these codes is mandatory.

Among the important items of fire protection equipment installed in a building are the detection, alarm, and communication systems. These systems are primarily intended to provide the indication and warning of abnormal conditions, the summoning of appropriate aid, and the control of occupancy facilities to enhance protection of life.

The provision for early warning of fire, accompanied by notification of appropriate authorities, is a key element of a fire protection program. Where people are involved, protective signaling assumes even greater importance. Fire detection without warning is not enough. People must be alerted to the existence of an emergency in order to take appropriate action.

Different occupancies require different types of signaling and alarm systems, and it is best to check the occupancy chapter for specifics. A health care facility, for instance, may require a more elaborate signaling system than an office building.

§10.6.1   Automatic Sprinklers and Other Extinguishing Equipment

Once a fire is under way, the most effective safeguard against loss of life is through the use of automatic sprinklers. Their value is psychological as well as physical, in that they give a sense of security to the occupants of a building and tend to minimize the possibility of panic in case of fire. There is no case, in the NFPA records of more than 100,000 fires in sprinklered buildings, in which water from automatic sprinklers in any way contributed to panic.

Requirements in the Code for automatic sprinklers have been carefully based on the sprinkler experience record, and show that a sprinkler system is the most effective device, when installed and maintained properly, for protecting and safeguarding against loss of life and property.

Claims that water in a sprinkler system can be heated to scalding temperatures and literally sprayed onto occupants, causing injury and panic, should be discounted. Other misconceptions about sprinkler systems, including occupants of a room being drowned, electrocuted or scalded by steam, have not been shown by fire record data to exist and should also be discounted.

NFPA 13, Standard for the Installation of Sprinkler Systems, is the so-called "bible" for sprinkler systems insofar as design, installation, and character and adequacy of water supply are concerned.

NFPA 13A, Recommended Practice for the Care and Maintenance of Sprinkler Systems, gives detailed information on maintenance procedures.

This Code also recognizes the fact that some quantity of smoke may be produced before fire is extinguished by automatic sprinklers, and that any smoke may create a panic hazard even though there may be no actual danger.

When buildings are equipped with sprinklers, there are usually Code provisions permitting an increase in travel distance to exits, and some possible reduction in fire resistance requirements, although not below a specified level.

§10.7   Summary

As mentioned earlier, Chapters 8-30 of the Code are the occupancy chapters. Most occupancies are discussed in two chapters each — one for new structures and one for existing structures. The specifics given are unique to the occupancy type, and are used to modify and supplement core chapters.

The user of the Code should keep in mind that each occupancy should be dealt with as a unique structure. For instance, health care occupancies will require smoke barriers rated differently from those used in educational occupancies. Individual chapters outline details for individual occupancies.

The last chapter of the Code, Chapter 31, discusses day-to-day operations that affect the level of life safety in various occupancies.

Prevention of "human or mechanical failure" puts the Code into the areas of care, operation, maintenance, and inspection of equipment. Further, since life safety depends upon proper human actions at the time of an emergency, drills and checklists of what is to be done at the time of the event become critical.

Chapter 31 of the Code also covers the aspects of a building's contents which have a bearing on life safety. These factors are covered in general for all occupancies, and are then reinforced for each specific occupancy within the chapter.

Finally, Chapter 31 focuses on the things that people — occupants, owners, tenants, and maintenance personnel — can do to complement the Code in achieving life safety. This allows for involvement of people in providing for their own safety and security, and includes exit drills and the locking up of combustible liquids.

A code suitable for enforcement purposes must be concise and without explanatory text. And, too, a code cannot be written to cover every situation that will be encountered; thus, it must be applied with judgment and an awareness of the rationale for the requirements that must be enforced. To meet this end, the Committee and NFPA have put together the *Life Safety Code Handbook*.

The Life Safety Code Handbook provides background information on the reasons for certain Code provisions. It also gives suggestions, through text and illustrations, on how some Code requirements can be implemented intelligently. But this commentary is only a combination of personal views of the Committee on Safety to Life, and should only be taken as such.

The Life Safety Code Handbook is especially helpful in explaining changes and additions to NFPA 101, and is revised and published immediately after each new edition of the Code.

CHAPTER 11

BUILDING OPERATIONS

§11.1   Electrical Systems

If properly designed, installed and maintained, electrical systems are both convenient and safe; otherwise they may be a source of both fire and personal injury. Electricity may cause a fire if it arcs or overheats electrical equipment; it may cause injury or death through shocks and burns.

To assist litigators in more fully understanding and appreciating the workings of an electrical system, this section describes the components and particular functions of an electrical system used in a building.

The basic circuit usually consists of five components: a power source; electrical conductors; overcurrent protection devices; some form of control device; and the load which will utilize the electrical current.

The power may be supplied by batteries, generators, or a public utility. The conductors may be made of copper or aluminum in the form of a cable, busbar or wire installed in a raceway system. A raceway is a tube for carrying and protecting electrical wires, cables or busbars. The conductors may also be in the form of copper or aluminum busbars in a busway. A busbar is a conductor forming a common junction between two or more electrical circuits. A busway is a grounded metal enclosure containing factory mounted, bare or insulated conductors.

The overcurrent protection device can be in the form of fuses or circuit breakers, the function of which is to open the circuit should an excessive amount of current, above which the conductors are rated, flow through the conductor. An excessive amount of current could cause overheating of electrical equipment.

A controlling device may be a switch, thermostat, float switch, motor controller, or any other type of device designed to control the energy delivered to the equipment utilizing the electricity.

The load is the piece of equipment utilized to convert the electrical pressure and energy that move the electrons through the wire into a form which will be of use. Such forms include lighting, heating, motors, or the mechanical motion used in devices such as elevators, washing machines and so forth.

Electrical energy is conducted from the public utilities through overhead type conductors. Overhead conductors are simply conductors that run overhead. The service drop picks up electrical energy at the public utility's transformer or power lines and runs to the building being served. The service drop includes overhead conductors from the last pole or aerial support to the service-entrance conductor at the building or other structure. There, the conductors, either in the form of cables or open conductors, are secured to the building with the proper clearance from the ground. The service drop conductors are supply conductors that extend from the street main or from transformers to the service equipment of the premises. Usually at this point the service entrance conductors are connected to the service drop conductors.

Where the service conductors are run underground, they are referred to as the service lateral. The service lateral consists of the underground service conductors, which run between the transformer or public utility's power lines to a junction box or meter located either inside or outside the building. From the junction box or meter, the conductors will terminate at a disconnect switch or will connect to the service entrance conductors. Under certain conditions, underground service installations may not have a service entrance conductor, because the service lateral may terminate directly at a disconnect switch.

In an overhead service system, the service entrance conductors are spliced to the service drop conductors, usually at the point of attachment on the building. The service entrance conductors then are run outside the building either by a cable or within a raceway system. In most dwelling installations, the service entrance conductors pass through a kilowatt-hour meter located either outside or inside the building. From the meter, they travel a short distance and terminate at a disconnect switch or distribution panel. Because the service entrance conductors are not provided with overcurrent protection, they are required to terminate at a disconnect switch in a readily accessible location near the point of entrance to the building.

These service conductors, in effect, now terminate in the service equipment. This equipment usually consists of a main disconnecting means — a distribution panel, or, in the case of larger installations, a switchboard. Located within the distribution panel or switchboard are the means for disconnecting the power as well as the overcurrent protection devices. These devices may be in the form of fuses or circuit breakers. This equipment must be accessible to the occupant of the building, except in the case of large multifamily residential structures, industrial buildings, or commercial installations. In such facilities, the service equipment is permitted to be behind a locked door when there is continuous building management supervision. This equipment will then be accessible to authorized personnel, such as an electrician or maintenance person familiar with the operation of the electrical system. The service equipment provides a main control and means of shutting off the electrical energy entering the premises.

Voltages usually associated with residential installations consist of 240/120 volts. In commercial or industrial installations, the voltage may consist of 208/120 volts, 480/277 volts, 240 volts, and 550 volts among others.

Another important area associated with the service equipment is the grounding electrode system. System and circuit conductors are intentionally grounded in order to limit the voltages due to lightning, line surges, or accidental contact of high voltage lines with a low voltage system. Systems and circuit conductors are solidly grounded in order to facilitate the operation of overcurrent protection devices should a ground fault occur.

A grounding electrode system may consist of a metal underground water pipe, metal frame of a building, a concrete-encased electrode consisting of one or more steel reinforcing bars or a bare copper conductor, or a ground ring encircling the building or structure in the form of a bare copper conductor. When grounding electrodes are not available on the premises, other types of electrodes must be used. Alternatives include electrodes of pipe or conduit, rods of steel or iron, copper-clad steel rods, or steel or copper plates buried in the ground.

NFPA 907M, Manual on the Investigation of Fires of Electrical Origin, outlines possible causes of electrical fires based on various fact patterns. When investigating a fire of this type, questions can and should be asked in order to correctly determine the cause. Was electrical power available in the area determined to be the fire's point of origin? Was any electrical equipment energized or operating at the time?

Inspection of the equipment for electrical damage may reveal melting, pitting, or beading of metal, and even melting of nonmetallic components. Equipment which may be fire damaged and which may have been the cause of the fire can often be distinguished by the severity of the heating and melting of the components in the equipment. Was the electrical power accidentally or intentionally made available at the scene? Were electrical wires or other conductive materials placed at the location intentionally, or did the fire occur accidentally, perhaps due to a failure in the wiring system?

Was the wiring of a temporary nature, installed during construction or remodeling and never removed, but still energized? Was the electrical power from a conventional supply system, such as a public utility or an on-site generating system used during the normal operations, or from an emergency or alternate supply system?

Many types of electrical equipment produce sparks under normal operating conditions. The litigator must be familiar with such electrical equipment pertinent to the case which produces arcing under normal operating conditions.

Switches used for the control of lighting, heating, motors, etc., produce arcing in the normal function for which they are utilized. Arcing usually occurs in a switch when it opens the ciruit of a system supplying equipment that uses components in the form of coiled conductors such as motors, transformers, and some heating elements. There can be a slight amount of arcing when the switch is closed; this is due primarily to contact bounce.

Normal arcing will be present at the point of contact between the brushes and commutator of a universal type motor. These motors are most commonly found in portable hand tools and appliances such as electric drills, electric saws, vacuum cleaners, blenders, and hair blow dryers among others. These motors are called universal motors because they will operate on alternating current as well as direct current. This fact is indicated on the nameplate of the appliance.

Other alternating-current, single-phase motors which may display arcing when starting up are of the split-phase type. These motors utilize a centrifugal switch which opens and disconnects the starting winding when the motor has reached a predetermined speed. Examples of this type are motors for washing machines, swimming pool pumps, and attic exhaust fans.

Normal arcing also occurs in overcurrent devices, such as fuses and circuit breakers. The arcing which takes place when the overcurrent device performs as designed is contained within the housing of the fuse or circuit breaker. Under normal conditions, there is no expulsion of hot metal particles and there is no fire hazard.

Arcing may also occur when a plug is removed from an electric receptacle where the equipment utilized is drawing current; for example, when the control switch is in the closed ("ON") position.

Abnormal arcing is usually indicated by certain damage which is detectable by careful observation. Some of the clues indicative of abnormal arcing are as follows:

- Broken or cracked electrical insulators.
- Carbon tracks on the insulator.
- Holes melted through metal enclosures.
- Hot spots or discoloration near electrical terminals or conductors.
- Melting or beading between two or more conductors.
- Melting or beading between conductors and metal raceways or enclosures.
- Overcurrent devices charred, melted or blown apart. (An explosion in an overcurrent device can occur when the available energy is greater than the interrupting current capacity of the fuse or circuit breaker.)
- Underground fault conditions, evidence of melted metal or splattering between couplings or connectors and the raceway and junction boxes.

Another avenue to pursue is improper installation or use of electrical equipment, such as:

- Overcurrent devices bypassed or disabled, either intentionally for arson purposes or because of the ignorance of the persons tampering with the devices. Overcurrent devices can be easily bypassed by the arsonist, who primarily intends to create an overload of the electrical system which will result in fire.

- Often, due to a lack of knowledge of the results of bypassing overcurrent devices, an individual will bypass a fuse as a "temporary" measure, but "temporary" sometimes ends up as permanent.

Weak links in a wiring system can be the splice and termination points. Overheating and damage of terminations, splices and the associated conductors and equipment occurs under the following conditions:

- Loose connections such as improper torque not in accordance with installation requirements.
- Terminals or splicing devices not suitable for the type of materials.
- Use of terminals too small for the size of the wire, where a number of strands are removed from the conductor to accommodate the terminal. This often happens when the installer uses a terminal designed for only one conductor, but wants to place two or three conductors in the terminating device; in such cases, some of the strands are removed and a potential hot spot is inadvertently incorporated in the system.
- Most terminals on overcurrent devices and receptacles utilize the heat-sink characteristics of the conductor to remove heat from the terminal.

Some splicing devices require application of an insulating material, such as varnished cambric, rubber, plastic, or friction tape. Such insulation must be applied to provide the insulating properties equivalent to the insulation on the conductors. Insulating material must also be suitable for the conditions of the environment, such as moisture, vibration, and temperature.

Splices must be contained within junction boxes or enclosures large enough to hold the conductors contained within them based upon the appropriate National Electrical Code tables. Where splices are made and jammed into too small an enclosure, the insulation on the splice or conductor may be damaged. Quite often a tapped splice is squeezed into the junction box and forced into the enclosure by the cover. The constant pressure of the splice or conductor against the cover will gradually cause the insulation to creep away from the pressure point, thereby reducing the thickness of the insulation. Eventually, the voltage stress exceeds the dielectric break-down voltage and a ground fault occurs, arcing and burning through the metal enclosure. The same thing can occur between two splices forced together when they are of different polarity.

The proper installation of the electrical conductor is another area requiring investigation. Circuit wiring is sized in accordance with the size of the load to be served, the required voltage, environmental conditions, and ambient temperature. The current drawn by the load must not exceed the ampacity of the conductor. Ampacity is based on the gauge of the wire and the temperature rating of the insulation.

Where the ambient temperature exceeds 30°C (86°F), correction factors must be applied that reduce the amount of current a conductor may safely carry without exceeding the temperature limitations of its insulation. Where the number of conductors in a raceway or cable exceeds three, or where the load operates continuously for three hours or more, the amount of current for which the conductor is rated in the ampacity tables must be reduced. (Refer to Article 310 of the National Electrical Code.)

The major enemy of the conductor and its insulation is heat. Damage to conductors during installation is hard to determine after the fire incident. Where the conductors short-circuit or ground out, such extensive damage occurs that it is hard to determine if the insulation was damaged before or during installation.

§11.1.1 Raceways and Cables

In an installation consisting of a metallic raceway, metallic cable or flexible metal conduit, failure of cable box connectors or locknuts to be properly secured or tightened can result in arcing at the loose connection, heating of the metal cables, and preventing proper operation of the overcurrent device.

Tight bends of small radii can damage insulation of the enclosed conductors and can also damage the armor. This can cause abrasion or cut into the conductor insulation, resulting in short circuits or ground faults.

The litigator should be familiar with the different types of fittings which are suitable for the proper connection of cables or raceways to an enclosure (box or cabinet). Use of the wrong type of fitting can result in damage to the cable, poor grounding continuity, and damage to conductor insulation.

Improperly supported or secured raceways and cables may result in damage to the raceway or cable, loss of grounding continuity, and cutting or abrasion of the conductor insulation. Lack of proper support may also result in the cable or raceway being too close to a deteriorating environment, such as heat, chemicals, or vibration.

§11.1.2 Lighting Fixtures

Due to the design of certain lighting fixtures, it is very easy to create fire hazards by misuse or by violating manufacturer's instructions. One of the major violations is overlamping — the use of light bulbs with a higher wattage than that recommended by the manufacturer or testing laboratory.

Overlamping may cause overheating of the lighting fixture and surrounding materials. Examples include the cardboard insulating shell which fits between the lampholder and the metal outer shell, and the Bakelite or plastic composition shell of the nonmetallic lampholder, as well as pyrolysis of the structure on which the lighting fixture is mounted.

Another area of concern is the overheating of the electrical branch circuit conductors and splices. Overheating causes deterioration of the conductor insulation, which has a certain temperature rating. When it is in an environment that exposes it to temperatures above that rating, the insulation loses its insulating qualities and may start to carbonize.

Depending upon the voltage, a small amount of current can leak through the insulation between conductors of different polarities. It also might leak between the hot conductor and a grounded material such as a light fixture (metallic), a junction box, metal raceways or metal-sheathed-type cables. The carbonized material is capable of conducting current, becoming red hot, and finally causing ignition of combustibles in the vicinity, including the conductor insulation.

Heat also is the worst enemy of terminations, connections and splices. Oxidation of the conductors occurs, increasing the contact resistance of the connections. This in turn increases the voltage drop at those locations, due to the heat generated by the current being forced through the resistance. The amount of heat generated in terms of watts is the current squared multiplied by the resistance.

The problem of overheating is further amplified by an improper initial installation, where the electrician had difficulty inserting the mounting screws for the canopy and discarded the fiberglass insulation pad. This insulation pad prevents excessive temperatures in the fixture canopy and protects the contents under the canopy from heating to excess of design values. Other problems of overheating include switches and receptacles with ratings insufficient for the loads, loose connections at terminals, and stranded conductors used for push-in, back-wired-type devices where solid conductors are required.

§11.2 Codes, Standards, and Guidelines

There are several reference sources for a litigator pursuing an electrical fire case. Some major references are listed below.

§11.2.1 The National Electrical Code (NEC) (NFPA/ANSI 70)

The NEC provides for the practical safeguarding of persons and structures and their contents from the hazards arising from the use of electricity. The NEC was first issued under its present name in 1897. It is revised every three years by the NEC Committee under the aegis of the National Fire Protection Association (NFPA), and combines experience and judgment of all parties interested in the safe use of electricity.

§11.2.2 Manual on the Investigation of Fires of Electrical Origin (NFPA 907M)

This manual, developed and published by NFPA, provides guidelines for examining possible causes of fires of electrical origin.

§11.2.3   National Electrical Safety Code (ANSI C2)

As interest in electrical safety increased in the United States, a need was felt for a code to cover the practices of public utilities and others when installing and maintaining overhead and underground electrical supply and communication lines. The National Electrical Safety Code was completed in 1916. Currently this code is published by the Institute of Electrical and Electronic Engineers (IEEE).

§11.2.4   Canadian Electrical Code

The Canadian Standards Association currently sponsors and publishes the Canadian Electrical Code. This Code, which establishes essential requirements and minimum standards for the installation and maintenance of electrical equipment, is offered for adoption and enforcement by electrical inspection departments throughout Canada. Its several parts are the Canadian equivalent of the NEC, the National Electrical Safety Code, and the standards of Underwriters Laboratories in the United States.

§11.2.5   The National Electrical Code Handbook

The NEC Handbook contains the entire text of the NEC supplemented by comments, diagrams, and illustrations that are intended to explain and clarify some of the intricate requirements of the NEC. The Handbook is published by the NFPA on a three-year cycle to coincide with the production of each edition of the NEC.

§11.2.6   Other NFPA Standards and Codes

Other pertinent NFPA standards and codes include the following: NFPA 56A, Standard for the Use of Inhalation Anesthetics (Flammable and Nonflammable); NFPA 70A, Electrical Code for One- and Two-Family Dwellings; NFPA 70B, Recommended Practice for Electrical Equipment Maintenance; NFPA 70E, Standard for Electrical Safety Requirements for Employee Workplaces; NFPA 75, Standard for the Protection of Electronic Computer/Data

Processing Equipment; NFPA 99, Standard for Health Care Facilities, Chapter 8; NFPA 79, Electrical Standard for Metalworking Machine Tools and Plastics Machinery; NFPA 493, Standard for Intrinsically Safe Apparatus and Associated Apparatus for Use in Class I, II, and III, Division I Hazardous Locations; and NFPA 496, Standard for Purged and Pressurized Enclosures for Electrical Equipment in Hazardous (Classified) Locations.

Three publications are available that list electrical appliances and materials that have been examined by UL. Devices and materials listed have been tested for actual field use in accordance with the NEC and against UL safety standards. These lists are published so that the names of the manufacturers of devices that meet test criteria may be readily obtainable.

The three publications are the *Hazardous Location Equipment Directory*, the *Electrical Appliance and Utilization Equipment Directory*, and the *Electrical Construction Materials Directory*. They are published annually by Underwriters Laboratories, with a supplement six months after the directory is issued. Certain types of electrical equipment are covered in other UL directories, e.g., the *Marine Products Directory*.

§11.3   Heating Systems and Related Appliances

Heat-producing systems and their related appliances are often cited as the sources of ignition. Because these systems require high temperatures generated either by fire or electrical resistance, because they may involve the use and storage of flammable fuels, and because their purpose is to heat a living environment, they are worth checking in any fire situation where the ignition source is subject to question. Heating systems have potential fire hazards that are unique among building operating systems.

Industrial furnaces use fire and high levels of heat for various manufacturing and finishing processes. Such use is beyond the scope of this book. Consideration here is focused on the general characteristics of the system used to heat the living environment of a building. If there is interest in the industrial use of fire, reference should be made to NFPA 86A, B, C, and D, standards for various types of industrial furnaces. Boiler furnaces, including firing by pulverized coal, oil and gas, are addressed in NFPA 85A, B, D, E, F, and G, providing the energy input is 12.5 million Btu per hour and above.

Because heating systems are made for the very purpose of using fire and high temperatures, ignition caused by a heating system usually results from the design, installation or misuse of the central burning unit itself; from a peripheral system, such as fuel storage; or from one of the attendant appliances, such as duct work heating vents.

§11.3.1   Fuels

In order for combustion to occur and maintain itself, there must be an adequate supply of fuel and air. These supplies of fuel and air must be continuous and controllable if a heating system is to function properly. Incomplete combustion — the failure of the air to fully oxidate the available fuel — may result from an improper fuel-air mixture, an insufficient quantity of air, or a temperature too low for complete combustion. Incomplete combustion is one of the principal producers of lethal carbon monoxide. Because of the distribution characteristics of a heating system, carbon monoxide can be spread throughout the building.

Air used by the heating system is usually referred to as primary or secondary. Primary air is air that is introduced through or with the fuel; secondary air is air otherwise supplied to the combustion zone.

Fuel for a heating system is usually wood, coal or some other solid fuel, natural gas or fuel oil. Wood is by far the most plentiful of the solid fuels and has had a great resurgence in popularity in recent years. A full discussion of the nature and burning characteristics of wood can be found in section 15 of Chapter 7 of this book.

Coal is the second most widely used solid fuel. The coal generally used for heating is one of four basic types: anthracite, bituminous, subbituminous and lignite. These four principal classifications and subclasses beneath them are based on the heating value and the amount of fixed carbon in the coal. The heating value is based on Btus. Anthracite has the highest amount of carbon, as much as 98 percent. Lignite has the least carbon content and the lowest heating value. The particular classifications along with some basic characteristics are shown in Table 11-1.

TABLE 11-1 Classification of Coals by Rank[A]

Class	Group	Fixed Carbon Limits, percent (Dry, Mineral-Matter-Free Basis)		Volatile Matter Limits, percent (Dry, Mineral-Matter-Free Basis)		Calorific Value Limits, Btu per pound (Moist,[B] Mineral-Matter-Free Basis)		Agglomerating Character
		Equal or Greater Than	Less Than	Greater Than	Equal or Less Than	Equal or Greater Than	Less Than	
I. Anthracitic	1. Meta-anthracite	98	...	...	2	...	...	nonagglomerating
	2. Anthracite	92	98	2	8	...	...	
	3. Semianthracite[C]	86	92	8	14	...	...	
II. Bituminous	1. Low volatile bituminous coal	78	86	14	22	...	...	
	2. Medium volatile bituminous coal	69	78	22	31	...	...	
	3. High volatile A bituminous coal	...	69	31	...	14,000[D]	...	commonly agglomerating[E]
	4. High volatile B bituminous coal	...	...	...	...	13,000[D]	14,000	
	5. High volatile C bituminous coal	...	...	...	...	11,500	13,000	
						10,500	11,500	agglomerating
III. Subbituminous	1. Subbituminous A coal	...	...	...	...	10,500	11,500	nonagglomerating
	2. Subbituminous B coal	...	...	...	...	9,500	10,500	
	3. Subbituminous C coal	...	...	...	...	8,300	9,500	
IV. Lignitic	1. Lignite A	...	...	...	...	6,300	8,300	nonagglomerating
	2. Lignite B	...	...	...	...	...	6,300	

[A] This classification does not include a few coals, principally nonbanded varieties, which have unusual physical and chemical properties and which come within the limits of fixed carbon or calorific value of the high-volatile bituminous and subbituminous ranks. All of these coals either contain less than 48% dry, mineral-matter-free carbon or have more than 15,500 moist, mineral-matter-free British thermal units per pound.

[B] Moist refers to coal containing its natural inherent moisture but not including visible water on the surface of the coal.

[C] If agglomerating, classify in low-volatile group of the bituminous class.

[D] Coals having 69% or more fixed carbon on the dry, mineral-matter-free basis shall be classified according to fixed carbon, regardless of calorific value.

[E] It is recognized that there may be nonagglomerating varieties in the groups of the bituminous class, and that there are notable exceptions in high volatile C bituminous group.

Reprinted, with permission, from the Annual Book of ASTM Standards. Copyright, American Society for Testing and Materials, 1916 Race Street, Philadelphia, PA 19103. ASTM D 388.

331

Coke is another product which is often used for the same purpose as coal. Coke is not a natural coal, rather it is a product of the destructive distillation of coal and is used in many manufacturing processes.

In order to utilize coal in a heating system, the coal has to be fed continuously into the burning chamber. There are three methods by which this is generally done — by hand; by mechanical methods that feed the coal from either above or below the point of combustion; or by pulverzing the coal before delivery into the furnace by conveyor. NFPA 85E, Standard for Prevention of Furnace Explosions in Pulverized Coal-Fired Multiple Burner Boiler Furnaces, and NFPA 85F, Standard for the Installation and Operation of Pulverized Fuel Systems, deal specifically with the particular combustion problems raised by pulverized coal systems. In any nonhand-powered fuel firing system, careful investigation of the storage, transportation, and final delivery system should be made. The possibility of spontaneous combustion and ignition at any stage in the process is something which must be investigated. The designs of hoppers, conveyors of the flat or screw type, and draft systems hold numerous potentials for unvented ignition.

When coal is burned in a firebox, oxygen in the air passing through the grate (primary air) unites with the carbon in the lower portion of the fuel bed, called the oxidation zone, to form carbon dioxide. Some of this carbon dioxide is then reduced to carbon monoxide in the upper part of the fuel bed, called the reduction zone. Gases liberated from the fuel bed are carbon monoxide, carbon dioxide, nitrogen, and some oxygen. Oxygen from the air admitted over the fuel bed (secondary air) combines with some of the carbon monoxide to form carbon dioxide. When fresh coal is applied to the fire, moisture is driven off as steam and the hydrocarbon gases are distilled, combined with oxygen, and burned above the grate in what is called the distillation zone.

Fuel oil is the most common type of liquid fuel for heating systems. In Chapter 8 the various types and grades of fuel oil and their properties were discussed. Reference should be made to Chapter 8 for a better understanding of the fuel oils.

Burning fuel oil is generally done by one of two methods: vaporization or atomization. In either case, air is supplied by either a natural or a mechanical draft. The way in which air is supplied is a characteristic which should be examined in any investigation. Both types of burners are basically methods by which the liquid fuel can be changed into a vapor or mist, and combine with air and ignite in the combustion chamber to provide the accompanying heat. Because the delivery of the fuel and air can be regulated to control the intensity of the combustion, the degree of heat also can be regulated. Vaporizing burners generally introduce fuel into a chamber below the point of ignition. They allow the heat of the combustion above the chamber to vaporize the fuel, then mix with air as the vapor rises to the point of continuous ignition. Pot-type and sleeve-type burners fall into this category. Atomizing burners use a pump that draws oil from a supply tank. Compressed air delivers the fuel under pressure to the nozzle at the combustion chamber. The fuel oil is sprayed as a fine mist which is ignited in the combustion chamber. This is the principle behind the popular gun-type burners which are in wide use.

Another method of classifying burners is based on their use, either residential or commercial-industrial. The residential type are those which have capacities of not more than 7 gph (gallons per hour) and are intended to be used with fuels not heavier than No. 2 fuel oil. Commercial-industrial type burners have larger capacities and generally burn heavier and less expensive grades of fuel.

There are few definitive and restrictive rules for the design of fuel oil burners. The basic underlying principle is for the fuel to be brought to the ignition point in a way that allows for easy ignition. Sufficient primary and secondary air to support ignition must be specifically introduced or allowed to enter the system prior to ignition.

Another widely used fuel is gas, which in most cases is either LP (liquefied petroleum) or natural gas. LP-gas has proven popular largely because of its portability and ease of use in many applications. Natural gas is popular because of its low comparative cost, nationwide transportation system through underground pipelines, and the fact that it burns cleanly leaving few noxious products of combustion, such as soot or odor.

The most common types of gas burners are the injection type, often known as the Bunsen burner; the luminous or yellow-flame burner; the catalytic burner; and the power burner. In the Bunsen burner the gas is injected into the air prior to combustion, while in the luminous burner the only air for burning is externally supplied at the point of combustion. The catalytic burner allows combustion at temperatures lower than normal, thus it has value in burning off unused gases which might not have fully burned. The power burner is similar to a Bunsen burner in that the air-gas mixture is supplied under pressure, thus increasing the intensity of the combustion.

Some gas systems use propane, propylene, butane, butylene or a mixture of these. These systems are not nearly as common as the LP and natural gas systems.

The necessary pre-mixing of air and fuel is very common among many of the different furnace systems. Explosions and implosions are among the chief hazards encountered in furnace systems. Chapter 7 discussed the basis of explosions. Implosions are negative pressures within the burner itself which may cause the walls of the burner to collapse. Three NFPA standards, 85B, 85D, and 85E, deal respectively with the prevention of furnace explosions in natural gas-fired, oil-fired, and pulverized coal-fired multiple burner boiler-furnaces. In the litigation of any explosion, these standards may prove applicable.

§11.3.2   Storage of Fuels

Due to the high potential for fire ignition, the storage of heating fuels (coal, oil, and gas) is an important point which should be researched by any litigator. However, the area of storage is broad and it is not really part of the building operating system. Reference should be made to other books for a better understanding.

Two applicable NFPA standards deal with the storage of fuel oil: NFPA 30, the Flammable and Combustible Liquids Code, and NFPA 31, the Standard for the Installation of Oil Burning Equipment. NFPA 30 covers, among other topics, tank storage, piping, valves and fittings, and container storage. NFPA 31, as the title indicates, deals with general requirements for heating equipment and accessories, and with storage of fuel oil as it directly pertains to the heating equipment.

Storage of gas is generally covered by NFPA 58, Standard for the

Storage and Handling of Liquefied Petroleum Gases; NFPA 59, Standard for the Storage and Handling of Liquefied Petroleum Gases at Utility Gas Plants; and NFPA 59A, Standard for the Production, Storage and Handling of Liquefied Natural Gas (LNG). Also, a major reference in this area is NFPA 54, the National Fuel Gas Code. In addition to these are standards developed by the American Petroleum Institute, the American Society for Testing and Materials, and the American Society of Mechanical Engineers. Numerous articles in the technical literature deal with precise questions concerning fuel storage and describe fire incidents involving storage or use of fuel.

§11.3.3 Heating Appliances

The heating of a building is usually accomplished by a central heating system along with attached appliances. There are, however, various types of individual room or area heaters. This section describes some of the more common heaters, including central system types and unit or space heaters.

In many central heating systems, boilers are used to heat water which is then circulated in order to radiate heat. Usually these are either low-pressure boilers that operate at pressures not exceeding 15 psi, or hot-water boilers that operate at greater pressures and temperatures not exceeding 250° F. Industrial and commercial boilers are often used for purposes other than strictly environmental heating, such as in manufacturing processes. The American Society of Mechanical Engineers' Boiler and Pressure Vessel Code contains requirements for construction and installation of industrial and commercial boilers. Fire problems from boilers usually result from their mountings or clearances from combustibles.

Warm-air furnaces are in use in many residences. These systems heat air and use fans, blowers and a duct system to circulate the heated air through the building. Circulation is sometimes left to gravity but it is usually boosted with a fan system. The duct work can cause a potential fire problem due to the buildup of heat. Other fire problems are usually based on clearances to combustibles surrounding the central heating unit or maintenance failures of the central unit or ducts. For residential use, automatic controls are necessary to limit the buildup of heat.

Three other kinds of furnaces are attached to a central fuel supply system. These are the pipeless furnace, the floor furnace, and the wall furnace. These are used in areas where heating is needed but where the entire building is too small to justify a central system. One pipeless furnace below the floor level is sufficient for a small structure. An area of a larger structure which has specific heating requirements can be served by a floor furnace or wall furnace. Control is a key problem with these types of furnaces because they should shut down when adequate heat is attained. Also, because these furnaces are often used for quick heating of a small area, individuals have created problems by using them to dry flammable clothing.

A fourth type of heater attached to the central fuel supply system is the duct furnace. These are small furnaces installed directly in ducts of warm air heating systems. Duct furnaces may be oil- or gas-fired, or, in some cases, electrical. They are used to increase the temperature of the air which might have cooled during circulation from the central warm air furnace.

Not all heating appliances are connected to a central fuel system. Free-standing area heaters are referred to as unit heaters when used for nonresidential purposes and room or space heaters when used in residences. Such heaters are usually straightforward in design, containing a heating source that is either gas- or oil-fired. With unit heaters, a fan or blower is used to spread the heated air. Unit heaters are either suspended from a wall or ceiling or floor-mounted. Room heaters are generally floor-mounted and circulate heated air by radiation. In either case, the venting of gases must be an integral part of the design since combustion takes place within the confines of the casing of the heater. In any case involving these types of heaters, careful attention should be given to the provisions for venting gases. Careful attention should also be given to the internal system, if any, to make sure it will shut down in case of overheating.

Residential room heaters have become popular in many areas, although there are some particular problems associated with them. There is also significant literature on room heaters which should be carefully reviewed in any litigation. Among the problems are improper installation, improper venting, construction too close to combustibles, insufficient insulation from flooring, improper duct work or connection to flues and chimneys, improper handling of ashes, use of the wrong fuel, creosote accumulation which can lead to fires in chimneys, and overfiring. Because room heaters are manufactured products, they can be tested, listed, and supplied with clear installation and operation instructions. In any case involving a residential room heater, the compliance with all manufacturer's instructions should be carefully checked, along with the full details of the actual installation. Flooring, clearance from combustibles, venting, and connections to chimneys must be checked. Table 11-2 gives basic requirements for floor mountings.

§11.3.4 Distribution

Every central heating system distributes the heated air or water through some configuration of pipes, ducts, or plenums. These may also serve the opposite purpose of returning to the central heating unit the secondary air needed for combustion. Major fire protection problems associated with pipes, ducts, or plenums involve the potential for these items to become overheated and ignite nearby combustibles.

Some buildings, particularly single-story residences lacking basements, may use a crawl space under the floor as a plenum for a supply of air. The ducts in the crawl space distribute the heated air to each room through registers cut in the floor or lower wall. In these cases, the use of combustible material for a plenum and the storage of combustible materials in the crawl space can be sources of fire.

Another hazard which should be investigated in any fire involving heating systems is the spread of fire. Heating ducts and plenums often spread through a large area of a building without any compartmentalization, providing easy travel for flames or smoke. Sophisticated systems include closing systems to prevent this spread of fire or the products of combustion, but residential properties are much less likely to be so protected.

TABLE 11-2   Floor Mountings for Heat-Producing Appliances

Type of Mounting	Required for the Following Types of Heaters and Furnaces
No Floor Protection: Combustible floors.*	Residential-type central furnaces so arranged that the fan chamber occupies the entire area beneath the firing chamber and forms a well-ventilated air space of not less than 18 in. in height between the firing chamber and the floor, with at least one metal baffle between the firing chamber and the floor. Low heat appliances in which flame and hot gases do not come in contact with the base, on legs which provide not less than 18 in. open space under the base of the appliance, with at least one sheet metal baffle between any burners and the floor. Other appliances for which there is evidence that they are designed for safe operation when installed on combustible floors.
Metal: A sheet of metal not less than No. 24 U.S. Gage, or other approved noncombustible material, laid over a combustible wood floor.*	Heating and cooking appliances set on legs or simulated legs which provide not less than 4 in. open space under the base. Ordinary residential stoves with legs. Residential ranges with legs. Residential room heaters with legs. Water heaters with legs. Laundry stoves with legs. Room heaters with legs.†

TABLE 11-2   Floor Mountings for Heat-Producing Appliances (Cont.)

Hollow Masonry:

   Hollow masonry not less than 4 in. in thickness laid with ends unsealed and joints matched in such a way as to provide free circulation of air through the masonry.

Downflow furnaces.

Heating furnaces and boilers in which flame and hot gases do not come in contact with the base:

   Floor mounted heating and cooking appliances.
   Residential stoves without legs.
   Residential ranges without legs.
   Room heaters without legs.
   Water heaters without legs.
   Laundry stoves without legs.
   Residential-type incinerators.
   Restaurant ranges on 4-in. legs.
   Other low heat appliances on 4-in. legs.
   Medium heat appliances on legs which provide not less than 24 in. open space under the base.

Hollow Masonry and Metal:

   Hollow masonry not less than 4 in. in thickness covered with a sheet of metal not less than No. 24 U.S. Gage, laid over a combustible floor. The masonry will be laid with ends unsealed and joints matched in such a way as to provide a free circulation of air from side to side through the masonry.

Two Courses Masonry and Plate:

   Two courses of 4-in. hollow clay tile covered with steel plate not less than 3/16 in. in thickness, laid over a combustible floor. The courses of tile will be laid at right angles with ends unsealed and joints matched in such a way as to provide a free circulation of air through the masonry courses.

Heating furnaces and boilers in which flame and hot gases come in contact with the base.
   Restaurant ranges.
   Other low heat appliances.

339

TABLE 11-2   Floor Mountings for Heat-Producing Appliances (Cont.)

Type of Mounting	Required for the Following Types of Heaters and Furnaces
Fire-Resistive Floors, Extending 6 in.: Floors of fire-resistive construction with noncombustible flooring and surface finish and with no combustible material against the underside thereof, or on fire-resistive slabs or arches having no combustible material against the underside thereof. Such construction will extend not less than 6 in. beyond the appliance on all sides, and where solid fuel is used, it will extend not less than 18 in. at the front or side where ashes are removed.	Floor mounted heating and cooking appliances. Residential-type room heaters. Residential-type water heaters.
Fire-Resistive Floors Extending 12 in.: Floors of fire-resistive construction with noncombustible flooring and surface finish and with no combustible material against the underside thereof, or on fire-resistive slabs or arches having no combustible material against the underside thereof. Such construction will extend not less than 12 in. beyond the appliance on all sides, and where solid fuel is used, it will extend not less than 18 in. at the front or side where ashes are removed.	Heating furnaces or boilers. Restaurant-type cooking appliances. Residential-type incinerators. Other low heat appliances.

TABLE 11-2   Floor Mountings for Heat-Producing Appliances (Cont.)

Fire-Resistive Floors Extending 3 ft:

Floors of fire-resistive construction with noncombustible flooring and surface finish and with no combustible material against the underside thereof, or on fire-resistive slabs or arches having no combustible material against the underside thereof. Such construction will extend not less than 3 ft beyond the appliance on all sides, and where solid fuel is used, it will extend not less than 8 ft at the front or side where ashes are removed.

Medium heat appliances and furnaces.

Fire-Resistive Floors Extending 10 ft:

Floors of fire-resistive construction with noncombustible flooring and surface finish and with no combustible material against the underside thereof. Such construction will extend not less than 10 ft beyond the appliance on all sides, and where solid fuel is used, it will extend not less than 30 ft at the front or side where hot products are removed.

High heat appliances and furnaces.

* Where an appliance is mounted on a combustible floor, and solid fuel is used or the appliance is a domestic type incinerator, a sheet of ¼-in. asbestos covered by a sheet of metal not less than No. 24 U.S. Gage will be required extending at least 18 in. from the appliance on the front or side where ashes are removed. (The sheet of asbestos may be omitted where the protection required under the appliance is a sheet of metal only.) For residential type incinerators the protection must also extend at least 12 in. beyond all other sides. If the appliance is installed with clearance less than 6 in. the protection for the floor should be carried to the wall.

† Floor protection for radiating type gas burning room heaters which make use of metal, asbestos or ceramic material to direct radiation to the front of the device should extend at least 36 in. in front when the heater is not of a type approved for installation on a combustible floor.

NFPA 90A, Standard for the Installation of Air Conditioning and Ventilating Systems, contains requirements for systems in spaces of less than 25,000 square feet, while NFPA 90B, Standard for the Installation of Warm Air Heating and Air Conditioning Systems, contains requirements for systems installed in spaces of more than 25,000 square feet. Reference should be made to these standards for requirements of these two systems.

Hot water heating systems have the same basic installation requirements near combustibles as do warm air systems. Because the water in these systems must be maintained at significant pressure, it can cause a serious problem, though not a fire problem, if the pipe ruptures.

§11.3.5   Installation

In discussing the various facets of heating systems, the importance of installation techniques has been repeatedly mentioned. The two principal fire dangers associated with heating systems are the danger of explosion during the combustion reaction and the danger of radiated heat raising temperatures of adjacent combustible materials to dangerous levels. Proper installation of the components within the system will go far to prevent an explosion during ignition. The proper installation of the whole system is the principal safeguard against dangerous temperature increases in adjacent combustibles. The remainder of this chapter explores some of the general considerations regarding safe installations.

There is an initial guiding consideration which must be taken into account when reviewing a fire scene where improper installation is suspected as an ignition cause. Wood and other combustibles may ignite at temperatures far below their usual ignition temperatures if they are continuously exposed to relatively moderate heat over long periods of time. A second major consideration is that, over time, heating systems can radiate heat through thick layers of solid material, such as concrete or brick, so as to cause damage to attached wooden beams or flat boards. The comprehension of these two principles will help the litigator understand fires that result from heat generated by heating systems.

At least three factors should be considered in any installation of a heating system: proper clearance from combustibles; the use of insulating materials; and air flow. Only the first factor can be quantified with easy accuracy prior to installation, thus making improper attention to clearances easiest to document and prove. Table 11-3, based on field experience and laboratory tests, shows minimum acceptable clearances for various types of heating devices under various conditions.

Insulating materials may be used to retard the possibility of fire ignition. The insulation is either attached to the heated surfaces themselves or placed between the heat source and any combustible surface. Note in Table 11-2 that sheet metal or masonry insulation can be used to reduce the clearances which otherwise would be needed. However, as with all insulation, the possibility of heat building up and transferring over time to adjacent combustible surfaces must be considered. Therefore, proper consideration must be given to air flow which dissipates heat.

Whenever heat is radiated from a source, it can be dissipated to some extent by air flowing between the source and a combustible surface. If insulation is used to protect surfaces against which heat may be radiated, the insulation should be built or installed in such a way as to allow for air flow which can assist in carrying away the heat. There are no formulas that mathematically show adequate air supply for this type of protection. Thus a lack of adequacy is a more difficult factor to prove in any litigation because it involves more than the size of clearances. However, air flow should always be considered in any fire growing out of heat transferred from a heating system.

Maintenance also should be considered when examining required clearances. Due to the potential buildup of combustible products and the need to control the combustion reaction, heating systems require periodic cleaning and maintenance. If a heating system is installed so that normal maintenance is made unduly difficult, the opportunity for such maintenance will be retarded and, of course, the conditions for ignition will increase.

Installation of heating systems should be according to the manufacturer's instructions, or the requirements of a listing from a testing laboratory, as well as following the local codes. In any case where installation may be a factor, the requirements should be fully ascertained and understood with relation to the necessity for and the reasons behind them.

**TABLE 11-3 Standard Installation Clearances, Inches, for Heat-Producing Appliances
(See Note 1)**

These clearances apply unless otherwise shown on listed appliances. Appliances should not be installed in alcoves or closets unless so listed. For installation on combustible floors, see Note 2.

Residential Type Appliances For Installation In Rooms Which Are Large (See Note 3)		Appliance				
		Above Top of Casing or Appliance	From Top and Sides of Warm-Air Bonnet or Plenum	From Front See Note 4	From Back	From Sides
Boilers and Water Heaters						
Steam Boilers—15 psi	⎧ Automatic Oil or Comb. Gas-Oil	6	—	24	6	6
Water Boilers—250°F	Automatic Gas	6	—	18	6	6
Water Heaters—200°F	Electric	6	—	18	6	6
All Water Walled or Jacketed						
Furnaces—Central						
Gravity, Upflow, Downflow, Horizontal and Duct. Warm-Air—250°F Max.	⎧ Automatic Oil or Comb. Gas-Oil	6[5]	6[5]	24	6	6
	Automatic Gas	6[5]	6[5]	18	6	6
	Electric	6[5]	6[5]	18	6	6
Furnaces—Floor						
For Mounting in Combustible Floors	⎧ Automatic Oil or Comb. Gas-Oil	36	—	12	12	12
	Automatic Gas	36	—	12	12	12
	Electric	36	—	12	12	12

344

TABLE 11-3 Standard Installation Clearances, Inches, for Heat-Producing Appliances (Cont.)
(See Note 1)

Appliance	Type	1	1	1	1 (Firing Side)	1 (Opp. Side)
Heat Exchanger						
Steam—15 psi Max.		—				
Hot Water—250°F Max.						
Room Heaters						
Circulating Type	Oil	36	24	—	12	12
Vented or Unvented	Gas	36	24	—	12	12
Radiant or Other Type	Oil	36	36	—	36	36
Vented or Unvented	Gas	36	36	—	18	18
	Gas with double metal or ceramic back	36	36	—	12	18
Radiators						
Steam or Hot Water	Gas	36	6	—	6	6
Ranges—Cooking Stoves		See Note 6				
Vented or Unvented	Oil	30	9	—	24	18
	Gas	30	6	—	6	6
	Electric	30	6	—	6	6
Clothes Dryers		See Note 9				
Listed Types	Gas	6	24	—	6	6
	Electric	6	24	—	0	0
Incinerators		See Note 9				
Domestic Types	—	36	48	—	36	36

345

TABLE 11-3 Standard Installation Clearances, Inches, for Heat-Producing Appliances (Cont.)
(See Note 1)

These clearances apply unless otherwise shown on listed appliances. Appliances should not be installed in alcoves or closets unless so listed. For installation on combustible floors, see Note 2.

Low Heat Appliances Any and All Physical Sizes Except As Noted		Appliance				
		Above Top of Casing or Appliance See Note 7	From Top and Sides of Warm-Air Bonnet or Plenum	From Front	From Back See Note 7	From Sides See Note 7
Boilers and Water Heaters						
100 cu ft or less						
Any psi Steam	All Fuels	18	—	48	18	18
50 psi or less						
Any Size	All Fuels	18	—	48	18	18
Unit Heaters						
Floor Mounted or Suspended—Any Size	Steam or Hot Water	1	—	—	1	1
Suspended—100 cu ft or less	Oil or Comb. Gas-Oil	6	—	24	18	18
Suspended—100 cu ft or less	Gas	6	—	18	18	18
Suspended—Over 100 cu ft	All Fuels	18	—	48	18	18
Floor Mounted Any Size	All Fuels	18	—	48	18	18
Ranges—Restaurant Type						
Floor Mounted	All Fuels	48	—	48	18	18
Other Low-Heat Industrial Appliances						
Floor Mounted or Suspended	All Fuels	18	18	48	18	18

TABLE 11-3 Standard Installation Clearances, Inches, for Heat-Producing Appliances (Cont.)
(See Note 1)

Commercial-Industrial Type Medium-Heat Appliances	Appliance				
	Above Top of Casing or Appliance See Note 8	From Top and Sides of Warm-Air Bonnet or Plenum	From Front	From Back See Note 8	From Sides See Note 8
Boilers and Water Heaters					
Over 50 psi } All Fuels	48	—	96	36	36
Over 100 cu ft }					
Other Medium-Heat Industrial Appliances					
All Sizes All Fuels	48	36	96	36	36
Incinerators					
All Sizes —	48	—	96	36	36
High-Heat Industrial Appliances					
All Sizes All Fuels	180	—	360	120	120

NOTES TO TABLE

1. Standard clearances may be reduced by affording protection to combustible material.
2. An appliance may be mounted on a combustible floor if the appliance is listed for installation on a combustible floor, or if the floor is protected in an approved manner. For details of protection reference may be made to the Code for the Installation of Heat-Producing Appliances, obtainable from the American Insurance Association (NBFU), 85 John Street, New York, NY 10038, or Part 6 of the National Building Code of Canada published by the National Research Council, Ottawa, Ontario, Canada.
3. Rooms which are large in comparison to the size of the appliance are those having a volume equal to at least 12 times the total volume of a furnace and at least 16 times the total volume of a boiler. If the actual ceiling height of a room is greater than 8 ft, the volume of a room shall be figured on the basis of a ceiling height of 8 ft.
4. The minimum dimension should be that necessary for servicing the appliance including access for normal maintenance, care, tube removal, etc.
5. For a listed oil, combination gas-oil, gas, or electric furnace this dimension may be 2 in. if the furnace limit control cannot be set higher than 250°F or this dimension may be 1 in. if the limit control cannot be set higher than 200°F.
6. To combustible material or metal cabinets. If the underside of such combustible material or metal cabinet is protected with asbestos millboard at least ¼ in. thick covered with sheet metal of not less than No. 28 gage the distance may be not less than 24 in.
7. If the appliance is encased in brick, the 18 in. clearance above and at sides and rear may be reduced to not less than 12 in.
8. If the appliance is encased in brick the clearance above may be not less than 36 in. and at sides and rear may be not less than 18 in.
9. Clearance above the charging door should not be less than 48 in.

§11.3.6   Conclusion

Heating systems, because of their inherent use of rapid combustion which requires fuel and produces heat and gases, have a strong potential for involvement in the ignition of a fire in any building. The ducts and air flow of each system can easily spread combustion products. Radiated heat over long periods of time can cause combustible materials to ignite at temperatures well below their normal ignition levels. There are numerous standards as well as jurisdictional requirements governing most aspects of a heating system and its installation. In any case where the heating system could be a factor, the litigator should see that these various standards and requirements are researched, understood, and used. In addition, the use of often unquantifiable conditions of good practice should be addressed in any fully-developed theory of recovery or defense.

§11.4   Air Conditioning and Ventilating Systems

Air conditioning, as defined by the American Society of Heating, Refrigerating and Air Conditioning Engineers, is the process of treating air so as to control simultaneously its temperature, humidity, cleanliness, and distribution to meet the requirements of the conditioned space.

Invariably, the use of air conditioning and ventilating systems, except for self-contained units, involves some use of ducts for air distribution. Likewise, the use of ducts invariably presents the possibility of spreading fire, fire gases, and smoke throughout the building or area served.

NFPA 90A, Standard for the Installation of Air Conditioning and Ventilating Systems, details the safeguards necessary to protect against the hazards of duct systems.

There are several types of air conditioning systems, including systems in which air is filtered or washed, heated, and humidified in the cold weather, and systems which cool and dehumidify the air in warm weather. A ventilating system supplies or removes air by natural or mechanical means to or from a space. The air may or may not be conditioned.

Fresh air intake ducts connect directly to the system's return duct. From here, the mixture of fresh and recirculating air passes through the air handling equipment. Air is subjected to filtration or cleaning, heating or cooling, and humidification or dehumidification. Conditioned air is then circulated throughout the area served via the duct system in a continuous process.

Fans, heaters, filters, and associated equipment should be located in a room cut off from the rest of the building area by walls, floors, and a floor-ceiling assembly with a minimum fire-resistance rating of one hour.

This location setup prevents a fire involving the equipment from immediately spreading to adjacent areas of the building. It also prevents access to equipment by unauthorized persons.

These rooms are ideally protected by automatic sprinkler equipment. At the very least, smoke or heat detectors should be present and arranged to initiate an alarm and shutdown of the air handling system.

Location of the fresh air intakes is critical in a building since fire, fire gases, or smoke originating outside the building can easily be drawn in and spread throughout the duct system.

Protection can be provided by the installation of fire doors or dampers at the intakes, controlled by fire and smoke detectors.

The possibility of sparks or other products of combustion from an exposed fire chimney or incinerator stack also needs to be considered. Smoke rises, so in most cases the lower the intakes, the less possibility of drawing in smoke. Screens of corrosion-resistant material to prevent matter from entering the system are also recommended. After a fire, check to see if screens were in place.

There are two basic sources of hazards in air cooling and heating equipment. The first is the electrical equipment; the second is the possible toxicity situation from the refrigerant.

Fire experience is generally good where cooling equipment is properly installed and maintained. Installation of electrical equipment should follow the manufacturer's instructions and NFPA 70, National Electrical Code.

Refrigerant poses a possible combustibility problem as well as the toxicity potential, but the greatest hazard is from explosion due to pressurization of the refrigerant.

Several recommendations for safety measures can be found in the Safety Code for Mechanical Refrigeration issued by the American Society of Heating, Refrigerating and Air Conditioning Engineers.

Air filters and cleaners are also important parts of the duct system and their condition should be checked. These filters and cleaners fall into three general categories: the fibrous media unit filters, the renewable media filters, and the electronic air filters. The first two are true filters; the third is a static precipitator.

The potential hazard of these filters is related to their function of removing entrained dust and other particulate matter from the air stream. Filters that become loaded with such matter become combustible. Toxic smoke and other products of combustion can be circulated throughout the building by the air handling system, thus posing a direct threat to life safety. There also exists a possiblity that the filter media may be combustible or may be coated with a combustible adhesive. Adequate protection to minimize the possibility of fire in the filters must be designed into and around the air conditioning system.

In a duct system, smoke detectors must be located in the main supply duct, downstream of the filters or air cleaner, to sense smoke or other products of combustion.

Detectors are interlocked so that when they are activated, the entire air conditioning system automatically shuts down. The detectors also control the operation of smoke dampers located in the main supply return ducts, thus isolating the entire air conditioning section of the system.

Because of the system make-up, usually two types of portable fire extinguishers — dry chemical and those used for electrical fires — are on hand.

Ducts are to an air conditioning system what pipes are to a water system — a means of distribution. By their very nature, ducts provide an excellent means for transporting deadly smoke and other products of combustion instead of breathable air in a fire situation. Without proper design and installation precautions, smoke, other products of combustion, heat, and even flame can spread throughout the area served by the duct system.

The usual result in such a situation is panic among the occupants.

Exit corridors used as plenums, lack of smoke detector-activated control equipment in the system, and lack of adequate smoke dampers in appropriate walls, ceilings, or partitions can lead to tragic situations.

In addition, at some point a duct will probably pass through a wall, partition, floor or ceiling which is designed specifically to provide fire resistance. Literally, a hole has been poked through a fire-rated design. If the installation has been made without proper regard for firestopping, a serious fire resistance problem could result. The Life Safety Code discusses proper fire resistance and ratings.

§11.4.1 Smoke Control

An in-depth discussion of smoke control is included in NFPA 101, Life Safety Code, and NFPA 90A, Standard for the Installation of Air Conditioning and Ventilating Systems. Below is a brief discussion of passive and active smoke control.

Passive smoke control recognizes the long-standing compartmentation concept which requires that fans be shut down and fire and smoke dampers in duct work be closed on detection of fire conditions. Active approaches to smoke control utilize the building's heating, ventilating, and air conditioning (HVAC) systems to create differential pressures; this prevents smoke migration from the fire area, and exhausts the products of combustion to the outside. When the active approach is used, both NFPA 101 and 90A permit the absence of smoke and fire dampers.

Smoke dampers are required in the air conditioning or ventilating ducts which pass through required smoke barrier partitions. The dampers operate automatically upon detection of smoke and must function so that smoke movement through the duct is halted. Basically, smoke dampers are intended as an interruption of air flow through the duct. NFPA 90A, Standard for the Installation of Air Conditioning and Ventilating Systems, permits the use of fire dampers for smoke control purposes. Recent developments include a UL standard (555S) for leakage rated dampers for smoke control systems.

When used in ducts, smoke detectors must include design features specifically for this use. Smoke detectors may activate equipment to shut down the air conditioning system or ventilating system, sound alarms, operate smoke control dampers, activate fire suppression systems, and/or initiate active smoke control functions.

Location of smoke detectors in the ducts is primarily intended to prevent smoke circulation throughout the entire area served, by shutting down the air handling system. They also prevent the smoke from circulating in the air filters or air cleaners.

§11.4.2   Active Smoke Control

Active smoke control has been in the limelight for some time now, especially in high-rise buildings. Attempts to completely confine smoke and other products of combustion are seldom successful. Smoke movement in high-rise buildings and its threat to life safety has been the object of in-depth investigation since the tragedies at the MGM Grand Hotel and the Las Vegas Hilton Hotel.

Active smoke control utilizes air conditioning or ventilating systems for smoke control and removal in case of fire. Smoke control systems must maintain safe exit routes with sufficient exiting time for building occupants to either leave or move to designated safe refuge areas. A primary concern in a fire area is the exhaust (removal) of smoke. The intention of these smoke control systems is to provide smoke-free routes for exiting the building.

A fan unit itself is not an undue hazard if properly installed and firmly supported on a rigid foundation. Units installed in ducts will need protective devices to cut off power before temperatures reach a point where smoke may be generated.

All air conditioning/ventilating systems need manual shutoffs for use in case of fire or other emergencies.

The maintenance and cleaning of these systems are of the utmost importance to safe operation of any air conditioning or ventilating system. Filters should be changed or cleaned as frequently as necessary.

Ducts, particularly on the return side of the system, are cleaned out periodically to prevent hazardous accumulations of combustible dust and lint. Evidence of any form of wiring or electrical equipment defect must be checked out and rectified at once.

Periodic testing of all fire protection devices, including fire suppression equipment, smoke control and fire dampers, and alarms should be conducted. In addition, all vibrations should be checked. The litigator should try to learn what test method was used and whether or not it was a standard test, conducted regularly.

§11.4.3 Air-Moving Equipment

Air-moving equipment (AME) includes all mechanical draft duct systems of both the pressure and exhaust types for removal of dust, vapors, and waste material, and for the conveying of materials.

Hazards of such a system lie in the probability of ignition of flammable materials or vapors by sources such as sparks generated by fans or other solid foreign materials (rocks, metals, etc.) or by overheating of fan bearings. Then such systems could inadvertently spread fires through the building.

Air-moving equipment systems do play an important role in fire protection. Material that has not been removed could result in an explosion of vapor and air mixtures. A flash fire due to such things as the accumulation of lint and general poor housekeeping conditions could result.

To reduce the hazard of fire, fans should be of noncombustible construction, be able to be shut down by remote control in case of fire, and be accessible for maintenance. They should also be structurally sound enough to resist wear and overcome distortion and misalignment caused by structural weakness or overloading.

Details on the installation of open air-moving equipment are given in NFPA 91, Standard for the Installation of Blower and Exhaust Systems for Dust, Stock and Vapor Removal or Conveying; and NFPA 96, Standard for the Installation of Equipment for the Removal of Smoke and Grease-Laden Vapors from Commercial Cooking Equipment.

§11.4.4 Hazards of Specific Uses

Because flammable vapors ignite easily, precautions must be taken to eliminate ignition sources in ducts which carry them. Using a single system to exhaust flammable vapors and particles from spark-producing processes is obviously dangerous. It is important that vapors be withdrawn from the room or equipment in which they are generated and taken directly to the outside of the building.

The ventilation system should provide sufficient air movement to maintain the vapor concentration below the lower explosive limit in the area where vapors are being liberated. If vapors are toxic, it may be necessary to maintain concentrations well below those required for flammability.

§11.4.5   Ventilation   of   Corrosive   Vapors   and Fumes

If corrosive vapors must be exhausted, the degree of expected corrosion is the governing factor in the construction of the ducts. In some cases, a heavier-gauge metal is used, and in others protective coatings may be sufficient. Occasionally, however, the corrosiveness is so extreme that a special lining is required. Stainless steel and asbestos cement have been used successfully in some cases.

Automatic fire protection is also necessary at the hood, canopy, or intake of plastic duct systems. NFPA 91, Standard for the Installation of Blower and Exhaust Systems for Dust, Stock and Vapor Removal or Conveying, addresses the use of these plastic duct systems. Plastic lining may be used for vapors that are flammable as well as corrosive, provided that such lining has a flame spread rating of zero.

§11.4.6   Ventilation of Kitchen Cooking Equipment

Exhaust systems for restaurant equipment are troublesome because grease condenses inside the ducts. Grease accumulations may be ignited by sparks from the stove or, more often, by a small fire on the stove caused by overheating of cooking oil or fat in a deep-fat fryer or on a grill. If a duct did not accumulate grease, fires on stovetops could often be extinguished before causing appreciable damage. Fires occur frequently in frying because cooking oils and fats are heated to their flash points. Thus, oils and fats can reach their self-ignition temperatures by accidental overheating or can be ignited by spills on the stovetop.

Kitchen cooking equipment should have grease-removal devices and sufficient ducts to minimize grease accumulations on the equipment.

For sizable cooking installations, the fire extinguishing equipment should consist of an approved fixed pipe extinguishing system, supplemented by portable dry chemical extinguishers.

Where fixed extinguishing systems are installed, they should be located so they can be easily activated from the path of egress, and, when activated, automatically shut off all sources of fuel and heat to all equipment protected.

§11.4.7 Dust Collecting and Stock and Refuse Conveying Systems

These systems, which are discussed in NFPA 91, Standard for the Installation of Blower and Exhaust Systems for Dust, Stock and Vapor Removal or Conveying, consist of suction ducts and inlets, air-moving equipment, feeders, discharge ducts and outlets, collecting equipment, and vaults and other receptacles designed to collect powdered, ground, or finely divided material.

Collecting and conveying systems are often a very important part of the operations of a plant. When the material systems handle a combustible dust, there is danger of a fire or explosion. These factors make it a prime necessity that the design, construction, and operation of the systems conform to recognized standards.

§11.4.8 Separating and Collecting Equipment

Separating and collecting equipment is designed to withstand anticipated explosion pressures, allowance being made for explosion relief vents.

This equipment includes cyclones, condensors, wet-type collectors, cloth screen and stocking arresters, centrifugal collectors, and other devices used to separate solid material from the air stream in which it is carried.

Cloth screen or bag-type dust collectors used for collection of fine dusts present a special problem because of the use of a combustible fabric, even though the dust itself may be noncombustible. Automatic sprinkler protection may be needed where cloth dust collectors are important for the continuity of production or where, if ignited, they might expose other property to fire.

Equipment of large volume, such as bins and dust collectors, in which pulverized stock is stored or may accumulate, should be protected by automatic sprinklers. Automatic carbon dioxide, dry chemical, halogenated agent or water-spray extinguishing systems can be used effectively in dust collecting systems. Inert gas may be effectively used to create safe atmospheres in conveying systems.

§11.5   Fire Protection

The number of fires originating in and spreading through ducts justifies the installation of automatic extinguishing systems in ducts handling flammable or combustible materials. Such systems may be automatically or manually controlled.

If the material is extremely flammable, it may be advisable to install a deluge system actuated by heat-responsive devices. Care must be taken that the ductwork does not obstruct the automatic sprinklers so as to seriously interfere with the spray pattern. In many cases, sprinkler heads will have to be extended below the ductwork to provide satisfactory coverage, or side-mounted heads will be needed to augment the standard ceiling devices.

Portable fire extinguishers of appropriate type or small hoses with spray nozzles are helpful where fires may occur in ducts. This is true even where fixed-pipe extinguishing systems have been provided. Judicious placement of extinguishers and hoselines with respect to location of access panels will enable them to effectively complement the sprinkler system.

Explosion-relief vents prevent or minimize damage to duct systems that carry explosive mixtures. The vents should lead by the most direct practical route to the outside of the building. NFPA 68, Guide for Explosion Venting, will help when studying the design of explosion vents. The use of suppression and inerting systems is limited by the configuration of the equipment and the physical and chemical properties of the material being conveyed. Details concerning these systems are given in NFPA 69, Standard on Explosion Prevention Systems.

When flammable vapors, dust or other materials pass through a duct, static charges on the duct are generated. If a charge is allowed to accumulate on an electrically insulated portion of a duct system, it could discharge to an adjoining duct section — at a joint for example — and ignite the material being conveyed. For this reason, exhaust systems which carry flammable vapors, dusts, gases, or other materials should be electrically bonded and grounded. Methods of eliminating dangerous static charges are treated in detail in NFPA 77, Recommended Practice on Static Electricity.

A

access door A fire door smaller than conventional doors which provides access to utility shafts, chases, manways, plumbing, and various other concealed spaces and equipment.

air filter An in-line air cleaning device that eliminates some contamination from the air prior to its use. A Class 1 filter, when clean, does not contribute fuel when exposed to flame, and emits only negligible amounts of smoke. A Class 2 filter, when clean, burns moderately when attacked by flame, or emits moderate amounts of smoke, or both.

air flow The movement of air within a building; it is especially important on the fire floor, when air flow often follows a fire fighter's advance toward the fire; for purposes of ventilation, it is usually measured in cubic ft/min.

air handling system All of the fans, ducts, controls, dampers, filters, intakes, heating and cooling apparatus that remove, clean and recirculate the air in a structure.

air supported structure A structure, usually made of nonrigid materials, that has been pressurized internally above atmospheric pressure to keep the structure erect. (*See also* structure.)

all hands A working fire at which all units of the first alarm assignment are engaged in fire fighting; an "all hands" is frequently followed by a multiple alarm or the transfer of companies to cover the empty stations. (*See also* working fire.)

alternator An electric generator which produces alternating current for motor vehicles and can maintain battery charge at relatively low engine speeds.

ammeter An instrument for measuring the flow of electric current, usually in amperes, milliamperes, or microamperes.

ampacity The current-carrying capacity of an electric conductor, expressed in amperes.

ampere (amp) A standard unit for the amount of an electric current at a particular point in a circuit; it is the amount of current sent by one volt through a resistance of one ohm.

arcing The flashing occurring at electrical terminals when the circuit has been opened or closed.

area ignition Simultaneous or rapid successive ignition of a number of individual fires that are spaced to support each other and spread the fire quickly. (*See also* broadcast burning, center firing, simultaneous ignition.)

arson The crime of willfully burning a dwelling, building, structure, or other property, including one's own.

assembly occupancy A place of assembly including, but not limited to, a building or portions of a building used for gathering together fifty or more persons for such purposes as deliberation, worship, entertainment, amusement, or awaiting transportation. (*See also* place of assembly.)

atomic fission The splitting of the nucleus into two parts accompanied by the release of a large amount of energy. Fission reactions occur only with heavy elements such as uranium and plutonium. (*See also* chain reaction.)

authority having jurisdiction A term used in many standards and codes to refer to the organization, office, or individual responsible for "approving" equipment, procedures and construction in a town, county, city or state.

autoignition temperature The lowest temperature at which a flammable gas or vapor-air mixture will ignite without a spark or flame. Vapors and gases will spontaneously ignite at a lower temperature in oxygen than in air, and their autoignition temperature may be influenced by the presence of catalytic substances. (*See also* spontaneous ignition.)

automatic closing device A mechanism that can be fitted to a door which will cause the door to close if there is a fire. (*See also* automatic fire door, self-closing device.)

automatic dry sprinkler system A sprinkler system that has air under pressure in the pipes; when the sprinkler head fuses from the fire, the air escapes and allows water into the pipes and through the open sprinkler heads. (*also called* dry pipe system)

automatic fire door A door designed so that it will close automatically if there is a fire; the closing device is often a fusible link which melts in the presence of heat or activation of a smoke detector.

automatic sprinkler system A system of pipes with water under pressure that allows water to be discharged immediately when a sprinkler head operates. (*See also* wet-pipe sprinkler system.)

awning window A window with panes of glass about one foot high each and as long as the window's width. The panes may be opened by means of a cranking mechanism. (*See also* window, double-hung window, factory window, jalousie window.)

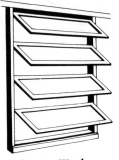

Awning Window

B

backdraft The explosion or rapid burning of heated gases that occurs when oxygen is introduced into a building that has not been properly ventilated and has a depleted supply of oxygen due to fire. (*See also* hot air explosion, smoke explosion.)

balloon construction Wood-frame construction in which the studs run two or more stories from the foundation to the eave line. The floor joists rest on ribbon boards nailed to the studs, leaving an unprotected opening running from the foundation to the attic. (*See also* platform construction.)

bearing wall A wall that supports floor or roof beams, girders, or other structural loads.

black powder A low explosive composed of a mixture of potassium nitrate, charcoal, and sulfur (sodium nitrate may be substituted for potassium nitrate), that is sensitive to heat, deflagrates rapidly, and does not detonate, but is a dangerous fire and explosion hazard.

BLEVE acronym Boiling Liquid-Expanding Vapor Explosion.

boiling liquid-expanding vapor explosion (BLEVE) A major container failure, in two or more pieces, at a moment when the contained liquid is at a temperature well above its boiling point at normal atmospheric pressure. If the liquid is flammable, an enormous explosion may result.

boiling point (bp) The temperature (which varies with the pressure and nature of the liquid) at which a liquid is rapidly converted to a vapor; normally it is reported at a pressure of one atmosphere.

boilover The expulsion of crude oil (or certain other liquids) from a burning tank; the light fractions of the crude oil burn off and produce a heat wave in the residue which, on reaching a water stratum, causes the violent expulsion of a portion of the tank's contents in the form of froth. (*See also* tank fire, frothover, slopover.)

British thermal unit (Btu) The amount of heat required to raise the temperature of one pound of water one degree Fahrenheit at atmospheric pressure.

broadcast burning An intentional burning in which fire is meant to spread over a specific area. (*See also* area ignition, center firing, simultaneous ignition, spot burning, edge firing.)

building 1. A structure that stands alone or is cut off from adjoining structures by fire walls, with all openings therein protected by approved fire doors. 2. Any structure used or intended to support or shelter any use or occupancy; the term is generally constructed as if followed by the words "or portion thereof." 3. A structure enclosed with walls and a roof, and having a defined height.

busbar A short conductor forming a common junction between two or more electrical circuits. (*also called* a bus)

business occupancies Occupancies used for the transaction of business other than that covered under mercantile occupancy; includes doctors' offices, dentists' offices, city halls, general offices, town halls, court houses, libraries and outpatient clinics among others. (*See also* mercantile occupancy, occupancy.)

C

Casement Window

casement window A window which is hinged on the side and swings outward when opened. (*See also* window.)

cause of fire The reason for ignition of an unfriendly fire, as recorded in fire service statistics.

ceiling The upper inside surface of a space, regardless of its height.

Celsius Unit of temperature named after the Swedish physicist Anders Celsius, based on the freezing and boiling temperatures of water; the freezing point of water is 0°C, and the boiling point is 100°C.

center firing A technique of broadcast burning in which fires are set in the center of an area to create a strong draft; additional fires are then set progressively nearer the outer control lines and, as the draft builds, it draws them toward the center. (*See also* area ignition, broadcast burning, simultaneous ignition.)

Centigrade (C) A measurement of temperature in which 0° is the freezing temperature and 100° is the boiling point of water; to convert Centigrade to Fahrenheit use the following formula: $F = (\%) C + 32$. (*See also* Celsius.)

chain reaction A self-sustaining series of chemical or nuclear reactions in which the products of the reaction contribute directly to the maintenance of the process. (*See also* atomic fission, fission.)

char 1. Carbonaceous matter formed by incomplete combustion of organic material such as wood. 2. To change into charcoal or carbon by pyrolysis. 3. To burn or scorch.

circuit 1. An electrical path of conducting components through which electrical current flows. 2. The conductor, or radio channel, and associated equipment used to transmit a fire alarm.

circuit breaker A device which opens a circuit when an electrical overload occurs.

Cleveland open cup apparatus A flash point testing device sometimes used in grading flammable liquids in transportation, under ASTM D92. (*See also* closed cup, open cup, Tagliabue closed cup, Tagliabue open cup.)

combustible 1. Capable of reacting with oxygen, and burning if ignited. 2. A material that will burn in air. 3. A material or structure that will burn.

combustible liquid Any liquid having a flash point, by a closed cup method, at or above 100° F and below 200° F and subdivided as follows: (1) Class II liquids — those with flash points at or above 100° F and below 140° F; (2) Class IIIA liquids — those with flash points at or above 140° F and below 200° F; (3) Class IIIB liquids — those with flash points at or above 200° F.

compartmentation A type of building design in which a building is divided into sections that can be closed off from each other so that there is resistance to fire spread beyond the area of origin; it is most common in highrise buildings and health care facilities.

conduction 1. The transmission of heat through or by means of a conductor by direct contact with a heated element. 2. The transmission of an electrical current.

conductor A substance that transmits electrical or thermal energy.

conduit 1. A trough or pipe containing and protecting electrical wires or cables. 2. A pipe or channel for conveying water.

convection The transfer of heat that occurs because of the mixing or circulation of heated fluid. (*also called* heat convection)

corrosive material Any solid, liquid, or gaseous substance that burns, irritates or destroys organic tissues, most notably the skin and, when taken internally, the lungs and stomach.

current The flow of electrons from one point to another, usually measured and expressed in amperes.

curtain wall An exterior nonloadbearing prefabricated wall, usually more than one story high supported by the structural frame, which protects the building interior from weather, noise, or fire. (*See also* panel wall.)

D

damage The total loss caused by fire, including indirect losses such as business interruption, loss of future production, and loss of grazing, wood products, wildlife habitat, recreation, and watershed values in forest, brush, or grass fires.

decomposition 1. Slow oxidation. 2. Burning without noticeable heat or light. 3. Chemical change of a single substance into two or more different substances.

decomposition explosion An explosion caused by a rapid exothermic reaction of a material, including some materials not ordinarily classed as explosives; it is usually accompanied by the release of large quantities of hot gases.

deflagration 1. Thermal decomposition that proceeds at less than some velocity, and may or may not develop hazardous pressures. 2. A rapid combustion that does not generate shock waves. 3. A burning that takes place at a flame speed below the velocity of sound in the unburned medium.

deluge sprinkler system A type of automatic sprinkler system, used to protect special risks. It consists of pipes and open sprinkler heads supplied through a valve connected to a detection system.

density The mass or quantity of matter of a substance per unit volume. (*See also* specific gravity.)

detonation 1. A thermal decomposition that occurs at supersonic velocity, accompanied by a shock wave in the decomposing material. 2. A burning that takes place at a flame speed above the velocity of sound in the unburned medium.

Double-Hung Window

double-hung window A type of window with two sashes at the center; the sashes open by sliding up and down. (*Also called* checkrail window; *see also* window.)

dry system A dry-pipe automatic sprinkler system that has air instead of water under pressure in its pip-

365

ing; dry systems are often installed in areas subject to freezing.

dwelling A building occupied with no more than two living units, each of which is occupied by a family; no more than three outsiders may rent rooms.

E

edge firing A technique of broadcast burning in which fires are set along the perimeter of an area and allowed to spread to the center.

educational occupancy A building in which six or more persons gather to receive instruction, such as a school, university, college and academy. (*See also* occupancy.)

egress A way out or exit.

electrical grounding A connection between a conductive body and the earth that eliminates the difference in potential between the object and ground.

electrical raceways Trenches in concrete floors, fitted with removable metal covers, where electrical connections and junction boxes are concealed.

electric switch A device for making, breaking or changing the connections in an electrical circuit.

endothermic reaction A process or change that absorbs heat and requires it for initiation and maintenance.

entrance An opening into a building.

exit 1. The portion of the means of egress that leads from the interior of a building or structure to the outside at ground level. 2. An area of refuge. (*See also* egress, fire escape.)

exit access Any portion of an evacuation path that leads to an exit.

exit discharge That portion of a means of egress between the termination of the exit and the exterior of the building at ground level.

explosion A sudden and violent release of energy from a material or compound as it decomposes, undergoes rapid chemical reaction, or changes from a solid to a liquid. (*See also* BLEVE.)

explosive atmosphere An atmosphere containing a mixture of a vapor or gas in any concentration within the explosive range.

explosive limits The range of concentration of a flammable gas or vapor (percent by volume in air) in which explosion can occur upon ignition in a confined area. (*See also* flammable limit.)

extra hazard Condition in which

the amount of combustibles or flammable liquids present is such that fires will be severe if they occur.

extra hazard occupancies Properties in which flash fires that open all the sprinklers in the area are a problem. In such occupancies, sprinkler spacing must be closer and pipe sizes larger than in other occupancies. (*See also* ordinary hazard occupancy, light hazard occupancy.)

F

faced wall A wall comprised of two different masonry materials, called wythes. The wythes are bonded together to act as one unit under load.

Factory Window

factory window A window that pivots out from the frame when it is opened. (*Also called* projected window; *see also* window.)

Fahrenheit (F) The temperature scale in which 212°F is the boiling point of water and 32°F is its freezing point.

fire Rapid self-sustaining oxidation accompanied by the evolution of varying intensities of heat and light. A Class A fire is one in ordinary combustibles; a Class B fire is one in flammable liquids, flammable gases, or grease; Class C fires are those involving energized electrical equipment; and Class D fires occur in combustible metals.

firebreak 1. A natural or constructed barrier that stops the spread of a fire or provides a control line from which to work. (*also called* a fire lane) 2. Urban areas in which decrepit buildings have been destroyed in order to reduce the chance of a conflagration.

firebug 1. An arsonist. 2. A person who enjoys attending fires, but does not set them. 3. A psychopathic fire setter.

fire code A set of rules and requirements whose purpose is to establish levels of fire protection considered adequate for procedures, practices, and equipment.

fire damper A device arranged to interrupt airflow automatically through a duct system, so as to restrict the passage of heat.

fire door A tested, listed, or approved door and frame assembly which prevents the spread of fire through a vertical or horizontal opening. (*See also* automatic fire door.)

fire escape A layman's term for any emergency exit from a building that is not in everyday use. (*See also* egress, exit.)

Fire Protection Handbook An extensive reference work covering virtually all phases of fire protection; it has been published since 1935 by the National Fire Protection Association.

fire raiser *Brit* An arsonist.

fire resistance rating The time, in minutes or hours, that materials or assemblies have withstood exposure to a fire, as established in accordance with the procedures of NFPA 251 and ASTM E119.

fire resistant construction *See* fire resistive construction.

fire resistive construction Construction in which the structural members, including walls, columns, floors, and roofs, are of noncombustible or limited-combustible materials, and have fire resistance ratings not less than those specified in NFPA 220; fire resistive construction has more ability to resist structural damage from fire than any other construction type.

fire retardant A surface or construction that will slow or limit the spread of fire.

fire separation 1. A floor or wall that meets specified fire endurance requirements as a barrier which prevents or retards fire spread. 2. A space or aisle between objects, such as goods in storage, buildings, or structures, that serves as a fire break and also as an area for fire fighting operations. (*See also* separation.)

fire signature Any product of combustion from a fire that can be used to detect or identify the fire.

fire stop An obstruction across an air passage or concealed space in a building to prevent fire from spreading.

firestopping The blocking off of concealed spaces in structures to prevent fire spread through walls and ceilings. (*also called* stopping)

fire tetrahedron A model of the four elements required by a fire: fuel, heat, oxygen, and uninhibited chain reaction; each side is contiguous with the other three. (*See also* fire triangle.)

fire tower 1. A smokeproof stair designed to limit or prevent penetration of heat, smoke, and gases into any part of a building. 2. A fire department training tower. 3. A fire department communications center. 4. A fire department water tower truck or an aerial ladder employed as a water tower. 5. A forest fire lookout tower. (*See also* smokeproof tower.)

fire triangle A three-sided figure representing three of the four factors necessary for combustion: oxygen, heat, and fuel; it has now been replaced by the fire tetrahedron.

fire wall 1. A wall constructed of solid masonry units faced on each side with brick or reinforced concrete used to subdivide a building or separate buildings to restrict the spread of fire; it begins at the foundation and extends through all stories to and above the roof, unless the roof is of fire-resistive or semi-fire-resistive construction, in which case the wall is carried up tightly against the underside of the roof slab. 2. A wall with adequate fire resistance used to subdivide buildings to restrict the spread of fire.

first degree burn A burn which causes only pain, redness, and swelling. (*See also* second degree burn, third degree burn.)

fission The splitting of an atomic nucleus by bombardment with neutrons from an external source, resulting in the release of large amounts of energy.

flame The burning gas or vapor of a fire, visible as a flickering light of various colors.

flame over The rapid spread of flame over one or more surfaces.

flame resistant A material or surface that does not propagate flame once the external source of flame is removed.

flame spread The propagation of a flame away from the source of ignition in a gas or across the surface of a liquid or a solid.

flammability The relative ease with which a fuel ignites and burns.

flammable Capable of being readily ignited.

flammable gas 1. A gas that will burn in the normal concentration of oxygen in air. 2. Any gas that will burn when mixed in any proportion with air, oxygen, or nitrous oxide.

flammable limit The highest or lowest concentration of a flammable gas or vapor in air that will explode or ignite. (*See also* explosive limit, lower flammable limit, upper flammable limit.)

flammable liquid Any liquid with a flash point below 100°F and a vapor pressure not exceeding 40 psi at 100°F. Class IA includes those with a flash point below 73°F, and a boiling point below 100°F. Class IB includes those with a flash point below 73°F and a boiling point at or above 100°F. Class IC includes those with a flash point at or above 70° F and below 100° F. (*See also* combustible liquid.)

flammable range The range between the upper and lower explosive or flammable limits of a flammable vapor or gas.

flammable vapor The vapor given off from a flammable liquid at and above its flash point.

flashback The jump of a flame from an ignition source across a distance to a supply of flammable liquid.

flashover The stage of a fire at which all surfaces and objects are heated to their ignition temperatures and flame breaks out almost all at once over the entire surface.

flash point The minimum temperature at which a liquid gives off sufficient vapor to form an ignitible mixture with air; it is usually determined by a closed cup test. (*See also* flammable liquid.)

flue gas Combustion products vented up a chimney from a fireplace or other combustion chamber.

forest fire A fire in a forest or other wildland. A Class A fire covers ¼ acre or less; Class B covers

fires more than ¼ acre but less than 10 acres; a Class C fire involves 10 acres or more but less than 100 acres; Class D includes fires of 100 acres or more but less than 300 acres; a Class E fire is one of 300 acres or more but less than 1,000 acres; Class F fires are those in 1,000 acres or more but less than 5,000 acres; and Class G fires are those of 5,000 acres or more.

frothover The overflow of a hot oil from a tank when water at the bottom of the tank boils. (*See also* boilover.)

fuel 1. Any substance that produces heat through combustion. 2. A substance that reacts with oxygen, or with the oxygen yielded by an oxidizer, to produce combustion.

fuel gas 1. Acetylene, hydrogen, natural gas, LP-gas, methylacetylene-propadiene, and some other liquefied or nonliquefied flammable gases that are stable because of composition, or because of the conditions of storage and use. 2. Manufactured gas, natural gas, undiluted liquefied petroleum gas (vapor phase only), liquefied petroleum gas-air mixture, or mixtures of these gases that would ignite in the presence of oxygen.

fuel load The expected maximum

amount of combustible material in a given fire area, usually expressed as weight of combustible material per square foot.

fuel oil Any liquid petroleum product burned in a furnace for the generation of heat, or used in an engine for the generation of power, except oils having a flash point below 100° F burned in cotton- or wood-wick burners.

fuse 1. An electrical safety device consisting of or including a wire or strip of fusible metal that melts and interrupts a circuit when the current exceeds a particular amperage. 2. To reduce to a liquid or plastic state by heat.

fusible link A connecting link of a low-melting alloy that holds an automatic sprinkler head in the closed position and melts at a predetermined temperature; it may also be used to hold a fire door or fire damper in the open position.

G

gable roof A double sloping roof that forms a gable at each end.

gas 1. The state of matter characterized by very low density and viscosity, comparatively great fluctuation as pressures and temperatures change, the ability to diffuse readily into other gases, and the ability to occupy with almost complete uniformity the whole of any container. 2. Gasoline.

gasoline A common flammable liquid mixture of volatile hydrocarbons used in internal combustion engines; it has an octane number of at least 56, and an explosive range of about 1.4 percent to 7.6 percent. The flash point and ignition temperature vary with the octane number.

general industrial occupancies Buildings in which ordinary and low hazard manufacturing operations are conducted; included are multistory buildings where floors are rented to different industrial tenants. (*See also* occupancy, high hazard industrial occupancy.)

gravity 1. The force of mutual attraction between masses. 2. The force that causes a body to accelerate while falling, usually expressed as 32.2 ft/sec^2. (*See also* specific gravity.)

ground A conducting connection, whether intentional or accidental, between an electrical circuit or equipment and the earth, or some conducting body that serves in place of the earth.

H

halogen Any one of the five elements — fluorine, chlorine, bromine, iodine, and astatine — that produces salt when combined with metals.

halon 1. Any one of several halogenated hydrocarbon compounds, two of which (bromotrifluoromethane and bromochlorodifluoromethane) are commonly used as extinguishing agents; they are inert to almost all chemicals, and resistant to both high and low temperatures.

halon extinguisher A fire extinguisher charged with a halogenated agent and an expellant; it is rated for use on flammable liquid and live electrical equipment fires.

hazardous material Any substance which, by reason of being explosive, flammable, poisonous, corrosive, oxidizing, irritating, or otherwise harmful, is likely to cause death or injury.

health care occupancies Those facilities used for the medical treatment or care of persons suffering from physical or mental illness, disease, and infirmity, and for the care of infants, convalescents, or aged persons.

heat Energy that is associated with and proportional to molecular motion, and that can be transferred from one body to another by radiation, conduction, and convection. (*See also* spontaneous heating.)

heat cramp A condition, less common than heat exhaustion, in which painful spasms of voluntary muscles, dilated pupils, and a weak pulse occur due to loss of body salt brought on by physical exertion in a hot environment and by profuse sweating.

heat exhaustion Distress characterized by headache, nausea, heavy perspiration, pale clammy skin, abnormal temperature, dilated pupils, and a general weakness brought on by overexposure to high temperature and a deficiency of salt. (*See also* heat cramp, heat stroke.)

heat flux The intensity of heat transfer across a surface expressed in watts/cm^2, joules/m^2, or Btu/in.^{2}/sec.

heat of combustion The amount of heat released during a substance's complete oxidation (combustion). Heat of combustion, commonly referred to as calorific or fuel value, depends upon the kinds and numbers of atoms in the molecule as well as upon their arrangement.

heat of vaporization The quantity of heat required to change a unit quantity of a liquid into a vapor. One pound of water at 212°F requires 972 Btu to vaporize. (*also called* latent heat of vaporization)

heat stroke A condition more serious than heat exhaustion, characterized by dry skin, high body temperature, and collapse due to long exposure to high temperature; if untreated, it may lead to delerium, convulsions, coma and even death. (*See also* heat exhaustion.)

heavy timber construction Construction in which nonbearing exterior walls, bearing walls, and bearing portions of walls are noncombustible, and columns, beams, and girders are heavy timber. Floors and roofs are of wood and there are no concealed spaces. (*also called* mill, plank-on-timber, and slow-burning construction)

high hazard industrial occupancy Those buildings of a factory or plant which contain extremely hazardous materials, processes, or contents; these include, among others, buildings in which large quantities of flammable liquids are present, explosive dusts are produced, and explosives are manufactured or stored.

hip roof A roof having sloping ends and sloping sides and equal angles on all sides.

horizontal exit A protected way of travel from one area of a building to another area in the same building or in an adjoining building on approximately the same level.

hot air explosion A backdraft explosion. (*See also* backdraft, smoke explosion.)

hydrocarbon An organic compound that consists exclusively of carbon and hydrogen and is derived principally from petroleum, coal tar, and vegetable sources.

hydrogen A gaseous element (H), with an ignition temperature of 752°F and explosive limits of 4 percent to 75 percent, that is highly flammable and explosive.

I

ignite To initiate combustion.

ignitible mixture 1. A vapor-air or dust-air mixture that can be ignited by a static spark. 2. A mixture that is capable of the propagation of flame away from the source of ignition when ignited.

ignition The point at which the heating of something becomes self perpetuating.

ignition temperature The minimum temperature required to initiate self-sustained combustion of a material or compound.

ignition time The time, in seconds, between the application of an ignition source to material and the instant self-sustained combustion begins.

incendiary 1. A fire believed to have been deliberately set. 2. An arsonist. 3. Of or relating to a device used to set an arson fire. (*See also* arson, firebug, fire raiser.)

incomplete combustion A process in which there is insufficient oxygen available for complete combustion; the result is a smoldering fire that produces smoke and carbon monoxide, and may be in danger of a backdraft.

industrial occupancy Factories of all kinds and properties devoted to operations such as processing, assembling, mixing, packaging, finishing or decorating, and repairing. (*See also* occupancy.)

inflammable A material that will burn. (*See also* flammable, nonflammable.)

ingress An entrance or the act of entering.

insulator An nonconductor of electricity, usually made of porcelain or glass, which encloses dangerous electrical equipment and prevents the leakage or passage of electricity.

interior finish The surface material of walls, fixed or moveable partitions, ceilings, and other exposed interior surfaces; the category includes plaster paneling, wood, paint, wallpaper, floor and ceiling tiles, and the like.

inversion A horizontal layer of air in which temperature increases as height increases.

ionization detector A detector that senses the presence of particulate combustion products and actuates an alarm.

J

Jalousie Window

jalousie window A window with many small panes, each about four inches wide and as long as the window, that opens and closes by means of a crank. (*See also* awning window, window.)

joule (J) A unit of work or energy equal to the force of 1 newton acting through a distance of 1 meter; 1 J = 10^7 ergs = 1 watt/sec.

K

kerosene A water-white oily liquid with a flash point of 100°F to 150°F, an ignition temperature of 410°F and an explosive range of 0.7 percent to 5 percent; it is moderately toxic if ingested or inhaled.

L

labeled Equipment or material to which has been attached a label, symbol, or other identifying mark of approval by a nationally recognized testing laboratory or inspection agency.

landing A flat platform at the head of a series of steps.

Lantern Roof

lantern roof A type of roof that is relatively rare but still exists in some older industrial buildings. Built on top of a roof, it is a structure with open or windowed walls to let in light and air.

lean mixture A mixture of air and gas that contains too much air for the amount of gas present to cause an explosion and is thus below the lower flammable limit.

lethal concentration That quantity of a substance administered by inhalation that is necessary to kill 50 percent of the test animals exposed to it within a specified time.

Life Safety Code NFPA 101, a standard containing provisions for building design whose purpose is to ensure that lives are protected should a fire occur; usually this entails providing a safe and accessible means of egress, but in health care occupancies it involves designing buildings which can be defended with occupants in place.

light hazard industrial occupancy An occupancy in which the potential rate of heat liberation is low, areas are subdivided, and a small number of sprinklers should be able to control any fires. (*See also* extra hazard occupancies, ordinary hazard industrial occupancies.)

liquefied natural gas (LNG) A cryogenic fluid composed predominantly of methane, but which may contain some butane, ethane, propane, nitrogen, or other components normally found in natural gas.

liquefied petroleum gas (LP-Gas or LPG) A colorless, nontoxic gas liquefied by compression; it has a flash point of -100° F, an ignition temperature of 800 to 1000° F, and is obtained as a byproduct in petroleum refining.

liquid A substance in which the molecules move freely so that the substance flows readily, but in which there is enough cohesion so that it does not expand indefinitely like a gas.

listed A device or product that has been tested, passed, and certified by a nationally recognized testing laboratory.

lower explosive limit (LEL) The minimum concentration of combustible gas or vapor in air that will ignite. (*also called* lower flammable limit)

lower flammable limit (LFL) *See* lower explosive limit.

low explosive An explosive that deflagrates or burns rather than detonates, such as propellants, certain primer mixtures, black powder, photoflash powders, and delay compositions.

M

Mansard Roof

mansard roof   A roof that has two slopes on all sides, with the lower slope steeper than the upper one.

means of egress   A safe, continuous, and unobstructed way of travel out of any building or structure; this includes the exit access, exit, and exit discharge.

mercantile occupancy   A building or any portion thereof used for the display, sale, or purchase of goods, wares, or merchandise.

metal   An element that forms positive ions when its compounds are in solution, and whose oxides form hydroxides rather than acids with water. About 75 percent of the elements are metals: most are crystalline solids with metallic luster, are conductors of electricity, and have high chemical reactivity.

mill-type construction   Construction in which bearing walls and exterior walls are noncombustible and have a fire resistance rating of 2 hours; columns, beams and girders are commonly heavy timber; and wood floors and roofs are built without concealed spaces. (*See also* heavy timber construction.)

minimum hourly fire-resistance rating   That degree of fire resistance deemed necessary by the authority having jurisdiction

multiple-death fire   A fire in which three or more fatalities occur.

N

National Building Code (NBC) A standard for building construction maintained and published by the American Insurance Association.

National Electrical Code (NEC) A standard for safe electrical installations prepared and revised every three years by a committee of the National Fire Protection Association.

National Fire Codes (NFC) A set of more than 250 fire protection codes, standards, manuals, and recommended practices prepared by committees of the National Fire Protection Association and published annually.

natural draft A current of air produced by the difference in the weight of a column of flue gases inside a chimney and a corresponding column of air of equal dimensions outside the chimney or vent.

natural gas A naturally occurring fuel gas consisting of mostly methane but which has some ethane, butane, propane, and nitrogen. It is colorless, and almost odorless unless a warning odor has been added, and has an ignition temperature of 900-1000° F, and an explosive range of 3.8 to 17 percent. It is used for fuel and cooking, for the synthesis of ammonia, and as a raw material for the petrochemical industry; as a liquid it is called liquefied natural gas (LNG).

nonbearing wall A wall which supports only its own weight.

noncombustible 1. A material that will not ignite and burn when subjected to a fire. 2. A material defined as such in NFPA 225. 3. A material classified as such by the Standard Method of Test for Noncombustibility of Elemental Materials, ASTM E136-73.

nonflammable 1. A liquid or gas that will not burn under the conditions set forth in the definition of flame resistant. (*See also* flame resistant.) 2. A liquid or gas that will not burn in 100 percent oxygen at pressures of 760 torr.

O

occupancy The use or intended use of a building, floor, or other part of a building. (*See also* assembly occupancy, business occupancies, educational occupancy, general industrial occupancies, health care occupancies, high hazard industrial occupancy, industrial occupancy, mercantile occupancy, penal occupancies, residential occupancy, special purpose industrial occupancies.)

occupant load The theoretical maximum number of persons that may occupy a building or an area in it at one time.

ohm (o) The amount of electrical resistance resulting from a voltage drop of one volt with the current of one ampere.

oil Any of the various greasy, combustible substances that are obtained from animal, vegetable, and mineral matter, liquid at ordinary temperatures, and soluble in certain organic solvents, though not in water.

open circuit 1. A break in an electrical circuit. 2. A fire alarm circuit which has no current except when a signal is being transmitted.

open cup A flash point testing device sometimes used in grading flammable liquids in transportation. (*See also* flash point, Cleveland open cup apparatus, Tagliabue open cup apparatus, Tagliabue closed cup tester, Pensky-Martens closed tester, setaflash closed cup tester.)

open cup test A method for testing the flashpoint of a flammable or combustible liquid.

open-joist construction Construction in which solid beams project below a ceiling surface more than 4 in. with intervals of 3 feet or less.

ordinary construction Building construction in which exterior bearing walls are made of noncombustible or limited-combustible materials and have minimum fire resistance ratings and stability; nonbearing exterior walls are of noncombustible or limited-combustible materials; and roofs, floors, and framing are wholly or partly of wood of smaller dimensions than heavy timber construction.

ordinary hazard industrial occupancies Industrial occupancies in which the processes, materials, and equipment are such that fires will probably burn fairly rapidly, or give off a considerable volume of smoke, but in which neither poisonous fumes nor explosions need be expected.

overcome   To be incapacitated by the effects of a fire, usually the smoke. (*See also* smoke inhalation.)

overcurrent   Electrical current which causes an excessive or dangerous temperature in the conductor or conductor insulation.

oxidation   Originally, a chemical reaction in which oxygen combines with other substances; it now indicates any chemical reaction in which electrons are transferred. Oxidation and reduction always occur simultaneously, and the substance which gains the electrons is called the oxidizing agent.

oxidizer   1. A substance that readily gives up oxygen without requiring an equivalent of another element in return. 2. A substance that contains an atom or atomic group that gains electrons such as oxygen, ozone, chlorine, hydrogen peroxide, nitric acid, metal oxides, chlorates, and permanganates. (*Also called* oxidizing agent; *see also* oxidation.)

oxygen (O)   A gaseous element present in the atmosphere in about 21 percent concentration; it is essential to combustion, is one of the four parts of the fire tetrahedron, and is the essential gas in respiration.

P

panelboard 1. A single panel or group of panels, accessible only from the front, from which lights, heat, or power circuits are controlled. 2. Thin laminated or pressed boards with a plastic facing or wood veneer on one side used for interior finish. (*also called* paneling)

panel heating A heating system in which the heating elements are concealed in the walls or ceilings.

panel wall A nonloadbearing wall made of panels fitted into steel or reinforced concrete framing members. (*See also* curtain wall.)

panic That sudden, unreasoning, and overwhelming fear that is experienced by people in the face of real or fancied danger.

panic hardware Special latches installed on exit doors so that the latch can be released and the door opened with a force of less than 15 pounds.

partition An interior space divider such as a wall.

party wall A wall common to two buildings, often owned or leased by both parties. (*also called* common wall, separating wall, zero lot line)

passageway A corridor, hallway, passage, or tunnel used for pedestrian traffic.

penal occupancies Places in which the occupants are under some degree of restraint or security, including jails, penal institutions, and reformatories. (*also called* correctional facilities)

Pensky-Martens closed tester A flashpoint testing device which tests liquids with a flashpoint above 200° F and certain other viscous or film-forming materials. (*See also* flashpoint, Cleveland open cup apparatus, Tagliabue open cup apparatus, Tagliabue closed cup tester, setaflash closed cup tester.)

perfect combustion Total combustion, in which the quantities of air and fuel present are so exact that they react completely. (*See also* stoichiometic air.)

photoelectric detector A fire detector with a photocell that either changes its electrical conductivity or produces an electrical potential when exposed to radiant energy; it also detects smoke by the reduction in the transmission of light through the smoke.

pilot ignition The ignition of a

material by radiation, in which a local high temperature ignition source is located in the stream of gases and volatiles issuing from the exposed material.

place of assembly A building or portion of one used by fifty or more persons for meeting or gathering. (*See also* assembly occupancy.)

platform construction Wood frame construction in which the first floor is built as a platform, i.e., the flooring is laid on the joists, and the frame for the first floor walls is erected on the first floor. The second story joists are then placed, the second story subfloor is laid, and the second floor walls are erected on the second story subfloor. In this method of construction there are inherent barriers to the spread of fire through the walls. (*Also called* western platform construction; *see also* balloon construction.)

polyvinyl chloride (PVC) A synthetic thermoplastic polymer which decomposes at 300°F, releasing toxic fumes of hydrogen chloride. It is often used for insulation on wiring.

portland cement A type of cement that hardens under water; it is made by burning a mixture of limestone and clay or similar materials.

post and beam construction Heavy-timber or mill-type construction.

pounds per square foot (PSF) A unit of pressure equivalent to 144 pounds per square inch.

pounds per square inch (psi) A unit of pressure; it is usually compared to atmospheric pressure (psig) or to pressure in a perfect vacuum (psia). (*See also* pound per square inch absolute.)

preaction sprinkler system A sprinkler system in which piping is dry and waterflow is actuated by a separate detection system.

propagation The spread of combustion through a solid, gas, or vapor, or the spread of a fire from one combustible to another combustible. (*See also* flame spread.)

psia *abbr* Pounds per square inch absolute.

pyrolysis Chemical decomposition caused by heat.

Q

quick response sprinkler A sprinkler head designed for rapid operation that has increased surface-to-mass areas or special actuators, i.e., metallic vane heat collectors or electronic squibs.

R

raceway A tube for carrying and protecting electrical wires, cables, or busbars.

radiant heat Heat energy carried by electromagnetic waves that can pass through gases without warming them, but that increases the temperature of solid and opaque objects. (*also called* radiated heat)

radiation 1. Energy that is sent forth, e.g., light, short radio, ultraviolet, and x-ray waves. 2. The transfer of energy, including heat, through visible light by electromagnetic waves. 3. Streams of high-speed atomic particles, e.g., alpha and beta particles and neutrons.

reflash The reignition of a flammable fuel by a hot object after flames have been extinguished. (*See also* flashback, rekindle.)

regression rate The burning rate of a solid or liquid, usually measured in centimeters per second measured perpendicular to the surface.

rekindle To reignite after extinguishment. (*See also* reflash.)

residential occupancy A place in which sleeping accommodations are provided for normal residential purposes; it includes all buildings designed to provide sleeping accommodations, i.e., hotels, motels, apartments, dormitories, lodging or rooming houses, and 1- and 2-family dwellings. (*See also* occupancy.)

resistance That property of a conductor by which it opposes the flow of an electric current, resulting in the generation of heat in the conducting material; the measure of resistance is usually expressed in ohms.

rheostat A device used for varying the resistance of an electrical circuit without interrupting the circuit.

rich mixture A fuel and oxidizer mixture having more than the stoichiometric concentration of fuel. (*See also* lean mixture.)

roof The outside top covering of a building. (*See also* gable roof, hip roof, lantern roof, mansard roof.)

rooming house Living quarters in which there are separate sleeping rooms for not more than fifteen persons and no separate cooking facilities for individual occupants.

S

safety glass Glass that has thin wires or plastic embedded in it or placed between laminates; if broken, it produces no flying splinters. (*See also* tempered glass.)

SCBA *abbr* self-contained breathing apparatus.

secondary exit An alternate exit.

second degree burn A burn in which the skin is blistered. (*See also* first degree burn, third degree burn.)

self closing Equipped with an approved device that will ensure closing after having been opened. (*See also* automatic closing device, self-closing device, self-closing door.)

self-closing device A mechanism that ensures that a door or other closure, when opened, will return to the closed position. (*See also* automatic closing device, self-closing, self-closing door.)

self-closing door A door which, when opened, will return to the closed position.

self-contained breathing apparatus Equipment that is worn by fire fighters to provide respiratory protection in a hazardous environment; it consists of a facepiece, a regulating or control device, an air or oxygen supply, a harness assembly, and, sometimes, filters.

separated storage Storage in the same fire area, but separated by as much space as practicable, or by intervening storage, from incompatible materials. (*See also* firebreak, fire separation, separate fire division, separation.)

separate fire division A portion of a building cut off from all other portions of the building by fire walls, fire doors. (*See also* fire separation, separation.)

separating wall *See* party wall.

separation 1. The spacing of buildings or materials to provide fire exposure protection. 2. A barrier to fire spread, such as a fire wall. (*See also* firebreak, fire separation, separated storage, separate fire division.)

service The conductors and equipment for delivering energy from the electrical supply system to the wiring system of the premise served.

service conductors Supply conductors that extend from the street main or from transformers to the service equipment of the premises.

service drop The overhead conductors from the last pole or aerial support to the service-entrance conductors at the building or other structure.

setaflash closed cup tester A flashpoint testing device used for testing aviation turbine fuels according to ASTM 3243, and for testing paints, enamels, lacquers, varnishes and related products and their components having flash points between 32°F and 230°F. (*See also* flash point, Cleveland open cup apparatus, Tagliabue open cup apparatus, Tagliabue closed cup tester, Pensky-Martens closed tester.)

short circuit Conduction of electrical current (usually inadvertent) made between points on a circuit between which the resistance is normally greater.

simultaneous ignition Ignition of a fire at many points at once; it is used in broadcast burning or backfiring to obtain a quick, hot, clean burn. (*See also* area ignition, broadcast burning, center firing.)

single-family dwelling A residential structure for one family.

slopover The overflow from a tank caused by frothing of burning oil when water or foam is applied. (*See also* boiling liquid-expanding vapor explosion, boilover, frothover.)

slow oxidation The oxidation of a material without the evolution of visible light, as in the rusting of iron. (*See also* oxidation.)

smoke The airborne solid and liquid particulates and gases evolved when a material undergoes combustion.

smoke control door A door designed to inhibit or act as a barrier to the spread of smoke in a building. (*also called* smoke stop door)

smoke damper A device to restrict the passage of smoke through a duct that operates automatically and is controlled by a smoke detector.

smoke density The proportion of solids in smoke.

smoke detector A detector that actuates if it senses visible or invisible particles of combustion. (*See also* ionization detector, photoelectric detector.)

smoke explosion An explosion of heated smoke and gases. (*Erroneously called* hot air explosion; *see also* backdraft.)

smoke inhalation Sickness or injury caused by the respiration of smoke and other products of combustion. (*Also called* smoke poisoning)

smoke partition Partitions or walls installed to divide a building into compartments and prevent the spread of smoke from a fire.

smokeproof stairway A stairway designed so that the passage of smoke and gases from a fire into it is limited.

smokeproof tower A continuous fire-resistive enclosure in a building that protects a stairway from fire or smoke. (*Also called* smoke tower, tower; *see also* fire tower.)

smoke shaft A continuous shaft, extending the full height of a building, with openings at each level and a fan at the top; during a fire, the dampers on the fire floor open and the fan vents the combustion products.

smolder To burn and smoke without flame.

Southern Building Code *See* Standard Building Code.

space heater A portable heating device used to heat small areas, some types of which use liquid fuel. Though these are prohibited in most jurisdictions, they are still found and have been the initial cause of many fatal fires.

spalling The deterioration of concrete due to fire exposure.

spark 1. A small, incandescent particle from burning wood or a glowing particle produced by metal grinding. 2. A localized instantaneous electrical discharge, accompanied by heat and light, between points at different voltages.

special purpose industrial occupancies As defined by the NFPA Life Safety Code, this term includes ordinary and low hazard manufacturing operations in buildings designed for and suitable only for particular types of operations, characterized by a relatively low density of employee population, with much of the area occupied by machinery or equipment. (*See also* occupancy, high hazard industrial occupancy.)

specific gravity (sp gr) The weight or mass of a given volume of a substance at a specified temperature, as compared to that of an equal volume of another substance. (*See also* density, gravity.)

spontaneous combustion *See* spontaneous heating, spontaneous ignition.

spontaneous heating Heating due to chemical or bacterial action in a combustible material. (*See also* spontaneous ignition.)

spontaneous ignition Ignition due to chemical reaction or bacterial action in which there is a slow oxidation of organic compounds until the material ignites; usually there is sufficient air for oxidation but insufficient ventilation to carry heat away as it is generated. (*Also called* self ignition; *see also* autoignition temperature.)

spontaneous ignition temperature *See* autoignition temperature.

spot burning A form of broadcast burning that starts in and is confined to accumulations of slash. Slash is the debris left after logging, pruning, thinning, or brush cutting, including logs, chunks, bark, branches, stumps, and broken understory trees or brush.

spray sprinkler An automatic sprinkler head designed to control fire by the spray principle of heat absorption. (*Also called* spray sprinkler head; *see also* sprinkler system.)

spread The extension of a fire.

sprinkler alarm A device that sounds an audible local alarm when sprinkler heads activate.

sprinkler block A device used to stop the flow of water from a particular sprinkler head without shutting the system down. (*See also* sprinkler tongs, sprinkler wedge.)

sprinkler connection A connection in which hose lines from a pumper are used to increase the pressure in a sprinkler system, usually a sprinkler siamese with two pumper inlet connections. (*Also called* fire department connection)

sprinklered Equipped with a sprinkler system.

sprinkler head A waterflow device in a sprinkler system, consisting of a threaded nipple that connects the head to a water pipe, a fusible link or other releasing mechanism held in place, and a deflector that breaks up the water into a spray.

sprinkler spacing The distribution of sprinkler heads to provide adequate protection for a given hazard.

sprinkler stopper *See* sprinkler tongs, sprinkler wedge.

sprinkler system A system of water pipes and spaced sprinkler heads installed in a structure to control and extinguish fires; it uses a suitable water supply, such as a gravity tank, fire pump, reservoir, pressure tank, or connections to city mains and usually has a controlling valve and an alarm that signals when the system is actuated. (*See also* automatic sprinkler system, automatic dry sprinkler system, deluge sprinkler system, dry system, preaction sprinkler system, spray sprinkler, sprinkler alarm, sprinkler block, sprinkler connection, sprinklered, sprinkler head, sprinkler spacing, sprinkler stopper, wet-pipe sprinkler system.)

sprinkler tongs A tool used to stop the flow of water from a sprinkler head without shutting down the system. (*Also called* sprinkler block, sprinkler stopper; *see also* sprinkler wedge.)

sprinkler wedge A tapered, wedge-shaped wooden block used to stop the flow of water from a sprinkler head without shutting down the system. (*Also called* sprinkler block, sprinkler stopper, sprinkler tongs.)

stack effect The air or smoke movement or migration through a tall building due to pressure differentials caused by temperature.

standard A document containing requirements and specifications, such as for building construction or fire protection. (*See also* fire code.)

Standard Building Code The model building code written and maintained by the Southern Building Code Congress International.

static electricity Charges of electricity accumulated on nonconducting materials, usually by friction.

Steiner tunnel test A method of testing the flame spread of interior finishes. It is described in ASTM E84, NFPA 255, and UL 723.

stoichiometric air The chemically correct amount of air required for complete combustion of a given quantity of a specific fuel. (*See also* perfect combustion.)

stratification The rising or settling of layers of smoke, according to density or weight, with the heaviest layer on the bottom; smoke layers usually collect from the ceiling down.

structure   An assembly of materials constructed to serve a specific purpose. (*See also* air supported structure.)

sublimation   Evaporation or release of vapors from a solid without going through the liquid phase.

suppress a fire   To extinguish a fire or confine it to an area burning within fixed boundaries.

suppression   All actions taken to extinguish a fire, from the time of its discovery; fire extinguishment.

surface burning   *See* flame spread.

surface flame spread   *See* flame spread.

suspicious fire   A fire of undetermined cause suspected to be arson.

T

Tagliabue closed cup tester A flashpoint tester used for all flammable liquids except certain viscose or film-forming materials with a flash point at or below 200°F; details are given in ASTM D56. (*Also called* Tag cup test, Tag tester, Tagliabue tester; *see also* Tagliabue open cup apparatus, Cleveland open cup apparatus, Pensky-Martens closed tester, flash point.)

Tagliabue open cup apparatus A flashpoint tester sometimes used in grading flammable liquids in transportation according to ASTM D1310. (*Also called* Tagliabue tester, Tag cup tester, tag tester; *see also* flash point, Tagliabue closed cup tester, Cleveland open cup apparatus, and Pensky-Martens closed tester.)

tank fire 1. A fire involving a flammable liquid storage tank. (*See also* boiling liquid-expanding vapor explosion (BLEVE), boilover.) 2. A small fire that can be extinguished with the water from a booster tank.

temperature conversion The conversion between the various temperature scales is: $C = \frac{5}{9} \times (\text{temp. F} - 32)$; $F = (\frac{9}{5} \times \text{temp. C}) + 32$; K = $C + 273.15$; and $R = F + 459.67$, where C is degrees Celsius, F is degrees Fahrenheit, K is degrees Kelvin, and R is degrees Rankine.

tempered glass Glass heat-treated in a special process so that it has a better resistance to bending, impact, or thermal shock than ordinary glass of the same thickness; when broken it falls into small, regular fragments instead of large, random-sized shards.

theoretical air *See* stoichiometric air.

thermal column A column of smoke and gases given off by fire because of the expansion and rise of heated gases from displacement by cooler air; from it the magnitude and intensity of a fire can often be judged. (*Also called* convection column, thermal updraft)

thermal conductivity The transmission of heat through a solid or liquid.

thermal convection *See* convection.

thermal decomposition The breakup of materials into other compounds as a result of heat. (*Also called* thermal degradation; *see also* pyrolysis)

thermal radiation   The emission of radiant energy waves from a heated body.

thermal updraft   *See* thermal column.

thermocouple   A temperature-measuring device composed of two dissimilar conductors connected at their ends. Heat causes a different voltage in each of the two conductors and voltage difference is proportional to the temperature of the material measured.

third degree burn   A flesh burn in which charring occurs. (*See also* first degree burn, second degree burn.)

total loss   1. The complete loss of an insured property, where there is nothing of value to salvage. 2. A loss exceeding the maximum amount written in an insurance policy. 3. The complete destruction of a property.

trailer   1. The rear end of a semi-trailer vehicle designed for towing by a tractor or power unit, such as the aerial ladder trailer, that carries the fire fighting equipment of a tractor-drawn aerial truck assembly. 2. A trail of combustible materials used by an arsonist to speed the spread of fire through a structure.

transformer   A device that raises or lowers the voltage of alternating current.

tunnel test   The popular name for a standard surface burning test of building materials, the details of which are given in ASTM E84. (*See also* Steiner tunnel test.)

U

UEL *abbr* Upper explosive limit.

UFL *abbr* Upper flammable limit.

UL label A label affixed to a product designating that it, or a prototype of it, has been tested by Underwriters Laboratories and found to be safe for general use.

under control The stage of a fire at which it has been contained and partially extinguished, and authorities are confident that it can be completely extinguished; some overhaul may begin at this point. (*Also called* tapped out)

Uniform Building Code (UBC) The building code that is maintained and published by the International Conference of Building Officials (ICBO); it contains construction specifications and includes provisions for materials, processes, and design.

Uniform Fire Code A fire code published by the International Conference of Building Officials and the Western Fire Chiefs Association.

unit of exit width The width necessary (22 in.) for the orderly movement of a single line of people along a passageway or through an exit during an emergency.

unprotected opening A vertical or horizontal opening through a floor, wall, or other partition in a building that allows the passage of smoke, heat, and flame.

unstable material Any material which will vigorously polymerize, decompose, condense, become self-reactive, or undergo other violent chemical changes.

upper explosive limit (UEL) The maximum concentration of vapor or gas in air above which flame propagation does not occur. (*Also called* upper flammable limit; *see also* flammable limit.)

upper flammable limit (UFL) The highest concentration of flammable vapor in air that will burn with a flame. (*Also called* upper explosive limit; *see also* flammable limit.)

utility gas Natural gas, manufactured gas, liquefied petroleum gas-air mixtures, or a mixture of any of these.

V

valve A gate in a passage used to regulate the flow of liquid, air, gas, loose material, etc.

vapor A substance in its gaseous state, particularly one that is liquid or solid at ordinary temperatures.

vapor density (vd) The weight of a given volume of a gas or vapor, as compared with the weight of the same volume of dry air.

vaporizing liquid A liquid extinguishing agent, usually carbon tetrachloride or chlorobromomethane, used on flammable liquid and electrical fires, that vaporizes and forms a vapor blanket heavier than air that chemically extinguishes the fire. Vaporizing liquid extinguishers are no longer approved for installation. (*See also* halon extinguisher.)

vapor pressure The pressure exerted at any given temperature by a vapor, either by itself or in a mixture of gases, measured at the surface of an evaporating liquid.

vent An opening for the release or dissipation of fluids, such as gases, fumes, smoke, etc.

volt (v) A unit of electrical potential that causes a current of one ampere to flow through a resistance of one ohm.

W

wall An upright structure of brick, concrete, wood or other similar materials serving for enclosure, division, or support, such as one of the enclosing sides of a building or room.

wall furnace A self-contained, vented heater installed in a building, mobile home, or recreational vehicle.

water A colorless, odorless, tasteless liquid (H_2O) used universally for extinguishing fire; its allotropic forms are ice (solid) and steam (vapor). The freezing point of water is $32°F$ and the boiling point is $212°F$; it is effective as a cooling agent because it has the largest latent heat of vaporization of all common materials, 970 Btu per lb.

watt (w) A unit of electrical power obtained by multiplying amperage times volts.

wet-pipe sprinkler system An automatic sprinkler system in which the pipes are kept filled with water. (*See also* automatic sprinkler system.)

wet water Water to which a wetting agent has been added in order to increase its penetration.

window An opening in a building, vehicle, or container for letting in light or air, or both; normally it has a pane or panes of glass, and can be opened. (*See also* awning window, casement window, double-hung window, jalousie window, factory window.)

wired glass Window glass with an embedded wire mesh to improve fire resistance and make it shatter-resistant.

wood A combustible hard fibrous substance, basically cellulose, that makes up the most part of the stem and branches of trees or shrubs.

wood frame construction A type of construction in which exterior walls, bearing walls, partitions, floors, roofs, and their supports are made wholly or partly of wood and other combustible materials, when the construction does not qualify as heavy timber construction or ordinary construction.

wood panel fire test A test for evaluating the effectiveness of Class A fire extinguishers; it consists of solid square wood-panel backings to which are applied two horizontal sections of furring strips spaced apart and away from the panel by vertical furring strips. This provides a large vertical surface area of wood subject to combustion.

wood roof   A roof made of wooden shingles or shakes.

wood stove   A stove designed to burn wood or coal that is usually used as a radiant heater for single rooms.

working fire   1. A fire that requires fire fighting activity by most or all of the fire department personnel assigned to the alarm. (*Also called* job, worker; *see also* all hands.) 2. A one-alarm blaze.

XYZ

yard hydrant A hydrant located in a private water distribution system; seldom does it have a fire department pumper connection.

BIBLIOGRAPHY

Brannigan, Francis L., *Building Construction for the Fire Service*, 2nd Edition, National Fire Protection Association, Quincy, MA, 1982.

Bugbee, Percy, *Principles of Fire Protection*, National Fire Protection Association, Quincy, MA, 1978.

Gold, David T., *Fire Brigade Training Manual*, National Fire Protection Association, Quincy, MA, 1982.

Handley, William, Editor, *Industrial Safety Handbook*, McGraw-Hill Book Co., Ltd, London, England, 1977.

McKinnon, Gordon, et al, *Fire Protection Handbook*, 15th Edition, National Fire Protection Association, Quincy, MA, 1981.

NFPA, *Fire Protection Guide on Hazardous Materials*, 7th Edition, National Fire Protection Association, Quincy, MA, 1978.

Plenner, Robert G., *Fire Loss Control*, Marcel Dekker, New York, NY, 1979.

Tuve, Richard L., *Principles of Fire Protection Chemistry*, National Fire Protection Association, Quincy, MA, 1978.

Smith v. Hobart Manufacturing Co., 302 F.2d 570 (3rd Cir. 1962)

Smith v. Insurance Company of North America, 213 F.Supp. 675 (M.D.Tenn. 1963)

Somerset County Mutual Fire Insurance Co. v. Usaw, 112 Pa. 80, 4 A.355 (1886)

South v. A.B. Chance Co., 635 P.2d 728 (Wash. 1981)

South Austin Drive-In Theatre v. Thomison, 421 S.W.2d 933 (Tex.Civ.App. 1967)

Speyer, Inc. v. Humble Oil & Refining Co., 403 F.2d 766 (3rd Cir. 1968), *cert. denied*, 394 U.S. 1015 (1969)

Standard Oil Co. v. Midgett, 116 F.2d 562 (4th Cir. 1941)

State v. Dean, 307 N.W.2d 628 (Wis. 1981), *rev'g, State v. Stanislawski*, 216 N.W.2d 8 (Wis. 1974)

State v. Stanislawski, 216 N.W.2d 8 (Wis. 1974)

State v. Trimble, 362 P.2d 788 (N.M. 1961)

State v. Valdez, 371 P.2d 894 (1962)

State Stove Manufacturing v. Hodges, 189 So.2d 113 (Miss. 1966), *cert. denied*, 386 U.S. 912 (1967)

Stein v. Girard Fire Insurance Co., 259 F.2d 764 (7th Cir. 1958)

Steinmetz v. Bradbury Corp., 618 F.2d 21 (8th Cir. 1980)

Summers v. Tice, 33 Cal.2d 80 (1948)

Sun Valley Airlines, Inc. v. Avco-Lycoming Corp., 411 F.Supp. 598 (D.C. Idaho 1976)

Swindle v. Maryland Casualty Co., 251 So.2d 787 (La.App. 1971)

Talley v. City Tank Corp., 279 S.E.2d 264 (Ga.App. 1981)

Taylor v. ROA Motors, Inc., 152 S.E.2d 631 (Ga.App. 1966)

Technical Chemical Co. v. Jacobs, 480 S.W.2d 602 (Tex. 1972)

Terpestra v. Niagra Fire Insurance Co., 256 N.E.2d 536 (N.Y. 1970)

Texas Farm Bureau Insurance Company v. Baker, 596 S.W.2d 639 (Tex.Civ.App. 1980)

Thibault v. Sears, Roebuck & Co., 395 A.2d 843 (N.H. 1978)

Turcotte v. Ford Motor Co., 494 F.2d 173 (1st Cir. 1974)

United States ex rel. Szocki v. Cavell, 156 F.Supp. 79 (W.D.Pa. 1957)

United States v. Burch, 294 F.2d 1 (5th Cir. 1961)

United States v. Turner, 528 F.2d 143 (9th Cir. 1975), *cert. denied*, 423 U.S. 996 (1975)

Universal Underwriters Insurance Co. v. Security Industries, Inc., 391 F.Supp. 326 (W.D.Wash. 1974)

Vockie v. General Motors Corp., 66 F.R.D. 57 (E.D.Pa.), *aff'd*, 523 F.2d 1052 (3d Cir. 1975)

Warners Furniture, Inc. v. Commercial Union Insurance Co., 349 N.E.2d 616 (Ill.App. 1976)

Watertown Fire Insurance Co. v. Grehan, 74 Ga. 642 (1885)

Webster v. Heim, 399 N.E.2d 690 (Ill.App. 1980)

Weiner v. Aetna Insurance Co. of Hartford, 259 N.W.507 (Neb. 1953)

Werner v. Upjohn Co., 628 F.2d 848 (4th Cir. 1980), *cert. denied*, 449 U.S. 1080 (1981)

Wilson v. Dake Corp., 497 F.Supp. 1339 (E.D.Tenn. 1980)

Winkler v. American Safety Equipment Corp., 604 P.2d 693 (Colo.App. 1979), *rev'd on other grounds, American Safety Equipment Corp. v. Winkler*, 640 P.2d 216 (Colo. 1982)

Witchita City Lines, Inc. v. Packett, 295 S.W.2d 894 (Tex. 1956)

Wojciechowski v. Long Airdox Div. of Marmon Group, Inc., 488 F.2d 1111 (3rd Cir. 1973)

Woodruff v. Wilson Oil Co., 382 N.E.2d 1009 (Ind.App. 1978)

W.T. Grant Co. v. United States Fidelity & Guaranty Co., 421 A.2d 357 (Pa.Super. 1980)